Karl Schroeder, Charles H. Carter

A manual of midwifery

Including the pregnancy and the puerperal state

Karl Schroeder, Charles H. Carter

A manual of midwifery
Including the pregnancy and the puerperal state

ISBN/EAN: 9783743357518

Manufactured in Europe, USA, Canada, Australia, Japa

Cover: Foto ©berggeist007 / pixelio.de

Manufactured and distributed by brebook publishing software (www.brebook.com)

Karl Schroeder, Charles H. Carter

A manual of midwifery

A

MANUAL OF MIDWIFERY

INCLUDING THE

PATHOLOGY OF PREGNANCY AND THE PUERPERAL STATE

A

MANUAL OF MIDWIFERY

INCLUDING THE

PATHOLOGY OF PREGNANCY AND THE PUERPERAL STATE

BY

DR KARL SCHROEDER

PROFESSOR OF MIDWIFERY AND DIRECTOR OF THE LYING-IN INSTITUTION IN THE UNIVERSITY OF ERLANGEN

TRANSLATED INTO ENGLISH FROM THE THIRD GERMAN EDITION

BY

CHARLES H. CARTER, B.A., M.D., B.S. LOND.

MEMBER OF THE ROYAL COLLEGE OF PHYSICIANS, LONDON,
AND PHYSICIAN-ACCOUCHEUR TO ST. GEORGE'S, HANOVER SQUARE, DISPENSARY

WITH TWENTY-SIX ENGRAVINGS ON WOOD

NEW YORK:
D. APPLETON AND COMPANY,
549 & 551 BROADWAY.
1873.

PREFACE

The Translator feels that no apology is needed in offering to the Profession a translation of Schroeder's 'Manual of Midwifery.' The work is well known in Germany, and extensively used as a text-book; it has already reached a third edition within the short space of two years.

It is hoped that the present translation will meet the want, which has been long felt in this country, of a manual of midwifery embracing the latest scientific researches on the subject.

The Translator has to thank his friend Dr Berkart for the great assistance he has rendered him in the translation, which has considerably lightened an otherwise very arduous undertaking.

CHARLES H. CARTER.

8, Old Cavendish Street, W.

PREFACE

TO THE

THIRD GERMAN EDITION

The Third Edition, which has unexpectedly become necessary within so short a time, is for the most part unchanged; some parts have been greatly altered, and some new matter has been added. The Author trusts that in this new form the book may show that the kind reception with which the former Editions have been received has aroused him to fresh and diligent research.

SCHROEDER.

Erlangen.

CONTENTS

PART I

PRELIMINARY OBSERVATIONS

PART II

THE PHYSIOLOGY OF PREGNANCY

PART III

PHYSIOLOGY OF PARTURITION

PART IV

THE PHYSIOLOGY OF THE LYING-IN STATE

PART V

THE PATHOLOGY OF PREGNANCY

PART VI

GENERAL PATHOLOGY AND THERAPEUTICS OF PARTURITION

PART VII

SPECIAL PATHOLOGY OF PARTURITION

PART VIII

THE PATHOLOGY OF CHILDBED

MANUAL OF MIDWIFERY

PART I

PRELIMINARY OBSERVATIONS

We presume that the student has a perfect acquaintance with the purely anatomical relations of the bony pelvis, as well as of the soft parts lining and covering it, and shall therefore only speak of the female pelvis with regard to its function as the canal of parturition.

CHAPTER I

THE BONY PELVIS

A.—THE CAVITY AND DIAMETERS OF THE PELVIS

The pelvis is divided by the ilio-pectineal line into an upper and a lower portion. The former is called the "false," the latter the "true" pelvis.

The false pelvis has on three sides only bony parietes; its anterior wall is formed by the integuments of the abdomen. It is of obstetric importance only so far as from deviations of some of its normal measurements inferences may be drawn as to abnormalities also of the true pelvis. This fact is of considerable importance, inasmuch as the false pelvis is much more accessible to external instrumental mensuration than the true.

The following measurements will have to be considered on this point:

a. The distance between the two anterior superior spines of the ilium in the bony pelvis, which is on an average . . . . . . 23 cm. or 9 inches.

b. The greatest distance between the crests of the ilium, measuring . . . . 25 cm. or 9·84 inches.

These measurements are the average distances between the above-mentioned points of the bony pelvis. They differ somewhat from the measurements obtained on the living woman by means of calipers.

In the chapters on the pathology of parturition and on the contracted pelvis this subject will be more fully entered into.

The small or true pelvis forms a canal, which runs downwards and backwards. The posterior wall is considerably higher than the anterior, which is constituted by the symphysis pubis. The dimensions of the canal vary in its course, and it is therefore necessary to consider separately different sections of it.

The "inlet" is formed behind by the promontory and the upper and inner margins of the wings of the sacrum, laterally by the ilio-pectineal line, and in front by the crests and the symphysis of the pubic bones. It has the following diameters:

a. The antero-posterior or conjugate, which is the shortest line connecting the promontory of the sacrum and the symphysis of the pubis, measures . . . . . . 11 cm. or 4·33 inches.

b. The transverse diameter, between the most distant points of the ilio-pectineal line 13½ cm. or 5·3 inches.

c. The two oblique diameters, between the sacro-iliac synchondrosis of one side and the ilio-pubic tuberosity of the other. The right diameter goes from the right sacro-iliac synchondrosis to the left ilio-pubic tuberosity, and the left *vice versa*; each measures 12¾ cm. or 5 inches.

The ilio-pubic tuberosity marks the inner margin of the iliacus and psoas muscles, and is always distinctly recognisable. The pubo-iliac synostosis, which has frequently been mentioned as the end of these diameters, is somewhat outside of it, and cannot be recognised in the fully developed pelvis.

For the more exact recognition of a certain class of faulty pelves, a knowledge of the distance between the sacral promontory and the region above the cotyloid cavity is indispensable. This, the sacro-cotyloid diameter, measures in the normal pelvis 8¾ to 9 cm. or 3·4 to 3·5 inches.

The pelvic cavity varies in width in different parts.

Thus, the widest part lies in a plane which passes through the middle of the symphysis the highest points of the acetabula and the points of junction between the second and third sacral vertebræ. The following diameters are distinguished:

a. The antero-posterior, from the middle of the symphysis to the upper margin of the third sacral vertebra . . . . . 12¾ cm. or 4·7 inches.

b. The transverse, between the highest points of the acetabula in the erect posture . 12½ cm. or 4·6 inches.

The narrowest part is in a plane which goes through the inferior angle of the sacrum the spines of the ischium and the apex of the pubic arch. The diameters are—

a. The antero-posterior, from the inferior angle of the sacrum to the apex of the pubic arch 11¼ cm. or 4·25 inches.

b. The transverse, between the spines of the ischium . . . . . . 10½ cm. or 3·9 inches.

The outlet consists of two triangles the bases of which coincide. Their common base is formed by the line joining the ischial tuberosities. The apex of the anterior triangle is at the vertex of the pubic arch, that of the posterior at the tip of the coccyx. The diameters are—

a. The antero-posterior, between the tip of the coccyx and the vertex of the pubic arch 9—9½ cm. or 3·3—3·5 inches.

During parturition, however, when the coccyx is pushed backwards, it may be prolonged for rather more than . 2 cm. or ¾ of an inch.

b. The transverse, between the middle of the ischial tuberosities, equals . . 11 cm. or 4 inches.

The statements of the normal measurements of the pelvic cavity are very varying, since the individual deviations of the canal are very considerable. The distances given above are obtained from the exact measurements of fifty pelves. The exact result of these measurements is on an average the following:

	Ant. pos.	Transv.	Right obl.	Left obl.	R. s. c. d.	L. s. c. d.
Inlet of the cavity...	10·97 cm.	13·41 cm.	12·69 cm.	12·53 cm.	8·71 cm.	8·86 cm.
Widest part of the cavity..............	12·63	12·41				
Narrowest part of the cavity..............	11·39	10·37	...	...	...	...
Outlet of the cavity	...	11·07	...	...	...	...

Nine out of the fifty pelves had the following average measurements:

Inlet of the cavity...	11·07	13·57	12·85	12·81	8·92	9·02
Widest part of the cavity..............	12·88	12·63			...	
Narrowest part of the cavity..............	11·7	10·55	...	...	...	...
Outlet of the cavity	...	10·85	...	...	...	...

Schweighäuser has called attention to the fact that the right oblique diameter is considerably larger than the left, whilst with the sacro-cotyloid diameter the reverse is the case. This slight oblique deviation, which probably is a physiological one, may be easily explained by the greater use of the right lower extremity. The weight of the trunk falling especially upon that, the acetabulum is pressed towards that side of the promontory, whilst the right sacro-cotyloid and the left oblique diameters are somewhat shortened.

The oblique diameters of the pelvic cavity and outlet have been omitted, because they depend partly upon the thickness of the soft parts; they do not present any constant measurement.

Till the second half of the sixteenth century obstetricians were entirely ignorant of the conditions of the normal pelvis. The opinion generally prevailed that, during parturition, the joints of the pelvis, more especially that of the symphysis, were separated, and that thus the necessary room was obtained for the passage of the fœtus; also that every pelvis was, in itself, too narrow to allow delivery. Andreas Vesalius (1543) was the first who combated this opinion, and gave an excellent anatomical description of the normal pelvis. He was followed by his pupils Realdus Columbus (1559) and Julius Cæsar Aurantius (1587), whilst Ambrose Paré (1573), and especially Severinus Pinæus, revived the old teaching of the separation of the joint of the pubic bones during parturition. Heinrich van Deventer (1701), the great Dutch obstetrician, called especial attention to the great importance in midwifery of acquaintance with the pelvis, and commences his 'Neues Hebammenlicht' with the description of it. Whilst the Englishman Smellie (1751) exhaustively described the pelvis, and was the first to give its correct and accurate measurements, considering it as a whole from an obstetrical point of view, his great rival in France, Levret (1747), made statements which do not accord with the actual conditions. The Dutchman Johann Hurvé (1735) had, however, before this, referred to the measurements of the pelvis. G. W. Stein the elder (1770) followed his teacher Levret, but avoided his errors, and Stein the nephew (1803) described the pelvic spaces, to which description little can be added even at the present time. Luschka corrected the erroneous opinion about the articulation of the pelvic bones, which had until then been considered as constituted by solid cartilaginous wedges. He has shown, moreover, that the sacro-iliac sychondrosis and the pubic symphysis are really joints.

B.—THE PROPORTION EXISTING BETWEEN THE SINGLE DIAMETERS OF THE PELVIS

Since the bones of the pelvis, as well as those of the other parts of the body, present very considerable individual differences in shape and size, it is greatly desirable to have a relative method of mensuration, by which the proportion of one of the diameters to all the others can be shown. Thus, from the given measurements pelves of different sizes can be compared with one another with regard to their shape. The conjugate diameter, the one obstetrically of the greatest importance, is taken as the standard, and all the others reduced to that standard. Now, it is easy to calculate the relative value of the sizes from the given absolute measurements. In making these calculations the coccyx is to be disregarded, because, as a movable part, it does not give any constant measurements. The diameters of the pelvic cavity are taken in the widest parts (the distance between the ischial spines is especially to be measured). The antero-posterior diameter of the outlet will be the distance between the symphysis pubis and the inferior angle of the sacrum. The following numbers will then be obtained—the length of the conjugate being equal to 100:

	Ant. p. d.	Trans. d.	Obl. d.	Isch. sp.
Inlet	100·	122·7	115·9	...
Cavity ...	115·9	113·6	...	95·5
Outlet ...	104·5	100·	...	...

Litzmann, who first applied this method of mensuration, has

obtained somewhat different results. This is probably due to the fact that in the pelves measured by him in Kiel the transverse and oblique diameters greatly preponderated over the antero-posterior. The relative measurements given by him are—

	Ant. p. d.	Trans. d.	Obl. d.	Isch. sp.
Inlet	100·	129·2	120·	...
Cavity ...	119·	115·1	...	96·
Outlet ...	105·	115·4	...	...

C.—INCLINATION OF THE PELVIS AND DIRECTION OF ITS CAVITY

The inclination of the pelvis in the upright position varies, according to Hermann Meyer, even in the same individual, and this variation especially depends upon the degree of rotation or abduction of the thighs. It is the least (40—45°) in very slight rotation inwards, or in moderate abduction of the thighs. By close approximation of the knees, by extreme abduction, as well as by great rotation inwards and outwards, the inclination of the pelvis can be increased up to 100°. In the usual erect posture it is on an average 54·5°.

The dried pelvis is, according to the Brothers Weber, in its normal position in relation to the horizontal line when it is so held that its acetabular cavities look directly downwards; according to H. Meyer the anterior superior spines of the ilium and the pubic tuberosities ought to be in the same vertical plane.

The direction of the entire pelvic canal is best determined by a line (Fig. 1, *c*, *d*) which is supposed to connect the middle points of the various sections of the pelvic canal. This line is called the axis of the pelvis, or the "guiding line." It is invariable only to the apex of the sacrum; there the direction is changed owing to the mobility of the coccyx. Since the sacrum runs almost directly backwards to the third sacral vertebra, if we assume the symphysis also to be straight, the pelvic axis forms a straight line as far as that point. The axis makes an angle with the inlet of about 90°, and its prolongation upwards would approximately pass through the umbilicus; the further course of the pelvic axis, corresponding to the direction of the lower part of the pelvic cavity, is that of a curved line.

FIG. 1.

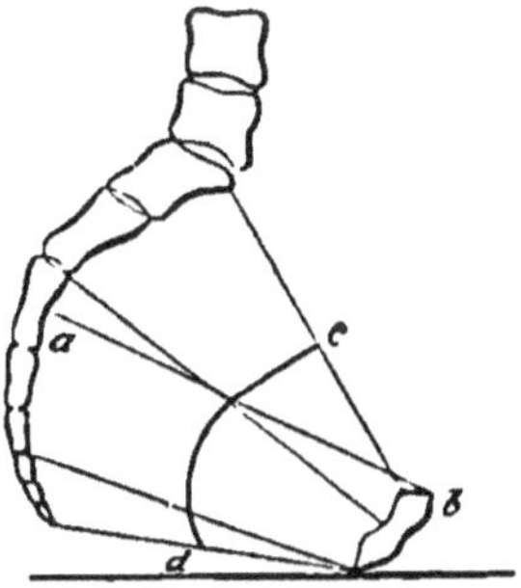

The angle which is made by the symphysis pubis with the conjugate diameter is of importance in the mechanism of parturition, as well as in the internal mensuration of the pelvis. In the normal pelvis it measures about 100°.

D.—DIFFERENCES BETWEEN THE MALE AND FEMALE PELVIS

Irrespective of the greater density and the more powerful development of the bones of the male in general, the female pelvis is especially characterised by the shallowness and width of its canal. The iliac bones are less perpendicular, the inlet more spacious, and the cavity funnel-shaped in the male, becomes wider in the female towards the outlet, partly on account of the greater recession of the sacrum and coccyx, partly on account of the greater distance between the ischial spines and the bending outwards of the lower edge of the rami of the pubic arch. The pubic arch, which in the male is an angle of 70—75°, is in the female one of 90—100°. The acetabula are farther apart, and are directed more forwards.

These differences of the female pelvis are caused by the development of the genital organs, which are situated within the true pelvis. By the growth of these the pelvis increases, especially in width. The outward contortion also of the pubic rami is due to the development of the corpora cavernosa, which extend along them.

As a proof that the due development of the pelvic cavity depends upon the female genital organs we may cite the instances of women with arrested mental and physical development; those with undeveloped genital organs always show a general narrowness of the pelvis. Roberts ('Journ. de l'Expérience,' 1843, No. 293, p. 99) observed in castrated females amongst the Hindoos uncommonly narrow pubic arches. On the other hand, a pelvis in the Anatomical Museum of Bonn, in which a double uterus had developed, shows a transverse diameter at the brim of 16 cm., or 6·3 inches.

E.—DIFFERENCES OF THE PELVIS IN INDIVIDUALS AND RACES

The individual differences of the female pelvis are very great. We rarely meet with pelves which, after a minute examination, can be pronounced perfectly regular. They are almost always asymmetrical.

The shape of the inlet furnishes the best means of recognising the differences. Stein junior and M. J. Weber distinguish four forms of the inlet:

1. That of the obtuse heart of the playing cards (Weber's oval and round oval).
2. That of an ellipse with a greater transverse diameter (this corresponds to Weber's third form, the quadriangular).
3. A rounded form (Weber's second original form).
4. That of an ellipse with a greater conjugate diameter (Weber's wedge-shaped pelvis).

The pelves of the different races are chiefly distinguished by the proportion between the conjugate and the transverse diameters. Corresponding to Stein's fourth form are the pelves of the Bushwomen and of the Malays; in these the conjugate is usually as large as, and only rarely larger than, the transverse diameter. The pelves of the aborigines of America and Australia are almost round, whilst the shape of the pelvis of the African negress resembles that of the Caucasian race. The pelvis of the latter is especially characterised by the greater dimensions of its cavity; the transverse diameter, especially, is very large. This is best exemplified in English women, and then, according to Litzmann's measurements, in the women of Holstein.

F.—THE PELVIS AT BIRTH AND THE CHANGES IT UNDERGOES

The pelvis of the new-born child differs essentially from that of the adult. In it the sexual difference is not marked. The breadth of the auricular surfaces of the sacrum is very small compared to that of the vertebræ. Its sacrum is almost quite straight (only at the fourth vertebra there is a slight curve); its anterior surface is more concave transversely, is inclined less forwards, and is less deeply pressed down between the iliac bones. The horizontal rami of the pubic bones are disproportionately short, and the pubic arch is an acute angle in both sexes. The iliac bones stand more upright, and the distance between the iliac spines is almost the same as that between the iliac crests. There is no transverse extension of the pelvis, and, consequently, at the inlet the conjugate is almost as large as the transverse diameter, exceptionally even larger. The walls of the true pelvis converge downwards, so that all the diameters, but especially the transverse, become smaller.

It is of great importance to be well acquainted with the causes by means of which the pelvis of the new-born child is converted into that of the adult with all the material differences of which we have spoken above. For the same causes, acting too powerfully on the pathologically softened bones of the pelvis, produce highly important pathological results in the shape of the pelvis. First of all, the shape of the pelvis is altered if all the parts do not grow in the same proportion. In the girl the sacrum, and proportionately the auricular surfaces (wings), grows the most in breadth, and the development of the pubic bones also considerably increases. In this way the roominess of the female pelvis becomes conspicuous.

The pressure of the weight of the trunk is the most important cause of the changes which take place up to puberty.

The old notion, that the sacrum is placed between the iliac bones like the keystone of a vault, is untenable, because the anterior (lower) surface of the sacrum is broader than the posterior. The very strong ilio-sacral ligaments fix the sacrum in its place between the iliac bones. By the pressure of the weight of the trunk the sacrum is forced deeper down into the pelvis; but since the centre of gravity of the trunk falls in front of the *point d'appui* of the sacrum, this turns so around its axis that the promontory descends deeper into the pelvis, and the inferior sacral angle, were the shape of the bone to remain unaltered, would look backwards. The lower portion of the sacrum, however, being fixed by the sacro-iliac and the sacro-sciatic ligaments, its deviation backwards is impeded, and the sacrum forms a curve from above downwards, most distinct at the third sacral vertebra. The transverse concavity is, however, diminished, because the pressure of the trunk forces the body of the bone, with which it is still connected by cartilages, somewhat out of its lateral position. At the same time each body of the sacral vertebræ is more compressed behind, so that its anterior surface is higher than the posterior.

The more deeply the upper portion of the sacrum descends into the pelvis, the greater the traction becomes which the sacro-iliac ligaments exert on the posterior superior spinal processes of the ilia. The iliac bones approach each other, so that they would gape at the symphysis if that were divided and the resistance absent on the part of the acetabula. But since the iliac bones are closely connected to each other, and since in the acetabula a counter-pressure is exerted against the weight of the trunk, there are two forces to act on the ends of the iliac bones: behind, the tension of the sacro-iliac ligaments; in front, the symphysis and the pressure of the femora; and therefore the still flexible bones are bent over their place of articulation with the sacrum and at the weakest spot, which is near to the auricular surface of the sacrum. In this way the greater transverse extension of the pelvis arises, and, therefore, the transverse diameters of the adult female pelvis greatly preponderate. In the accompanying woodcut (Fig. 2) the effect of the

FIG. 2.

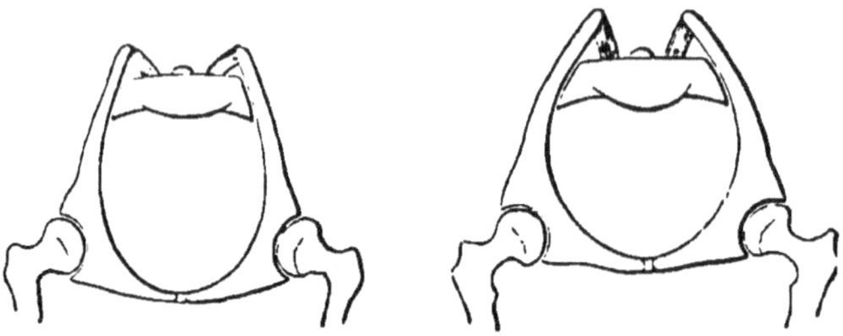

pressure of the trunk on the sacrum and the shape of the pelvis resulting from it are diagramatically represented.

That the pressure of the trunk, even without the counter-pressure of the femora, by the traction which each side exerts on the other through the symphysis, suffices to produce the transverse extension of the pelvis, is shown by the interesting pelvis of Eva Lank, forty years of age, examined by Holst and figured by Förster. She had no lower extremities, and there were also no acetabula to the pelvis; but since she could stand, or rather sit on the pelvis, the pressure of the trunk was sufficient to produce the usual changes in the pelvis. The true pelvis is of normal size and somewhat flattened. This *amelus* shows that the pelvic outlet is at least widened by sitting when through the normal development of the genital organs the pubic arch itself is not too narrow. The transverse diameter of the outlet measured 15 cm., or 5·9 inches, and the antero-posterior to the coccyx, which was attached by bone to the sacrum and directed backwards, measured 13½ cm., or 5·3 inches. The iliac bones were so bent over their axis from behind forwards that the distance between the crests of the ilium was only 20¼ cm., or 7·9 inches.

G.—THE TRUE PELVIS, WITH ITS SOFT PARTS

The shape of the bony pelvis is greatly altered by the soft parts situated within and around it.

Close to the inlet the powerful belly of the ilio-psoas muscle descends on each side of the promontory over the sacro-iliac synchondrosis and between the anterior superior spine of the ilium and the ilio-pubic tubercle to be inserted into the small trochanter of the femur.

In the pelvic cavity the large notches between the sacrum and the ischium are divided by the sciatic ligaments into an upper, larger, and roundish, and into a lower, smaller, and triangular space. The upper opening is filled up by the pyriform muscle, which takes its origin from the sacrum; the lower partly by the obturator muscles, which at their origin cover the obturator foramen.

It is at the outlet that the soft parts produce the greatest changes. By the coccygei, levatores ani, and transversales perinæi muscles imbedded in the manifold layers of the strong pelvic fascia forming the elastic floor of the pelvis, the outlet is so altered that it looks downwards and forwards, instead of downwards and backwards as in the bony pelvis.

It will be seen that the conjugate diameter, the one most important from an obstetric point of view, is least of all altered by the soft parts. But here also the proportion of the fœtal head to the measurements obtained from a dried pelvis does not perfectly correspond with the conditions in the living subject. The peritoneum, with the subjacent connective tissue, passes in front of the sacral promontory down into the pelvis, and out of it over the posterior surface of the uterus; behind the anterior wall of the pelvis also the urethra and bladder, surrounded by connective tissue, are situated. The head also enters the conjugate almost always covered by the uterus; and the double thickness of the uterine walls has to be added, and so makes the proportion between the fœtal head and the antero-posterior diameter of the pelvis still more unfavorable.

These conditions ought in every case to be taken into consideration before forming an estimate of the size of the pelvis of the living woman.

Literature.—Andr. Vesalii, Bruxell. de hum. corp. fabr. libr. septem. Basil, 1543.—Realdi Columbi, Crem. in almo Gymn. Rom. anat. celeb. de re anatom., libri XV. Venet., 1559.—Henr. a Deventer, operat. chir. nov. lumen. exh. obstetr. Ludg. Bat., 1701.—W. Smellie, A Treatise on the Theory and Practice of Midwifery. London, 1752.—G. W. Stein d. j., Lehre der Geburtshülfe. Th. 1. Elberfeld, 1825.—Schwegel, Monatsschr. f. Geb. u. Fr., B. 18, Suppl., p. 67.—H. Luschka, die Anatomie des menschlichen Beckens. Tübingen, 1864.—G. W. Steine d. ä., Theoretische Anl. zur Geburtsh. 1770. Kap. 2.—v. Ritgen, Gemeins. deutsche *Z.* f. G., B. I, H. 1, p. 17.—C. C. Th. Litzmann, die Formen des Beckens, &c. Berlin, 1861. F. C. Naegele, das weibliche Becken. Carlsruhe, 1825.—W. u. E. Weber, Mechanik der menschl. Gehwerkzeuge. Göttingen, 1836.—Krause, Handbuch der menschl. Anatomie, 2 Aufl., Hannover, 1841, p. 327.—H. Meyer, Archiv für Anat. und Phys., 1861, pp. 137—178.—Hegar, Arch. f. Gynaek, Bd. I, p. 193. G. W. Stein d. j., Neue Zeitschr. f. Geb., B. 12, p. 345.—Litzmann, die Formen des Beckens, § 5. G. W. Stein, Lehre der Geburtshülfe, Th. 1, § 53 und Taf. 1.—M. J. Weber, die Lehre von den Ur- und Raçenformen der Schädel und Becken der Menschen. Düsseld., 1830.—H. F. Kilian, die Geburt des Kindeskopfes, Bonn, 1830, p. 60 seq.—G. Vrolik s. Froriep's geb. Dem., Heft. VII, Taf. 27—30.—Joulin, Mém. sur le bassin, &c. Archives génér., 1864, II, p. 5.—C. Martin, M. f., G. B. 28, p. 23.—O. v. Franque, Scanzoni's Beiträge, B. VI, p. 163. De Fremery, Diss. i. de mutat. fig. pelvis, &c. Lugd. Batav., 1793.—Freund, M. F. G. B. 13, p. 202.—C. Th. Litzmann, die Formen des Beckens, § 5.—M. Duncan, Researches in Obstetrics, Edinburgh, 1868, p. 78 seq. and p. 95 seq.

PART II

THE PHYSIOLOGY OF PREGNANCY

CHAPTER I

THE OVUM AND ITS DEVELOPMENT

A.—OVULATION AND FECUNDATION OF THE OVUM

Of the thousands of ova which are found imbedded within the Graafian follicles in the new-born female, comparatively few come to be expelled from the ovary. As the openings of the tubular glands, from which the Graafian follicles are developed by constriction of a portion of the gland, are situated towards the periphery of the ovary, the most developed follicles of the new-born female will be found more towards the centre, and this condition will remain the same up to puberty. Not until the other organs of the body have almost reached the highest degree of maturity does a new phase of development begin in the organs of generation. This is seen in the ovaries by the enlargement of the Graafian follicles. According to Pflüger their slow but continued growth causes a constant irritation of the terminations of the nerves which are imbedded in the rigid stroma. This irritation, however, is so slight that it is not sufficient to set up immediately a reflex action, but at periodic intervals the sum of those irritations becomes so great that reflex action takes place in the form of a considerable arterial congestion of the genital organs. This suddenly increased afflux of blood has a double effect. Not only is that Graafian follicle which is the farthest advanced in development ruptured by the increased intrafollicular pressure, but from the vessels of the mucous membrane also hæmorrhage takes place upon the free surface of the uterus. The escape of the ovum from the follicle and the menstrual hæmorrhage are both consequences of one and the same cause, namely, of the pressure which the growing follicle exerts upon the terminations of the nerves contained in the ovarian stroma. It is therefore this pressure which causes the periodical reflex action—congestion of the genital organs.

Growth, rupture, and involution of the Graafian follicle take place in the following way:

The Graafian follicle, by its gradual growth, advances from the centre to the periphery of the ovary. At this time it consists of a distinct connective tissue covering—the *theca folliculi*,—and its inner surface has a stratified epithelium—the *membrana granulosa*. The epithelial cells collect at one point (more towards the centre of the ovary, according to Schrön, 'Zeitschrift für Wissensch. Zoologie,' vol. xii) in a greater number, and surround the ovum. The rest of the cavity of the follicle is filled with a serous fluid—*liquor folliculi*. The intrafollicular pressure, increasing with each menstrual hyperæmia, makes thinner and thinner the weakest point of the follicular sheath, that is, the one formed by the peripheral layers of the ovary, until it finally ruptures during a menstrual hyperæmia, and the contents of the follicle are poured out into the abdominal cavity.

The gradual thinning and final rupture of the follicular wall very easily take place, since, as was first shown by Pflüger ('Ueber die Eierstöcke der Säugethiere und des Menschen,' Leipzig, 1863, p. 31, plate ii, fig. 10), the so-called tunica albuginea of the ovary is no membrane proper, but only the external layer of the stroma. Also the surface of the ovary is not covered by a perfect serous membrane, but only by a single layer of epithelium. Waldeyer ('Sitzungsb. d. schl. Ges. für vaterl. Cultur,' 11th October, 1867, and 'Eierstock und Ei.,' Leipzig, 1870, p. 5) and Küster ('Centralblatt für die Med. Wissensch.,' 1868, No. 49) positively state that the peritoneum does not cover the ovaries, but that they project free into the abdominal cavity through an opening in the peritoneum, and are covered only by the epithelium of a mucous membrane, the prolongation of which forms the epithelial lining of the tubular glands of Pflüger.

The change occurring in a Graafian follicle after its rupture (*vide* Spielberg, 'M. f. G.,' B. 26, p. 7; His, 'Schultze's Archiv,' B. 1, p. 181: Waldeyer, loc. cit., p. 94; and Starjansky, 'Virchow's Archiv,' 1870, B. 51, p. 486) varies according to whether the ovum is fecundated and continues to develop, or remains sterile and perishes. In the first case the formation of a true corpus luteum takes place, according to Waldeyer, in the following way:—As a sequence of the general nutritive irritation of the organs of generation, the cells of the membrana granulosa greatly proliferate, and produce by their simultaneous breaking down a yellow, granular yelk mass. At the same time proliferation begins in the walls of the follicle, so that new vessels are pushed forwards into the cavity of the follicle, which, surrounded by migratory colourless blood-corpuscles, enter the yelk mass. The larger vessels form greater projections, and thus cause the pretty regular folds of the corpus luteum. In the interior of the follicle at times, but by no means constantly, a small quantity of blood is poured out, which passes through the usual changes. In the third and fourth months of pregnancy the true corpus luteum has reached its highest stage of development; it is then red, and looks somewhat like muscle. From this time the retrogressive metamorphosis begins. As the circulation begins to slacken in the abundant narrow capillaries, the cells are only imperfectly nourished; they consequently undergo fatty degeneration, and thus produce the yellow colour of the follicular contents. The fat is reabsorbed and the connective-tissue stroma of the granulations undergoes cicatricial contraction. Finally, all that remains of the corpus luteum is a slight depression on the surface of the ovary, a real cicatrix, the pigmentation of which is due to the blood stagnating in the vessels.

But if the ovum is not impregnated it perishes, and a so-called false corpus luteum is formed. Here the process of granulation is much less active, and the new-formed cells early undergo the fatty degeneration.

The ovum lies, as was stated above, imbedded in an accumulation of the cells of the membrana granulosa—the so-called *cumulus proligerus.* It has a diameter of one eighth to one tenth of a line, and to the naked eye is only visible as a very small white dot. Under the microscope it is seen to consist of the following parts:

The membrane appears as a clear, rather thick ring—*the zona pellucida.* The cell contents—*yelk, vitellus*—is a finely granular but turbid plasma, and lying excentrically in it is the nucleus of the ovum-cell—*the germinal vesicle, vesicula germinativa*—together with the nucleolus, *the germinal spot—macula germinativa.*

By the rupture of the Graafian follicle the ovum, together with the adhering cells of the membrana granulosa, escapes to the free surface of the ovary, that is, into the peritoneal cavity. The reception of the ovum by the Fallopian tube takes place in the following way:

The cilia of the Fallopian tube, vibrating in a direction towards the uterus, cause a continuous current in the fluid which is always present upon the peritoneum in the neighbourhood of the abdominal opening of the Fallopian tubes. This current is able to carry along with it very small substances, such as the ovum, and in this way the ovum generally enters the wide abdominal opening of the oviduct of its own side.

The further advance of the ovum in the external third of the oviduct is solely caused by the vibrations of the cilia, because the tube is here far too wide for the contraction of its muscular elements to act upon the small ovum. But in the middle portion the oviduct becomes so narrow that the contraction of the layers of its circular fibres can grasp and push it forward. It travels along the tube very slowly (according to Bischoff it requires eight to ten days in the dog, in man certainly a much longer time). Should the ovum not be fecundated during its passage into the uterus, it perishes there without any further development.

If the ovum is fecundated it adheres to the uterine mucous membrane, which had been wounded by the menstruation and undergoes a series of very important changes. The act of fecundation, that is, the entrance of the spermatic filaments into the ovum, takes place, in the great majority of cases, undoubtedly within the Fallopian tubes, rarely in the ovary, or even in the uterine cavity. A micropyle, which has been met with for the entrance of the spermatic filaments into the ova of many of the invertebrata and of some fishes, has not yet been observed in the ova of mammalia. Pflüger, however, has seen, in the ovum of the cat, a number of the cells of the membrana granulosa sticking in the zona pellucida like a peg, so that after their expulsion an opening must arise in it. As far as we know at present, the spermatic filaments penetrate the zona pellucida and are lost in it. The disappearance of the germinal vesicle and the germinal spot is the first sign that impregnation has taken place. At the same time the yelk somewhat contracts, and in its interior a nucleus and nucleolus are seen. This is the first segmentation sphere, from which the others proceed by subdivision.

The mode of origin of the embryo and the development of the individual organs belong to the subject of embryology. But the knowledge of the development of the membranes and of the conditions of the embryo in each month of pregnancy is of such immediate practical importance to the obstetrician that it cannot be omitted here.

B.—THE MEMBRANES OF THE HUMAN FŒTUS

The fœtus in utero is enveloped in three membranes, which not only vary in their mode of origin, but which also remain distinct throughout the whole time of the intra-uterine life of the fœtus.

The external of the three, the membrana decidua, is supplied by the maternal parts, whilst the other two, the chorion and amnion, spring from the ovum itself.

1. *The Decidua*

When the fecundated ovum has entered the uterus it becomes inoculated in the wounded uterine mucous membrane, as Pflüger calls it. Whilst with an unimpregnated ovum the catamenial hyperæmia of the genital organs rapidly passes off, the fecundated ovum produces an intense irritation of the uterus. Its mucous membrane is thickened by an enormous proliferation and entirely covers the still very small ovum, so that it is quite embedded in the uterine mucous membrane. The growing mucous membrane itself is the decidua, and it is distinguished into the decidua vera, that portion which continuously lines the inner surface of the uterus, and into the decidua reflexa, that portion which has covered the ovum. The portion of the decidua vera upon which the ovum rests is called the decidua serotina.

According to Friedländer's researches, the formation of the decidua gives the following microscopic appearances:

In the earliest period of pregnancy the decidua consists, in its internal portions towards the ovum, of the greatly proliferating connective tissue of the mucous membrane. This layer, formed by the greatly enlarged and closely packed connective-tissue-cells, is perforated by the directly ascending ends of the uterine glands. These are more developed in the external portions of the decidua. The epithelium of the glands, which, according to Friedländer and Lott (Rollet, 'Unters. a. d. Phys. Instit. in George II,' Leipsig, 1871, p. 520), is originally ciliated, is even now distinctly cylindrical. Later on, two distinctly separated layers can be recognised. The inner, the cellular layer, consists towards the chorion of large round cells separated from each other by a scanty intercellular substance, which become partly spindle-shaped exteriorly. These decidua-cells are derived from the first-mentioned proliferating connective-tissue-cells (Friedländer, Hegar, Maier). The second layer, limited by the muscular coat of the uterus, consists of the flattened, compressed, and distorted gland-ducts, with one layer of epithelium. As long as the ovum remains small a little mucus separates the free surface of the decidua reflexa from the decidua vera. By the increasing growth of the ovum the vera and reflexa come into the most intimate connection. The reflexa undergoes very material changes, due to the great extension to which that small piece of mucous membrane is subjected. Externally (towards the uterus) it is smooth and without epithelium. In the third month, when it is still one third to one half of a line thick, it is entirely devoid of vessels, but it contains large, round, and spindle-shaped decidua-cells, like those of the vera, only that those of the decidua reflexa are much earlier filled up by fine molecules, and at the end of pregnancy they have almost entirely undergone fatty degeneration. From the sixth month both decidua form one thin membrane, and can only be separated at a few places for a short distance.

The thickness of the decidua of the expelled ovum varies greatly. It depends partly upon the great difference in the proliferation of the mucous membrane and partly upon how much of it has been retained within the uterus. Sometimes the ovum is expelled covered almost exclusively by the very thin decidua reflexa, so that nearly the whole decidua vera has been retained. More frequently thicker pieces of the decidua vera are found attached to many places on the periphery of the ovum, whilst at others it is entirely wanting. At any rate the muscular coat of the uterus is never denuded by the expulsion of the ovum, but a portion of Friedländer's cellular layer and the whole glandular layer remain within the uterus. The thickest part of the decidua is in the vicinity of the placenta.

The changes of the decidua serotina will be mentioned in the description of the chorion.

Very excellent and natural drawings of the development of the decidua are given by William Hunter in his above-mentioned large work (Plate 34), which also from an artistic point of view must be considered as a masterpiece of the first rank. Hunter was unhappily not permitted to write the work explanatory of his plates. His brother John considered the decidua to be coagulated lymph exuded from the uterus, and from him and Matthew Baillie, the editor of Hunter's posthumous works, that opinion is derived which has lasted until very recently. Generally the decidua was considered an exudation from the uterus, covering externally its inner surface and also that of the openings of the Fallopian tubes. Starting from that supposition, it was thought that the ovum, escaping from the tube, pushed the membrane in front of it, and that by its growth the membrane was gradually thinned out. The invested portion of the membrane was therefore called decidua reflexa. If the ovum pushed the membrane in front of it the membrane ought to be absent from the place of its insertion. But as here, also, a decidua was found, it was supposed that it formed later, and was therefore called decidua serotina. E. H. Weber ("Zusatze zur lehre vom Bau u. v. d. Ver. der Geschlechtsorgane" in 'Abh. d. k. sächsisch Ak.,' 1846) and Sharpey (in the Eng. trans. of Müller's

'Physiology') found in the decidua reflexa the same glandular openings as in the vera, and concluded therefrom that the reflexa must also belong to the uterine mucous membrane: Sharpey then supposed that the ovum is embedded in a fold of the proliferating mucous membrane of the uterus, the decidua vera, and that the proliferation of this completely covers the ovum and closes over it. This view has lately been confirmed by the researches of Coste. As the ovum, when it enters the uterus, is at most one eighth of a line in diameter, such a proliferation may, of course, very easily take place.

2. *The Chorion*

From the zona pellucida of animals, and probably also of man, a primitive chorion is developed, with small structureless villi. This covering disappears very early, and is replaced by the true chorion. The latter consists of two parts, viz. an external epithelial layer, the *exochorion*, derived from the serous coat of the ovum, and a subjacent vascular layer of connective tissue, the *endochorion*, derived from the allantois. Around the whole circumference of the ovum villous proliferations spring up; at first these are solid, afterwards they become hollow. At the end of the third or the commencement of the fourth week, the connective-tissue layer, together with the vessels of the allantois, extends into them. Whilst the villi of the circumference of the ovum corresponding to the decidua reflexa (which never extend through the reflexa into the vera) do not develop any further and their vessels atrophy, those at the periphery, the part corresponding to the decidua serotina, continue to develop and ramify. The arrangement of the vessels is such that a branch of the umbilical artery enters each villus, and there breaks up into a capillary network, the blood of which passes from the villus by the vein. Accordingly, from the end of the second month the chorion is clearly separated into two parts; the one—chorion læve—forms a thin connective-tissue membrane, connected with the decidua reflexa by small, atrophied, non-vascular villi, the other—chorion frondosum—forms the placenta, a thick layer of exceedingly branched villi, provided with large vessels. This great proliferation is due to the epithelium of the existing villi continually forming new excrescences, into which the connective-tissue layer with its vessels enters. Thus at the end of pregnancy the placenta is, so to speak, nothing else but a thick felt of chorion villi. The decidua serotina, that piece of the uterine mucous membrane into which the villi proliferate, does by no means disappear. It has the same elements as the vera, namely, the glandular layer close to the muscular coat of the uterus, and the superficial cellular layer. The latter, the proliferations of which are even extended like a wedge between the cotyledons of the placenta (although never reaching up to the origin of the chorion villi), adheres partly to the expelled afterbirth. It can always be seen as a thin membrane, distinguished by its greyish-white colour from the red villi of the chorion, but can be separated from it only with great difficulty in small pieces. The deeper portion of the cellular layer and the whole glandular layer remain within the uterus. Besides the cells mentioned in the description of the decidua vera there are the so-called giant cells in the cellular layer of the serotina, containing many nuclei (twenty or more). They are seen also, but in small number, in the adjoining decidua vera. The serotina is perforated by the large maternal vessels, which lose in it their walls, with the exception of the very thin endothelium. The afferent vessels bring the blood through the serotina into large cavernous spaces, which are not lined and are situated between the villi of the chorion, and the veins take it back again. Therefore, in the fœtal placenta the blood is found circulating perfectly freely within the large cavernous spaces between the ramifications of the villi of the chorion. They are surrounded by the maternal blood, and this is separated from the fœtal blood only by the layer of epithelium and connective tissue of the fine villi. The maternal blood returns from the placenta partly by veins of the decidua serotina, which run into the deeper uterine veins, but chiefly by the so-called "marginal" vessel. This passes around the periphery of the placenta, and is a sinus with distinct walls; it belongs to the decidua vera, from which this blood, received into it by many radicles from the placenta, again flows by numerous canals into the veins of the deeper portion of the decidua vera and to those of the muscular walls of the uterus. It rarely surrounds the placenta as a complete circle, and is frequently interrupted in one or more places.

The minute structure of the placenta, especially the relation of the villi of the chorion to the maternal placenta, has not yet been quite satisfactorily ascertained. According to the researches of Jassinsky, the above-described conditions are met with in part, but in the maternal portion of the placenta villi are found with a double layer of epithelium. The external layer, as well as the single, thick, and *knobby* villi, filled with detritus, into which no blood-vessels enter, is considered by Jassinsky the remains of the uterine glands, into which the villi of the chorion had partly grown. Ercolani is of opinion that in the human female a new glandular organ develops during gestation from the decidua serotina; that this extends itself between the villi of the chorion and surrounds all of them with a sheath, consisting of a structureless membrane and one layer of epithelium. This sheath lines also the maternal sinus, and the villi of the chorion do not dip immediately into the maternal blood, but are in contact only with a liquid (uterine milk) secreted by the cells of the glandular organ which serves for the nutrition of the embryo. According to Friedländer's researches above quoted, the villi of the chorion only penetrate to the cellular layer, whilst beneath it there is still the glandular layer.

Langhans observed that some villi of the chorion, branches about 1 mm. thick, as well as fine terminal filaments, by losing their epithelium, and terminating with a budlike enlargement, become so firmly connected with the maternal placenta that the connection cannot be loosened by powerful traction, but that the maternal structure always tears.

The formation of the placenta varies in the different orders of Mammalia. In Quadrumana, Carnivora, and Rodentia, the relation is similar to that in man, that is, the villi of the chorion grow at the placental insertion into the uterine mucous membrane, and this connection is so intimate that at least a part of the mucous membrane is expelled with the ovum; the decidua reflexa belongs exclusively to man.

3. *The Amnion*

The embryo at first lies flat upon the yelk, but as soon as its ventral walls begin to form it is placed in a depression of the embryonic spot. The external layer of the embryo is slightly curved in its whole extent, so that around the embryo a wall-like elevation is seen, which, according to the position, is distinguished into a cephalic, a caudal, and a lateral sheath. Now, this elevation grows by simple cellular hyperplasia towards an ideal point situated above the back of the embryo, and when union has occurred the cephalic, the caudal, and the lateral sheaths having united with one another, the embryo lies in a sac, which, commencing at the widely gaping abdominal walls, is closed above the back of the embryo. This sac—*the amnion*—develops in man not only from the epidermic, but also from the cutaneous layer, so that it forms a prolongation of the whole skin, and especially of the skin of the abdomen of the embryo. Very soon the amnion entirely separates from the external membrane, the serous coat with which it was originally connected at the place of union. After its formation into a sac the amnion lies close to the back of the embryo, but gradually, depending upon the closure of the abdominal cavity, it surrounds a larger part of the embryo. By the secretion of the liquor amnii the space between the embryo and the amnion also enlarges, and after the formation of the umbilicus the whole embryo lies enclosed in the amnion. Thus the abdominal walls of the embryo form immediately the umbilicus, and thence they again enter into the formation of the amnion enclosing the embryo. At all times the amnion is non-vascular, and consists of one layer of pavement epithelium situated towards the cavity corresponding with the epidermis, and of an external, striped, fibrous basement substance in which are found both spindle-shaped cells and others ramified and starlike with a long nucleus. This connective-tissue layer enters into Wharton's jelly of the umbilical cord and corresponds with the cutis of the abdomen of the fœtus.

In the middle of pregnancy the amnion is already in close apposition with the chorion, but even in the mature ovum it may easily be separated. Between the two membranes there is a very small quantity of a gelatinous tissue. This is the remains of an albuminous liquid, which, when the chorion and uterus were not in apposition, existed between the two membranes (tunica media of Bischoff).

Two important structures of the human ovum are still to be mentioned, viz. the vitelline sac and the allantois.

4. *The Vitelline Sac or the Umbilical Vesicle*

Before the abdominal cavity begins to develop, the inner germinal layer forms a large cavity, which, as soon as the abdominal plates grow towards each other, is divided into two large portions. That portion of the germinal layer situated within the abdomen of the embryo develops into the intestines, and the other outside the abdomen becomes the vitelline sac. Just as the intestines are derived from two layers, namely, the inner germinal or intestinal glandular layer and the intestinal fibrous layer, the inferior lamella of the divided lateral plates, so also the vitelline sac is formed of an internal epithelial and an external connective-tissue layer, which contains the omphalo-mesenteric vessels of the first fœtal vascular system. The more the abdominal cavity closes the more the intestinal canal and vitelline sac are separated from each other, and their communication becomes restricted. But the vitelline vesicle does not grow, its vessels atrophy, and finally the intestinal canal is connected only by a narrow duct, which later also atrophies—ductus omphalo-entericus—with the atrophied vitelline vesicle. According to Schultze, this can almost always be seen (140 times in 150 cases) in the mature human afterbirth, as a whitish vesicle situated between the amnion and the chorion, to which sometimes a white cord derived from the funis goes, namely, the ductus omphalo-entericus.

5. *The Allantois*

Before the amniotic cavity has perfectly closed, two slight elevations are formed at the margin of the entrance to the cavity containing the pelvic intestines. These two elevations, solid originally, soon become blended into one, and a cavity forms in it. The vesicle thus formed is provided with a pedicle, and is very early supplied with vessels from the ends of both primitive aortæ. The vesicle connected by the hollow pedicle with the intestine passes between the amnion and the vitelline sac to the periphery of the ovum, and supplies the epithelial layer of the chorion with a connective tissue, a vascular substratum. In describing the chorion it was stated that the vessels of the decidua reflexa atrophy, except at the place for the placenta where they continue to develop abundantly, and become the agents for the exchange of blood between the mother and the fœtus. The allantois, therefore, has the important function of carrying the two umbilical arteries to the periphery of the ovum, to allow of the formation of the placenta. The part of the allantois enclosed in the abdominal cavity then becomes the urinary bladder and the urachus, which is afterwards obliterated, and forms the middle ligament of the bladder.

In the accompanying diagrams we give a concise graphic description of the gradual process of the formation of the membranes of the ovum.

FIG. 3.

FIG. 4.

Fig. 5.

Fig. 6.

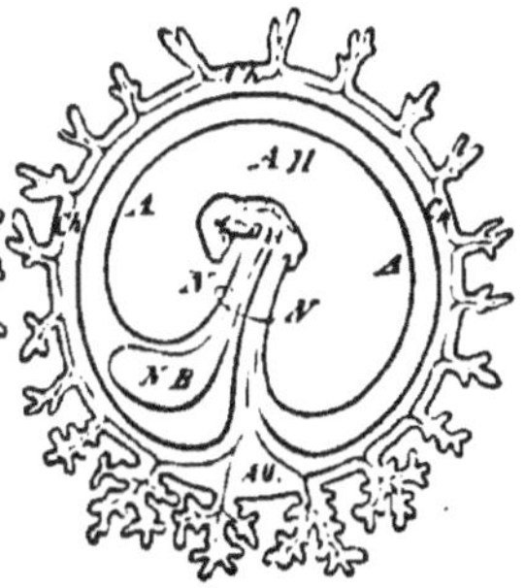

The letter A = amnion.
„ A H = amniotic cavity.
„ N B = umbilical vesicle.
„ D H = intestinal cavity.
„ All = allantois.
„ Ch = chorion.
„ N = umbilical cord.

In Fig. 3 the folds of the amnion are represented growing to meet one another.

In Fig. 4 they have met, and the allantois buds out. In both figures the primitive chorion, with the structureless villi, which afterwards disappear, are still visible.

In Fig. 5 the amnion is already separated from the serous layer, the epithelial layer of the permanent chorion, from which hollow villi develop, and the allantois has reached the periphery.

In Fig. 6 the vessels of the allantois have grown around the whole periphery of the ovum, and extend into all the villi of the chorion. They are, however, more developed at the future position of the placenta, whilst the vessels in the other places atrophy. The umbilical cord, scarcely indicated in Fig. 5, is in Fig. 6 already distinctly recognisable.

If we finally consider the ovum and all its membranes, as seen at the termination of pregnancy, we shall find at the periphery of the ovum the maternal membranes, which at the placental insertion consist of the decidua serotina, and at other parts varying in thickness, of a membrane formed by the blending of the decidua vera and reflexa. With the latter the chorion is connected by scanty delicate non-vascular villi, whilst at the place of attachment of the ovum the chorion forms, by its enormous proliferation, almost the whole mass of the placenta. Beneath the chorion lies the amnion, which can be easily separated. It is the innermost membrane of the ovum which, commencing at the abdominal walls of the fœtus, passes over the greatly developed and spirally twisted umbilical cord, and thence over the back of the child, where it unites and forms a closed sac. Between the chorion and the amnion is found a thin layer of albuminous substance without organization; this is the remains of a large quantity of serum, which formerly separated the two membranes.

Within the cavity of the amnion the fœtus is surrounded by the so-called 'embryo water' (liquor amnii). This is a serous fluid, of a slightly alkaline reaction, and contains the thrown-off epidermic scales and the lanugo of the fœtus. The liquor amnii is of very varying specific gravity, 1002—1028, contains some albumen (more so in the first months than afterwards), various salts, urea, and kreatinin. Urea, however, is not constantly present, at least in the early period of pregnancy. The quantity of the liquor amnii also greatly varies within physiological limits. At the end of gestation it is on an average from one to two pounds. In rare cases the liquor amnii is discoloured and fetid though the fœtus be perfectly healthy.

A very interesting question, and one which has received very different answers, is the following—Is the liquor amnii immediately derived from the mother, or is it a secretion of the fœtus? Many a pathological experience would go to prove that from the maternal vessels liquid may transude into the cavity of the amnion. There are cases where the embryo has very early atrophied, or has entirely disappeared, and nevertheless a quantity of liquid, corresponding to the age of the ovum (not of the fœtus), has been found. There are also diseases of the mother which, leading to serous transudations in other parts of the body, are often attended by hydro-amnion, that is, an abnormal increase of liquor amnii. Sometimes even in a stunted non-dropsical fœtus a greatly hypertrophied placenta and hydro-amnion are found. From these facts it is certain that the liquid in the ovum can be derived from the mother. On the other hand, the lanugo and the epidermic scales show that there are substances in the liquor amnii which are derived from the fœtus. From the regular occurrence of urea and other nitrogenous products of tissue metamorphosis (after the first months), it is probable that the excretion of the kidneys is emptied into the liquor amnii. It is also proved by the fact that in occlusion of the excretory ducts of the urinary apparatus the excretion may accumulate in the parts of the urinary system, and lead to considerable impediment in parturition, or even to rupture. It is not likely that the fœtus excretes any considerable quantity of fluid by its skin, although at an earlier period it is very easily permeable, and from the great quantity of water in the liquor amnii the contrary would be rather expected. Normally the excretions of the intestinal canal are absent from the liquor amnii. The regular occurrence of epidermic scales and of lanugo in the intestinal canal would show that the fœtus swallows liquor amnii, though it does not serve as nutrition.

The placenta is a spongy body, about one inch thick and more than six inches in diameter, weighing more than one pound. The convex portion is separated by deep furrows into single lobes, the so-called cotyledons. This is due to the villi of the chorion not everywhere adhering with the same degree of firmness. The uterine surface of the placenta is covered by a whitish-grey membrane, the maternal placenta, the upper portion of the decidua serotina, which sends wedge-shaped processes between the individual cotyledons. Its fœtal surface is covered by the amnion, and therefore smooth; beneath the amnion are the ramifying vessels of the umbilical cord.

In the majority of cases the placenta is situated on the anterior or posterior wall. If, as an exception, it is placed on one of the sides, it occupies more frequently the right than the left.

The umbilical cord varies in length; it is on an average about fifty cm., or twenty inches, of the thickness of the little finger, and almost always twisted spirally, the turns of which are most frequently (beginning at the fœtus) to the left, rarely to the right. This regular spiral twisting is probably due to the unequal length of the two arteries—v. Andreæ, 'Nabelschnurwindung,' &c., D. i, Königsberg, 1870. The umbilical cord is enclosed by a sheath of the amnion, and contains, besides Wharton's jelly, a gelatinous, embryonal connective tissue, two umbilical arteries going to the placenta and one umbilical vein coming from it. The vessels sometimes form short coils, which, together with an accumulation of jelly, constitute the so-called false knots. Real knots occur very rarely, and are caused by the fœtus passing through a loop of the umbilical cord, which is gradually tightened.

The insertion of the umbilical cord into the placenta is rarely exactly, but most frequently approximately, in the middle of the placenta—*insertio centralis;* sometimes, however, it is inserted close to the margin—*insertio marginalis.* In some cases it is inserted somewhat distant from the edge of the placenta into the membranes themselves—*insertio velamentosa.*

C.—THE FŒTUS

1. *The Fœtus in each Month of Gestation*

The duration of pregnancy in the human female is, as a rule, from 270 to 280 days. This time is most suitably divided into ten so-called gestation months, of four weeks each. As it is of great importance that an obstetrician should be able to determine the age of an expelled fœtus from its appearance, we shall consider its development progressively during each month of gestation.

First month.—The youngest human ova which have been subjected to careful examination were of the second week. Two only have been described by Thomson, one of which was not quite normal, as it, no doubt as frequently occurs, had continued to grow after the death of the contained fœtus. The other, which Thomson estimated to be twelve or thirteen days old, had a length of three lines; the embryo in it was only one line long. Probably the amnion had already formed, since the back of the embryo was attached to the external membrane, but the allantois was still absent. Coste has very accurately described an ovum of the third week, which was six lines long. The intestinal cavity of the embryo, which had a length of two lines, communicated by a very large opening with the vitelline sac, and an umbilical cord did not yet exist. The nutrition of the ovum was evidently still maintained by the vitelline sac, and this was provided with two arteries and two veins. The allantois had already brought the vessels to the periphery of the ovum, but these had not yet spread into the hollow villi of the serous coat. Similar embryos, with the commencement of a very short and broad umbilical cord, have been described by John Müller and R. Wagner.

A great many ova at the end of the first month have been described. The ovum of that period is about the size of a pigeon's egg; the embryo itself is 1 cm., or $\frac{1}{3}$ of an inch, long, and is already larger than the umbilical vesicle, and resembles the embryos of other animals. It is much curved, has still the branchial arches, a distinct tail, and only a trace of extremities. The umbilical cord is quite short, and the opening in the abdominal walls not very large. The accumulating liquor amnii commences to separate the body of the embryo from the amnion.

Second month.—In this period the ovum undergoes very great changes. It acquires the size of a hen's egg; the embryo is from 2½ to 3 cm., or ·98—1·2 inches long, and weighs about 4 gram., or 62 grains. The embryo has already its permanent shape, the extremities have distinctly separated into their three portions, the umbilical cord is longer, and the intestine still extends into it. In the clavicle and inferior maxillary bone the first centres of ossification appear.

Third month.—The ovum is of the size of a goose's egg; the embryo is from 7 to 9 cm., or 2·7 to 3·5 inches, long, and weighs from 5 to 20 gram., or 77·5 to 310 grains. The intestines are retracted from the umbilical opening, and centres of ossification are found in most of the bones; the fingers and toes with their nails are distinctly discernible, and the external genitals begin to assume the sexual distinction.

Fourth month.—The fœtus is from 10 to 17 cm., or 6·6 inches, long, and weighs as much as 120 gram., or 1860 grains. The sex can now be distinctly recognised.

Fifth month.—The fœtus is from 18 to 27 cm., or 7 to 10·5 inches, long, and weighs on an average 284 gram., or 4402 grains. The skin is less transparent. Hair is growing on the head, and lanugo over the whole body.

Sixth month.—With a length of from 28 to 34 cm., 11 to 13 inches, the fœtus has a medium weight of 634 gram., or 9827 grains. Fat begins to be deposited in the subcutaneous connective tissue, but it is still small in amount so that the skin is as yet quite wrinkled. The head is disproportionately large; the fontanelles and sutures are widely opened. A fœtus born at this time makes inspiratory movements, moves its limbs, but dies in a very short time.

Seventh month.—The fœtus is from 35 to 38 cm., or 13·7 to 15 inches, long, and weighs 1218 gram., or 42 oz. The eyelids are separated, the whole body is on account of the scanty development of fat, thin, the skin red and covered with vernix caseosa. A fœtus born between the twenty-fourth and the twenty-eighth week sometimes moves the limbs with some strength, but cries with a feeble voice, and even though very careful attention is paid to it, almost always dies in the first few hours, or at the most days, after birth.

Eighth month.—The fœtus is from 39 to 41 cm., or 15 to 17 inches, long, and weighs on an average 1567 gram., or 49 oz. The pupillary membrane is disappearing, and the external skin maintains its red colour. The fœtus is still thin, and has a senile aspect. However, children born at this time (twenty-eighth to thirty-second week) may under favorable circumstances be kept alive, but they very frequently die, and from slight causes.

Ninth month.—The length of the fœtus is 42 to 44 cm., or 16 to 17 inches, its medium weight 1971 gram., or 62 oz. The great development of fat contributes to give the body a rounded shape, and the face loses its wrinkled appearance. Children born in the thirty-second and thirty-sixth week show a far greater mortality than those born at term, but under favorable circumstances they are, as a rule, kept alive.

Tenth month.—In the first weeks of the tenth month the fœtus is from 45 to 47 cm., or 17·5 to 18·5 inches, long, and weighs 2334 gram., or 82 ozs. The lanugo gradually disappears, but is still almost everywhere distinctly visible, and most so on the shoulders. The nails have not yet reached the tips of the fingers, the cartilages of the ears and nose are soft, the skin is still red, but smooth and full.

Towards the end of the tenth month the fœtus acquires all the characteristics of the mature child and can no longer be distinguished from it.

2. *The Fœtus at Term*

The characteristic marks of a mature fœtus are the following:—It is, on an average, 51·2 cm., or 20 inches, long, and weighs 3275 gram., or $6\frac{1}{2}$ lbs. The skin is white; the fine lanugo is visible only on the shoulders, and has disappeared from every other part; the child is more or less covered by vernix caseosa (a whitish grease, formed of thrown-off epithelium, fine lanugo, and the secretion of the sebaceous glands). The hair of the head is usually dark, from one to one and a half inches long; the cartilages of the ear and nose feel hard; the nails also are usually hard, horny, and project at least in the upper extremities over the tips of the fingers. The umbilical cord is inserted somewhat below the middle of the body. In males the testicles can be felt in the thick wrinkled scrotum; in females, the labia majora are closely approximated; often, however, the labia minora are visible. The cranial bones are hard, they lie close to each other (the sutures are narrow), and the ossific centre of the lower epiphysis of the femur is about $\frac{1}{2}$ cm., or $\frac{1}{5}$ inch, in its greatest diameter. The child cries immediately after birth with a loud and powerful voice, and actively moves the extremities. It also passes urine and meconium. The latter is of a blackish or brownish-green colour, and consists of mucus, intestinal epithelium, bile, epidermic scales, and lanugo (Förster, 'Wien. Med. W.,' 1858, No. 32).

The head of the child, being its most voluminous and most unyielding part, is of special importance in the mechanism of parturition, and therefore requires a more detailed consideration.

The face appears to be very small in proportion to the cranium. This latter is formed by the two frontal bones, the two parietal bones, the squamous portion of the occipital bone, and at the sides by the temporal bones and the large wings of the sphenoid. All these bones have not grown to one another, but are separated from each other by spaces, the so-called sutures. The following sutures are distinguished:

1. The frontal suture, between the two frontal bones.
2. The sagittal suture, between the two parietal bones.
3. The coronal sutures, one on each side between a parietal and a frontal bone.
4. The lambdoidal sutures, one on each side between an occipital

and a parietal bone. The serrated temporal sutures, by which the squamous portion of the temporal bone is connected with the parietal bone on each side, cannot be felt on heads covered by the soft parts, the temporal muscle especially covering them.

At the place where the two coronal, the frontal, and the sagittal sutures meet, the bones present greatly rounded-off angles, and a large space is formed in the bony walls of the head, which is called the large fontanelle. It has the shape of a trapezium; its angle, formed by the meeting of the two frontal bones, is much more acute than that between the two parietal. The small fontanelle does not consist of a true membranous gap, but its place is marked only by the meeting of the sagittal suture with the lambdoidal sutures. At the ends of the lambdoidal sutures, close to the mastoid portions of the temporal bones, there are two distinct intervals in the cranium, the lateral fontanelles—*font. Gasserii.*

To determine the size of the fœtal head, the following measurements are taken:

1. The antero-posterior diameter, between the glabella and the most prominent point of the occiput, 4·62 inches, or $11\frac{3}{4}$ cm.

2. The larger transverse diameter; the greatest distance in a transverse direction, measures 3·64 inches, or $9\frac{1}{4}$ cm.

3. The smaller transverse diameter, the largest transverse distance between the two coronal sutures, 3·14 inches, or 8 cm.

4. The oblique or large diameter, from the chin to the most distant point of the cranium close to the small fontanelle, measures 5·31 inches, or $13\frac{1}{2}$ cm.

5. The vertical diameter, between the vertex and occiput, which in the living child cannot be accurately measured, from 3·74 to 3·93 inches, or $9\frac{1}{2}$ to 10 cm. The circumference of the head is $34\frac{1}{4}$ cm., or 13·58 inches.

There are, however, considerable individual deviations from the above-given average measurements; generally the heads of male children are a little larger than those of female, and the children of older pluriparæ have greater measurements than those of young primiparæ.

3. *The Nutrition and Circulation of the Fœtus*

The nutrition of the fecundated ovum is at the first carried on by mere osmosis. The nutritive irritation, which the ovum embedded in the uterus exerts, causes an increased supply of nutritive material and the structureless primitive villi of the chorion enlarge, as Kiwish says, "the surface for absorption, and are more permeable also for liquids on account of their delicate formation. The activity thus roused immediately converts a part of the absorbed fluid in the ovum into cells, that is, the embryo, as it were, crystallizes out of it."

As soon as the primitive vascular system is developed the omphalo-enteric vessels absorb the nutritive material contained in the vitelline vesicle and bring it to the embryo.

The most important change, however, takes place when the allantois has brought the fœtal vessels to the serous membrane which becomes the exochorion. Their great development at the placental insertion and the immediate prolongation of the villi of the chorion into the maternal sinuses allow of such an extensive interchange of the gases and liquids of the two bloods that the respiration and nutrition of the fœtus become thereby possible.

[It is not exactly known what kind of interchange this is; however, it appears almost certain that the red blood-corpuscles of the fœtus give off carbonic acid and receive oxygen, and that the plasma of the fœtal blood so exchanges its constituents with those of the mother's blood that the final products of the tissue metamorphosis of the fœtus enter the blood of the mother, and the highly organized combinations which the maternal blood has derived from the chyle enter the fœtal blood. An immediate and direct mixing of the two kinds of blood nowhere occurs; they are separated by the epithelium lining the villi of the chorion. Reitz ('Centralblatt f. d. Med. Wissens.,' No. 41, 1868, p. 655) after injecting a gravid rabbit with cinnabar found the colouring particle also in the blood of the fœtus, and especially in the capillaries of the pia mater. According to Cohnheim's observations on the migration of the white blood-corpuscles from the vessels, such a passage, undoubtedly connected with the white blood-corpuscles, is easily intelligible, and consequently cells of the maternal blood would also enter into the circulation of the fœtus.]

The nutrition of the fœtus is almost exclusively carried on by the placenta, and not by the liquor amnii which chiefly contains products of a retrogressive metamorphosis and only very little albumen. As regards the respiration of the fœtus, Pflüger correctly draws attention to the less amount of movement (heat and work) performed by the fœtus, and the consequently smaller quantity of oxygen required by it then as compared with extra-uterine life. The fœtus is suspended in a medium the heat of which corresponds to about that of its own blood, does not receive into its intestinal canal nor into its lungs cold substances to be heated therein, nor does it lose heat by radiation or by evaporation, either at the surface of its body or through the lungs. The work of the muscles, too, is proportionately small. Active movements are easily executed in a liquid the specific gravity of which is the same as that of its own body. The respiratory muscles are not in action, and the heart alone is powerfully at work. Nevertheless the fœtus does respire—that is to say, there is a consumption of oxygen; we conclude that this is so from the known fact that an interruption of the fœtal circulation, when the placenta is not replaced by the lungs, always causes death, and this occurs at a time when the want of nourishment cannot possibly prove fatal; and again on examining the bodies of those who have died from such a cause there are the unmistakable signs of death from asphyxia.

It has also been proved that the fœtus makes respiratory movements when its communication with the placenta is interrupted (Vesal); and, on the contrary, the new-born child ceases to inspire, becomes apnœic, when oxygen is supplied to it in another way than by the lungs (Mayow).

From this we may conclude that the fœtus inspires as soon as it feels the want of oxygen, and that during intra-uterine life oxygen is supplied by the placenta. Pflüger has also experimentally shown from the colour of the blood that the fœtus consumes oxygen.

The circulation of the blood in the fœtus takes place in the following way:—The umbilical arteries, the principal terminations of the iliac arteries, convey the blood through the umbilical cord into the placenta, and in the chorion villi the inter-changes with the maternal blood occur. The umbilical vein collects the blood suitable for the nutrition and respiration of the fœtus from the placenta, and brings it through the umbilical cord to the liver (for the sake of brevity we shall call this blood arterial, and that which has circulated in the fœtus venous, although neither of them exactly corresponds to the arterial and venous blood of extra-uterine life). A part of it circulates through the liver, another portion passes through the ductus venosus Arantii directly into the inferior vena cava, and both these mixed enter the right auricle, together with the venous blood returning from the lower half of the body. At an earlier period of fœtal life the inferior vena cava opens opposite the septum of both auricles into the left as well as into the right auricle, but its blood chiefly enters the left on account of the greater development of the Eustachian valve on the right side of the septum. In the second half of fœtal life the Eustachian valve becomes smaller, whilst, through increased development of the valve of the foramen ovale, the blood current of the inferior vena cava is more and more exclusively directed into the right auricle. At that time, therefore, its blood is partly mixed with that of the superior vena cava, so that mixed blood enters the right ventricle as well as the left auricle. The greater part, however, of the venous blood, which returns by the superior vena cava from the upper half of the body, enters the right ventricle and is hence forced into the pulmonary artery. But since a pulmonary circulation (in the true signification of the word) does not yet exist, only about half of the blood from the right ventricle enters the pulmonary artery, whilst the other half is poured through the ductus Botalli into the descending aorta. This latter contains blood from three different sources. The left auricle receives only a relatively small quantity of blood from the yet undeveloped pulmonary veins. It obtains its chief supply through the foramen ovale, through which, as stated above, the current of the inferior vena cava passes; this is chiefly arterial blood, though not quite unmixed. Hence this blood passes through the left ventricle into the first portion of the aorta, so that the carotid and subclavian arteries are supplied chiefly with arterial blood. Beyond the origin of those arteries the more venous blood of the right ventricle, and chiefly derived from the superior vena cava, enters through the ductus Botalli into the descending aorta. Therefore the umbilical vessels and the lower half of the trunk are supplied with more venous blood than arterial, chiefly by the force of the right ventricle.

These conditions, however, are at once altered at the birth of the child. By the first respiratory movements the lungs are expanded and thereby the pulmonary arteries dilated. Hence the blood of the right ventricle is exclusively forced into the pulmonary artery. In consequence of this a far larger quantity of blood, after being arterialised in the lungs, is returned to the left auricle. The pressure of the blood in the left auricle is now increased; at the same time, through the cessation of the placental circulation, less blood enters from the inferior vena cava into the right auricle, and thus the pressure of the blood in the latter is lessened, so that the pressure in both the right and left auricles is equal. If, however, that of the left auricle predominate, the passage of its blood into the right auricle is prevented by the peculiar formation of the valve of the foramen ovale (experiments of Kehrer, loc. cit., p. 98).

We have seen, also, above that the pressure in the descending aorta, which controls the whole placental circulation, is due to the contractions of the right ventricle, which forces the blood through the ductus Botalli. When the pulmonary circulation commences, suddenly a new and large channel is opened for the current of the blood of the right ventricle, the pressure in the right heart diminishes, and the ductus Botalli is narrowed by the contractions of its own walls. The walls come together and unite without the formation of a thrombosis. This takes place with greater certainty since the blood pressure in the left heart, which might keep the duct pervious, then only increases, after that in the right heart has diminished. Exceptionally only the ductus Botalli remains pervious, and this is especially the case when in pulmonary atelectasis the pressure in the right heart is not at all or only inconsiderably diminished. Since in extra-uterine life the pressure in the left heart very soon becomes greater than that in the right, we find, in children who have lived for some time, the aortic end of the duct wider than the pulmonary. With the closure of the ductus Botalli the force of the current in the descending aorta and in its branches must be suddenly very considerably lessened, since it had been maintained principally by the right heart. Were the placenta even to remain in connection with the child, the long placental circulation could not be sufficiently supplied. Consequently thrombi form in the two umbilical arteries, whilst the umbilical vein is commonly only greatly contracted. Even the rapidly increased pressure in the left heart is unable to maintain the slackened placental circulation. In this way, at the birth of the child, the circulation is established as it is observed throughout the whole of extra-uterine life. The ductus venosus Arantii contracts like the umbilical vein with the cessation of the placental circulation. As a rule, the ductus Botalli is obliterated as soon as the pulmonary circulation is established, whilst the foramen ovale sometimes remains open for a longer period, without there being, for the reasons given above, any intercommunication between the blood of the two auricles.

4. *Presentation, Position, and Attitude of the Fœtus within the Uterus*

By the presentation (*situs*) of the fœtus is meant the relation of its longitudinal axis to that of the uterus. If both approximately coincide the fœtus is said to be placed longitudinally, if the two axes cross each other it is said to be situated transversely, and, according to whether the cephalic or pelvic end presents, cephalic or pelvic presentation.

By the position (*positio*) of the child is indicated the various relations which, after the situation has been determined, a certain part of the fœtus (*e. g.* the back) has to certain sides of the uterine walls. If the fœtus is longitudinally situated the position of the fœtus with the back directed to the left uterine wall is called the first, and that with the back directed to the right uterine wall the second position. A subdivision is made, according to whether the back is more in front or behind, and the first subdivision is when the back is more in front, the second with the back more behind, so that, *e. g.*, the back to the left and in front would be the first subdivision of the first position.

By the attitude (*habitus*) of the fœtus within the uterus is understood the position of one part of the body in relation to another.

The attitude is the following:—The fœtus is bent over the abdominal surface, so that the whole vertebral column forms an arch with the concavity forwards. The chin rests upon the chest, the thighs drawn up to the abdomen. The legs are flexed, and the feet so stretched that the dorsum of the foot is drawn up to the leg, and the heel is the lowest. The arms are placed on the sides of the chest, and the forearms are crossed, or close to each other in front of the chest. In the space between the upper and lower extremities the umbilical cord is, as a rule, situated. This is the normal attitude, which, however, frequently varies during pregnancy, although these variations are inconsiderable and only temporary.

Only very rarely during the course of pregnancy are we able to prove, as continuing for any length of time, that the head is extended so that the occiput rests upon the back, a variation from the normal attitude which is oftener observed constituting a face presentation during parturition.

Whilst in the vast majority of cases the fœtus presents at birth by the head, yet during pregnancy this presentation is by no means constant, but subject to frequent changes, although even during pregnancy head presentations are by far the most frequent.

This change of presentation occurs more easily and more frequently during the early months of pregnancy. It is more often observed in pluriparæ than in primiparæ, in the former often just before delivery, whilst in primiparæ it very exceptionally occurs during the last three weeks of gestation. The heavier the child the less often are these changes of presentation met with; a contracted pelvis favours its occurrence.

Transverse presentations most frequently change into head presentations; the reverse is not quite so frequently, yet sufficiently often, met with. The change from a pelvic to a head presentation is likewise very frequent; the converse also is not rare.

The change from a pelvic to a transverse presentation and the converse is relatively seldom met with.

It happens far more frequently that the position of the child is changed. Since head presentations are the most frequent of all, it is natural that this should be most often observed in them, yet in pelvic presentations it is not a rare occurrence. The position is, as a rule, constant only when the head is firmly placed in the true pelvis.

Since the time of Hippocrates it has been commonly taught that up to the seventh month the fœtus is placed with its pelvic end downwards, but that it then suddenly turns so as to present by the head. This notion of the *culbute* (which also is found amongst the Chinese) was universally accepted up to the period of the great anatomists.

Realdus Columbus (1544), the pupil of A. Vesalius, was the first to speak against it. But only Smellie (1751), Solayris de Renhac (1771), and Baudelocque (1781), succeeded in entirely superseding that teaching, and from that time until very recently the opinion was prevalent that the fœtus is originally placed with the head downwards, and that this position is maintained unchanged during the whole period of gestation. Onymus also shared in that opinion ('D. m. i. de naturali fœtus in utero mat. situ.,' Lugd. Bat., 1743), and deserves to be mentioned because he appears to have been the first to observe the changes in the fœtal position by repeated examinations of pluriparæ, in whom the internal os was widely open. He found that in forty-three pregnant women in twenty-seven only the fœtal position remained unchanged up to birth, and he explained the normal head presentations as well as the other positions and their changes as due to the laws of gravity. Recently Scanzoni was the first to speak against the unchangeableness of the fœtal position during gestation, and he modified the old notion of the *culbute* in so far as he has stated that breech presentations are more frequent at an early period than towards the end of pregnancy, and that the change to the head presentation gradually takes place within the last months. Hecker then again drew attention to the subject, and showed by a series of observations that in the latter period of pregnancy the position and presentation of the fœtus are by no means as constant as they are generally supposed to be, that there are not only changes from a breech to a head position, but also the reverse. Similar observations, partly made at the same time as those of Hecker, were published in greater number by Credé, and later by Heyerdahl, Valenta, and Schröder. Lately Schultze, and after him Höning, have tried to elucidate this question by almost daily examinations. From the results they obtained, as well as from our own observations, the above data of the position and presentation of the fœtus have been formulated. The constancy of position in primiparæ during the last three weeks, which Schultze and Höning also (in part) found, we could not altogether affirm. Of 214 primiparæ examined by ourselves (including 4 cases with contracted pelvis) we found in 81 (and in one of these twice even) changes of position at that time. There was a change from a transverse to a head position six times, from a transverse to a breech position once, from a head to a transverse position once, and from a breech to a head position once.

Before we quit the subject of the position and presentation of the fœtus, we must first answer the question how it is that the head presents so disproportionately often. To explain this fact the most varied hypotheses have been adduced at all times, and it would lead us too far to give a detailed exposition of them. We would refer the reader for that purpose to the historical part of a paper by Cohnstein ('M. F. G.,' B. 31, p. 141). We shall only mention that in recent times two opinions have especially been contended for. The one, suggested by Aristotle, tries to explain the position of the fœtus by means of the laws of gravity; the other, supported by Simpson with a great deal of ingenuity, says that the fœtus makes reflex movements as long as it is in an uncomfortable position. Its most comfortable position, if its shape be compared with that of the uterus, is with the head downwards. If it is not in this position a greater pressure is exerted on it by the uterine walls. This pressure causes reflex movements, and they continue until the most comfortable position — head presentation — is found. Others (Credé, Kristeller) thought the head position was dependent upon uterine contractions.

To decide this question, it is first of all necessary to determine how the fœtus lies if solely under the influence of gravity, for the fœtus in the uterus is, as well as all other things, subjected to the laws of gravity. In very numerous experiments, made first by Duncan, and then by Veit (loc. cit., p. 256) and Höning (loc. cit., p. 93), in which a fœtus recently dead was allowed to swim in a balloon filled with salt water of a specific gravity equal to that of the fœtus, it was seen that the head lies lower than the breech, and that the right shoulder looks downwards. The fœtus lies in the position (only the right shoulder somewhat lower down) as represented in Fig. 7 after M. Duncan, (Obstetric Res., p. 26). This position

FIG. 7.

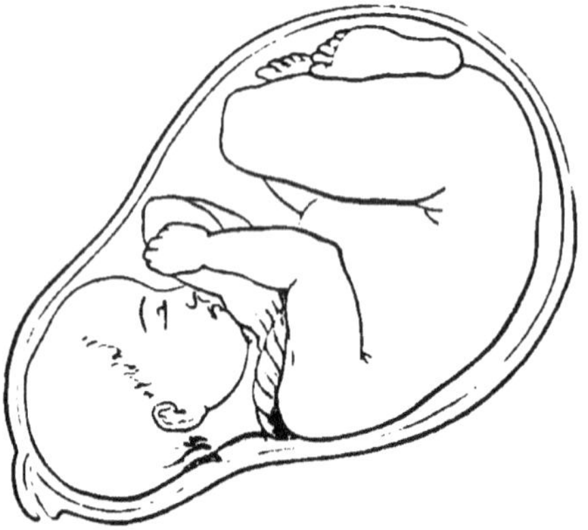

is doubtless due to the greater weight of the head and the liver. That Kehrer ('Beitrage,' &c., p. 109) did not find the centre of gravity of the fœtus nearer to the upper half of the body is no proof that the above explanation is wrong, and Duncan has shown ('Obst. Res.,' p. 22, Annotations) that the specific gravity of the head is greater than that of the decapitated trunk, and Kehrer himself has observed (l. c., p. 110) that the head has a greater tendency than the breech to assume the downward position. Poppel, however, has found the centre of gravity to be nearer to the cephalic end. However, the observations made by Veit on floating fœtuses have shown that the fœtus, if no other force act upon it, is placed with the head, not in the vertical line, but obliquely downwards.

Let us now consider the position of the uterus relatively to the horizon in the various positions of the woman. In the erect posture the axis of the uterus, which correctly enough may be said to coincide with the axis of the inlet, and with the usual pelvic inclination of 55°, makes with the horizon an angle of 35°. The uterus lies so obliquely to the horizon that it is not its internal os, but a spot in the anterior uterine wall, which occupies the lowest position. If the back of the fœtus lies in front, it follows that in the erect posture of the woman, since the head cannot deviate backwards, the head must lie in consequence of the laws of gravity alone upon the os uteri in the way shown in Fig. 7. If the back of the fœtus is quite to one side, the head must, according to the laws of gravity, deviate somewhat to the other side. But in the erect posture, since the fœtus floats with the right side lower down than the left, the back of the child must lie to the left and behind; consequently, if the position of the fœtus is solely determined by gravity, the head must deviate in the erect posture of the woman from the os uteri or inlet somewhat to the right.

If the woman lies on her back the axis of the uterus makes an angle with the horizon of about 55°. The uterus is now situated far more vertically than in the erect posture. Now, if gravity alone determines the position of the fœtus, the back must lie to the right and behind, and the head deviate somewhat to the left.

Although the statements just made are based only on experiments made by floating a fœtus in water, but otherwise constructed on theory, yet experience has actually shown that the fœtus is usually placed with the head somewhat deviated laterally from the inlet when the walls of the uterus are very soft and flaccid, when its shape is very changeable, and when the uterus influences the position of the fœtus as little as possible. But if the uterine walls are firmer, and if the transverse diameter is shortened by actual contractions, a lateral deviation now becomes impossible, and the head must consequently remain upon the brim. In primiparæ the uterine walls are of themselves much more firm and resisting, and in the last weeks already contractions occur. Consequently the head lies during the latter period of gestation on the brim, and usually even enters the true pelvis. In pluriparæ also, if the uterine walls are firm during pregnancy, or if slight contractions occur, the head equally lies on the brim. But if the walls are very flaccid the head, as a rule, deviates to one side, and assumes only the straight position when the first contractions occur which shorten the transverse diameter of the uterus.

In order to give a diagramatic representation of the usual condition, we have above only considered the positions when the woman is in the erect posture or lying horizontally on the back. In point of fact, however, a number of other positions have still to be considered. The recumbent posture, with the upper half of the body somewhat elevated, will still better explain the above-mentioned conditions, since by it the uterus is still more vertically placed. It is different, however, in the lateral posture. The fundus of an especially easily movable uterus may, in the lateral posture, fall so far to one side that it lies even lower down than the lower uterine segment, which by a lever action is drawn somewhat to the opposite side. Were the fœtus to follow the laws of gravity, its head would now fall into the fundus and a pelvic position would be established. Towards the end of gestation, in primiparæ at least, this is, as a rule, prevented, because the uterus grasps the fœtus so tightly that a change of position is impossible (as stated already, it is very rare in primiparæ during the last three weeks). In pluriparæ, as well as at an earlier period of pregnancy, it is probable that in this way at first transverse and then pelvic positions arise. It is, however, not denied that it may often be caused by active movements of the fœtus, and that it occasionally occurs under suitable conditions, as in the case of P. Muller, where, in a primipara with by no means flaccid uterine walls the fœtus turned six times in five days under active movements.

It would lead us too far to refute each objection individually which from many a side is made against the dependence of the position of the fœtus upon the laws of gravity. All we can say is that the same objections again and again appear, although Battlehner and Duncan have, now fifteen years ago, refuted the most weighty of them. It is no proof against this that in the earlier months, in living or recently dead fœtuses (in putrid fœtuses the centre of gravity changes according to Veit's experiments) the pelvic position is the more frequent, though here the laws of gravity ought especially to influence the presentation. We have seen that at an early period of pregnancy the tendency to changes in the fœtal position is very great, and all those conditions which are able to alter the position of the head will act more easily and more frequently. It must also be added that pelvic position of the immature fœtus often arises, according to Scanzoni, in the way of the so-called self-evolution from head positions with lateral deviation.

As regards the fœtal attitudes and their changes, it has already been shown that in the erect posture the back of the fœtus must be directed forwards and to the left, and in the recumbent posture backwards and to the right. The natural conclusion would be that, unless the pelvis or the uterine walls impede the movements of the fœtus, the presentation must change with each altered posture of the mother. Höning observed even directly that the first head position was converted into the second when the mother lay down, and *vice versâ*.

The cause of the normal attitude of the fœtus will simply be found in the peculiar direction in which the fœtus develops. At the earliest period of development the fœtus is strongly bent over its longitudinal axis, and preserves the curved attitude until other active causes modify it. Therefore, not the flexion of the head with the chin to the thorax, but its extension, as seen in face presentations, demands an explanation.

D.—MULTIPLE PREGNANCY

It is rather exceptional that the gravid uterus contains two or more fœtuses.

Such a multiple pregnancy may arise in several ways.

1. From one ovum several fœtuses develop. The ovum then contains either two germs which fecundated at the same time continue their simultaneous evolution, or one germ has by division given rise to the formation of several fœtuses.

2. One Graafian follicle may contain several ova, which after the rupture of the follicle are fecundated at the same time.

3. Several Graafian follicles, either of one or both ovaries, may simultaneously rupture during menstruation, and the expelled and fecundated ova develop together in the uterus.

The relations of the membranes also vary according to the different origins of the multiple pregnancy.

The decidua vera, which is nothing else but the proliferating mucous membrane of the uterus, must in all cases of multiple pregnancy necessarily be common to the several fœtuses, because the mother has only one uterus. The only exception is in cases of uterus bicornis or septus.

The decidua reflexa is simple when the several fœtuses are derived from one ovum, or when the several ova have been imbedded in the mucous membrane in close apposition. If the ova adhere to several places, each one is covered over by the proliferating mucous membrane, and thus obtains a separate reflexa.

The fœtuses have the same chorion if they are derived from the same ovum; those from different ova have always separate choria, which may, however, greatly atrophy at the place of contact of the two ova.

The amnion, which is formed neither from the mother nor from the ovum, but from the embryo itself, and is only the prolongation of the abdominal walls of the embryo, must consequently be multiple in several fœtuses. In rare cases two fœtuses are found lying in one amniotic cavity. Then the vessels of the umbilical cords of the two fœtuses are for some distance from the placenta covered by the same amnion, so that the cord with two sets of vessels at the placenta divides into two at some distance from it.

The cases of simple amnion may partly depend upon laceration and subsequent atrophy of the originally existing septum. But where the twin pregnancy is due to the division of the germ the full development of the two amnia may have been equally impeded by a close approximation of both embryoes, as under like circumstances double monsters are formed.

The placentæ also are originally always separate, since in all those cases each embryo forms its own allantois, which, independently of the other embryo, grows towards any spot in the periphery of the ovum. In different ova they may, but in a simple ovum they always must, be so close together that they more or less intimately coalesce. In the simple ovum there is, according to Hyrtl, always an anastomosis between the umbilical vessels of the embryoes.

Triplets also may form in the various ways mentioned above. Credé showed a triplet ovum with a common chorion; Pflüger, Grohé and Schrön have found three ova in one follicle, and Scharlau showed a triplet ovum, each of which had a separate chorion and also a separate placenta. More frequently triplets develop from two ova. Twins are most frequently derived from different ova (31 times in 126 cases from the same ovum, 95 times from different ova, Spaeth). Derived from the same ovum, the twins have consequently a common chorion, never different sexes, and frequently show a striking similarity of bodily formation and of mental aptitude. (Each of the two pairs of twins of Shakespeare's 'Comedy of Errors' is to be supposed to have arisen from one ovum.)

Twin pregnancy is not very rare. Amongst 13,000,000 births Veit found one twin pregnancy in 89 births; triplets were born once in 7910, and four once in 371,126 cases. Five at a birth are still more rare, and there is no authenticated case of a still greater number of fœtuses at one birth. In the greatest number (64 %) of all cases the twins have the same sex.

The weight and size of the twins are almost always beneath the average, even if they are born at term, which is commonly not the case. Three or four fœtusus are usually still more feebly developed and have very little chance of living.

Very frequently the fœtuses are unequally developed. It also occurs that one fœtus prematurely dies and is then flattened by the pressure of the other growing one and dries up. At birth this prematurely dead mummified and flattened twin fœtus (*fœtus papyraceus*) is expelled with the membranes. In some cases one of the twins is expelled prematurely, whilst the other remains and is born at term.

It is still a doubtful matter whether multiple pregnancy is effected by one coitus or whether several such acts are required, also with what interval between two coitus impregnation is still possible.

Superfecundation, that is, the impregnation of several ova of the same period of ovulation, is, at any rate, possible; there is no plausible reason why the possibility should be denied.

Superfœtation, that is, the impregnation of several ova of different periods of ovulation, is, as a rule, impossible, because after the impregnation of an ovum ovulation ceases until after delivery. Yet the question whether it is possible that during gestation ova may be expelled from the ovary is not definitely decided. The occurrence of such a superfœtation is not yet confirmed by any authenticated case; there is no absolute impossibility of such an event.

E.—THE DURATION OF PREGNANCY

1. The duration of pregnancy in the human female varies. In the great majority of cases it lasts from 265 to 280 days, on an average about 271 days. Since the day of the actual conception, that is, the day when the sperm has entered the ovum, is never accurately known, and the day of the fruitful intercourse only rarely so, we have, for determining the time of delivery, only the occurrence of the last menstruation. If from that day we count three months backwards, then add seven days, there will, as a rule, be an error of only a few days. Exceptionally there may be, in this way of calculating, an error of several weeks.

To calculate from the time when the mother perceived the first fœtal movements is much less accurate. Most frequently this takes place towards the twentieth week; according to Ahlfeld in primiparæ on an average at the 137th, in pluriparæ at the 130th day and, therefore, twenty to twenty-two weeks would have to be reckoned from that time. But the first fœtal movements are sometimes perceived before the eighteenth, in other cases after the twentieth week, so that this mode of reckoning is chiefly of value as a check.

Literature.—Valentin, Handbuch der Entwickelungsgeschichte des Menschen. Berlin. 1835.—R. Wagner, Lehrbuch der Physiologie. Abth. I. Leipzig, 1839, und Icones phys.—Bischoff, Entwickelungsgeschichte der Säugethiere und des Menschen. Leipzig, 1842.—Coste, Histoire gén. et part. du dével. d. corps org. 1847-1859.—Kölliker, Entwickelungsgeschichte d. Menschen und der höhern Thiere. Leipzig, 1861.—Bischoff, Beweis d. v. d. Begattung unabh. per. Reifung und Losl. der Eier der Säugethiere u. d. Menschen. Giessen, 1844.—Pflüger, Unters. aus dem physiol. Laborat. zu Bonn. Berlin, 1865, p. 53.—J. F. Lobstein, Essai sur la nutrition du fœtus. Strassb., 1802. Teutsch von Th. F. A. Kestner. Halle, 1804.—Th. Bischoff, Beiträge zur Lehre von den Eihüllen d. menschl. Foetus. Bonn, 1834. W. Hunter, Anatomia ut. hum. grav. illustr. Birm., 1774.—Robin, Mémoire, &c. Archives génér., Juillet, 1848, p. 265.—R. Wagner, Meckel's Archiv f. An. u. Phys., 1830, p. 73.—Hegar, M. f. G., B. 21. Suppl., p. 1.—Eigenbrodt u. Hegar, M. f. G., B. 22, p. 166.—Dohrn, M. f. G., B. 26, p. 120.—Friedländer, Phys. anat. Unters. üb. d. Uterus. Leipzig, 1870, p. 7, &c.—Winkler, Textur, &c., in d. Adnexen d. menschl. Eies. Jena, 1870.—Hegar u. Maier, Virchow's Archiv, 1871, B. 52, p. 161.—E. H. Weber, Hildebrandt's Handbuch der Anatomie. 4 Ausgabe, 1831, B. 4.—Seiler, Die Gebärm. u. d. Ei des Menschen. Dresden, 1832.—Schroeder van der Kolk, Verh. van het K. Nederlandsche Institut, 1851, p. 69.—Virchow, Ueber die Bildung d. Placenta. Ges. Abhandlungen. Frankfurt, 1856, p. 779.—Dohrn. M. f. G., B. 26, p. 119 u. 122 (vergl. Hegar, M. f. G., B. 29, p. 1).—Ernst Bidder, Holst's Beiträge zur Gyn. u. Geb. Tübingen, 1867, 2 H., p. 167.—Jassinsky, Virchow's Archiv, B. 40, p. 341 (vergl. Dohrn, M. f. G., B. 31, p. 219).—Ercolani, Delle glandule otricolari dell' utero, &c. Bologne, 1868 (s. das aus dem Französischen übersetzte Referat des Autors Amer. Journ. of Obst., B. II, p. 121, und das Referat von Hennig. M. f. Geb., B. 33, p. 236.—Langhans, Arch. f. Gyn., B. I, p. 317.—Winkler, Textur, &c., des m. Eies. Jena, 1870, p. 34.—Friedländer, Phys. anat. Unters. üb. d. Uterus. Leipzig, 1870.—Reitz, Stricker's Handb. d. Lehre von den Geweben. V, Leipzig, 1872, p. 1183.—Dohrn, M. f. G., B. 26, p. 116.—Winkler, Textur, &c., d. m. Eies. Jena, 1870.—B. S. Schultze, Das Nabelbäschen, ein constantes Gebilde in der Nachgeburt d. ausgetragenen Kindes. Leipzig, 1861.—Soemmering, Icones embryon. hum. Francof., 1778.—Ecker, Icones phys. Leipzig, 1851—1859. T. 25, 26 u. 27.—Hecker, Klinik der Geburtskunde. Leipzig, 1864, p. 22.—Ahlfeld, Arch. f. Gyn., B. II, p. 361.—Veit, M. f. G., B. 6, p. 104.—Siebold, M. f. G., B. 15, p. 337.—C. Martin, M. f. G., B. 30, p. 428.—Van Pelt, s. M. f. G., B. 16, p. 308.—E. Martin. M. f. G., B. 19, p. 76.—Stadfeldt, M. f. G., B. 22, p. 462.—Brummerstädt, Bericht aus der Rostocker Hebammenanstalt. Rostock, 1866, p. 46.—Frankenhaüser, M. f. G., B. 13, p. 170, und Jenaische Z. f. M. u N., 3 Bd., 2 u. 3 Heft.—Spiegelberg, M. f. G., B. 32, p. 276.—Schröder, Scanzoni's Beiträge z. Geb. u. Gyn., Bd. V, Heft 2.—Kulp, De pelvi obliqua. D. i, Berol., 1866, p. 10.—Hoth, Ueber die Veränd. der Kopfform Neugeborner, &c. D. i, Marburg, 1868.—Kiwisch, Geburtskunde. Erlangen, 1851. 1 Abth., p. 166.—Schwartz, Die vorzeitigen Athembewegungen. Leipzig, 1858.—Kehrer, Vergl. Phys., &c., p. 100.—Pflüger, Archiv f. d. ges. Phys., 1 Jahrg, 1 Heft, p. 59.—B. Schultze, Jenaische Z. f. M. u. N. 1868. B. 4, Heft 3 u. 4, und der Scheintod Neugeborener. Jena, 1871.—W. Hunter, Anat. ut hum. gr. tab. illustr. Birm., 1774. Plate VI.—J. Simpson, Edinburgh Monthly Journal, Jan., 1849, p. 423.—Battlehner, M. f. G., 1854, B. 4, p. 419.—M. Duncan, Edinburgh Medical and Surgical Journal, 1855, and Res. in Obst., 1868, p. 14 seq.—G. Veit, Scanzoni's Beiträge, B. IV, p. 279.—Hecker, Klinik d. Geb. Leipzig, 1851, B. I, p. 17, und 1864, B. II, p. 53.—Credé, Obs. de fœtus situ inter grav. Lipsiæ, 1862 u. 1864.—Heyerdahl, M. f. G., B. 23, p. 456.—Valenta, M. f. G., B. 25, p. 172.—Scanzoni,

Wiener med. Wochenschr., 1866, No. 1.—van Almelo u. Küneke, M. f. G. B. 29, p. 214.—Schroeder, Schwang., Geb. u. Wochenbett. Bonn, 1867, p. 21.—Schatz, Der Geburtsmechanismus d. Kopfendlagen. Leipzig, 1868, p. 35.—B. Schultze, Unters. über den Wechsel der Lage u. Stell. d. Kindes. Leipzig, 1868. —Poppel, M. f. G., B. 32, p. 321, und B. 23, p. 279.—Hoening, Scanzoni's Beiträge, Bd. VII, p. 36.—Fassbender, Berl. Beitr. z. Geb. u. Gyn., B. I, p. 41.—H. Meckel, Ueber die Verh. des Geschlechts, die Lebensfähigkeit u. s. w. Muller's Archiv, 3, 1850.—Veit, Beitragt zur geb. Statistik. M. F. G., B. 6, p. 126.—Spaeth, Studien über Zwillinge, Zeitschr. d. Ges. d. Aerzte zu Wien, 1860, No. 15 u. 16.—H. Ploss, Zur Zwillingsstatistik. Monatsblatt für med. Statistik u. s. w. Beilage zur deutschen Klinik, 1861, No. 1, p. 2.—Hyrtl, Die Blutgefässe der menschlichen Nachgeburt. Wien, 1870, p. 125.—Kleinwächter, Lehre von den Zwillingen. Prag, 1871.—Montgomery, Die Lehre von der menschl. Schwangerschaft, übersetzt von Schwann. Bonn, 1839, p. 297.—Berthold, Ueber das Gesetz der Schwangerschaftsdauer. Gottingen, 1844.—J. Reid, On Duration of Pregnancy in the Human Female. Lancet, 1850, vol. i, p. 438 u. 596, and vol. ii, p. 77.—G. Veit, Ueber die Dauer der Schwaung. Verh. d. Ges. f. Geb. in Berlin. 7 Heft, 1853, p. 102.—J. Simpson, Edinburgh Monthly Journal, July, 1853, u. Sel. Obst. Works, vol. i, p. 81.—M. Duncan, Edinburgh Medical Journal, Nov., 1856, p. 410 (s. M. f. G., B. 9, p. 379), and March, 1871, p. 788.—N. E. Ravn, Oom Svangerskabtidens Graendser. Kjoebenhavn, 1856 (s. M. f. G., B. 16, p. 238).—Elsässer, Henke's Zeitschr. fur Staatsarzneikunde. 37 Jahrg, 1857.—Schwegel, Zur Frage, &c. Wiener med. Wochenschrift, 1857, No. 44.—Hecker, Klinik d. Geb., 1861, p. 33.—Spiegelberg, M. f. G., B. 32, p. 270.—Ahlfeld, M. f. G., B. 34, p. 180 and 266.

CHAPTER II

THE CHANGES IN THE MATERNAL ORGANISM

A.—IN THE ORGANS OF GENERATION AND IN THOSE IN THEIR VICINITY

THE uterus is the seat of the greatest and most important changes which take place in the female organism during pregnancy. Whilst the virgin uterus weighs about one ounce, the organ towards the end of pregnancy has a weight of two pounds. This voluminous increase of the uterus is chiefly due to the hypertrophy and hyperplasia of its muscular fibres. They increase during gestation to eleven times their previous length and to double or even five times their original breadth. There is, at the same time, and especially in the internal layers of the uterine walls, a new formation of contractile fibre-cells. The connective tissue also between the muscular fibres increases and becomes looser. The blood-vessels also, grow abundantly, especially at the placental insertion, and multiply by new formation. Nerves and lymphatics equally participate in the enlargement to such an extent that the cervical ganglion, for example, which in the virgin condition is three quarters of an inch long and half an inch broad, acquires a length of two inches and a breadth of one and a half.

According to Luschka, the muscular fibres are arranged at the end of pregnancy in the following way:

The most superficial, a very thin layer, consists of a continuous stratum, covering the uterus like a cap. Its bundles extend to the Fallopian tubes the round ligaments and the ligaments of the ovaries. The sides of the uterus are not covered by this layer.

The middle layer, the thick mass of the muscular substance of the uterus, is made up of transverse and longitudinal bundles of fibres intimately interwoven, and perforated by a network of large veins. The transverse bundles have a varying course, individual bundles now penetrating deeply, now superficially placed. The longitudinal bundles are partly a prolongation of the transverse which have altered their course, partly of independent origin.

The innermost layer beneath the mucous membrane forms concentric rings around the three openings of the uterus, viz. the two Fallopian tubes and the os.

The peritoneal covering of the uterus is also stretched on account of the enlargement of the enveloped organ. The broad ligaments unfold more and more with the growing uterus, and towards the end of pregnancy the ovaries and the Fallopian tubes are closely applied to the uterus.

The mucous membrane of the gravid uterus also undergoes very important changes. These have already been mentioned whilst speaking of the formation of the decidua.

The enlargement of the uterus is, at any rate at the commencement of pregnancy, not due to the pressure of the growing ovum, but to an excentric hypertrophy arising in the uterus itself. For the ovum is, at first, too small to enlarge the uterus mechanically, and in extra-uterine pregnancy the same hypertrophy occurs, and also, if pregnancy goes on in one cornu of a double uterus the other cornu is also hypertrophied. At a later period the pressure of the growing ovum has a share in the enlargement, for then the size of the uterus corresponds to that of the ovum. The walls of the pregnant uterus at term are not thicker than in the virgin state, they scarcely measure $\frac{1}{2}$ cm., or $\frac{1}{6}$ in., in thickness.

The growing organ now rises into the false pelvis and pushes the intestines backwards and to the sides and the diaphragm somewhat upwards. Through the latter the position of the heart also is somewhat altered and the area of the heart's dulness somewhat enlarged. Only rarely the uterus lies straight in the abdominal cavity, and in most cases it lies somewhat to the right, its left lateral border being directed to the anterior abdominal wall; the reverse is far more rarely the case. Since, on account of the great inclination of the pelvis, the centre of gravity of the uterus falls in the erect posture of the woman far in front of the symphysis, it is the anterior abdominal wall rather than the symphysis which carries the weight of the uterus. Its position, therefore, in relation to the horizontal plane is accordingly determined by the more or less yielding abdominal walls. On the average, the angle which the gravid uterus makes with the horizontal plane is 35°, and in the half-recumbent and half-sitting posture, provided the abdominal walls are not too flaccid, the uterus is vertically placed.

The shape also of the uterus is not immaterially altered. Whilst the virgin uterus with the cervix has a shape similar to a calabash, it now becomes an ovoid organ by the exclusive growth of the body of the uterus, to which the small cervix adheres like an appendage. The slight curve over the anterior surface observable in the virgin state considerably increases with the growth of the uterus, and becomes a flexion, so that towards the end of pregnancy the body of the uterus and the cervix make an obtuse, sometimes even an acute, angle with the vertex directed backwards. Only at the commencement of strong contractions (which sometimes occur before the actual setting in of labour) is the external os forced asunder by the presenting fœtal part, so that, as a rule, during parturition alone, and rarely at the end of pregnancy, do the body and cervix form one cavity. The vagina also enlarges, though in a much less degree than the uterus. The muscular fibres of the vagina enlarge and multiply, and the mucous membrane hypertrophies so much that the vagina becomes longer and more roomy. Although the upper portion of the vagina is always drawn up (lengthened), yet a portion of the anterior wall is frequently found projecting at the vulva, and appears as a bluish-red swelling. The increased nutrition of the vagina is seen in the looseness and greater thickness of its tissue. The mucous membrane is of a bluish-red colour, the carunculæ are thicker, and the papillæ swell, in some cases so greatly that the whole vagina feels rough like a grater. The secretion of the mucous membrane becomes more abundant, creamy and is the habitat of the *Trichomonas vaginalis.* The vaginal portion of the uterus is somewhat shortened on account of the more elevated position of the uterus and of the swelling of the whole vagina. The dilatation of the anterior lip, which in primiparæ is always produced by the head pressing on it at the end of pregnancy, will be considered as we pass on.

The vulva also participates in the hypertrophy of the pelvic organs. The labia majora and minora are swollen, and thick blue veins become transparent through the mucous membrane, whilst the whole vulva is widely open.

The joints of the pelvic bones also become looser, *more moistened*, and somewhat more movable. Yet this increased mobility is insufficient to allow any great widening of the pelvic cavity. By the deposition of fat in the subcutaneous connective tissue the whole pelvic region obtains a rounder and fuller appearance. The abdominal walls are stretched by the growing uterus, which causes solution of continuity in the rete Malpighii. Sometimes the recti muscles are so widely separated that a kind of hernia along the linea alba occurs.

The pressure of the uterus gives rise also to disturbances on the part of the bladder and the rectum, in the former producing most frequently partial incontinence of urine, in the latter constipation. The same cause also produces varicose dilatations of the veins of the lower half of the body, œdema of the legs, and neuralgic pains in them, which, however, rarely last long.

The breasts are enlarged even as early as the second month, and become still more so in the fourth or fifth month. The sebaceous glands around the nipple enlarge, the areola becomes of a brown tinge, and from the gland itself a clear watery secretion can be pressed; this at times trickles out spontaneously.

B.—CHANGES IN THE BODY GENERALLY

Pregnancy is attended by various and profound changes of the whole body, so that functional disturbances arise, which in other individuals would be considered pathological. But here we regard them as a part of the physiological process, for they are not attended by grave symptoms or severe noxious consequences, and they completely disappear on delivery.

Such changes are especially observed in the blood. It is more watery, resembling that in chlorosis; the quantity of fibrin is increased, whilst that of albumen is diminished. The proportion between red and white blood-corpuscles is altered in favour of the latter. The circulation itself is disturbed in a variety of ways. This is shown by palpitations and vertigo. Congestion to the head leads to the formation of irregular plates of osseous substance (osteophytes) on the inner table of the skull-cap.

The urine is increased in quantity and more watery; the other constituents (urea) are not altered. Albumen is seldom found in the urine.

In the digestive organs functional disturbances are almost constantly met with. Vomiting is frequent in the first months of pregnancy. This comes on in the morning before food has been taken, rarely occurs after eating, and does not diminish the appetite. Salivation also is not a rare occurrence.

Pigmentations of the skin are commonly met with, not only of the areola and of the linea alba, but irregularly scattered patches over the whole body, and especially the face (chloasma uterinum). Exceptionally only, this is seen most frequently upon the chest. These patches are due to a fungoid growth (*Pityriasis versicolor*); usually the pigment is really deposited. Jeannin thinks that the chloasma uterinum is the consequence of the amenorrhœa which accompanies pregnancy.

The nervous system also is the seat of various disturbances, such as headache, toothache, hemerálopia, amblyopia, difficulty of hearing, altered taste, and most regularly mental disturbance. Women at other times serious now become gay; there is, however, more frequently mental depression, which may increase to pronounced melancholia.

By careful observations made in the last three months of pregnancy, Gassner has shown that the weight of the body greatly increases in normal cases (1500 to 2500 gram. in each month). This increase of weight is not exclusively due to the growth of the uterus and its contents, but to an increase of the whole body.

With regard to the changes of the thorax during pregnancy, Küchenmeister, Fabius, Wintrich, and Dohrn, by means of the cyrtometer, have shown that the capacity of the lungs does not diminish at this period. The thorax of pregnant women is less deep, but this decrease is balanced by an increase in breadth of the base of the thorax. In the puerperal state the thorax again becomes narrower but deeper.

Literature—W. Noortwyk, Uteri hum. grav. anat. et hist. Lugd. Bat., 1743.—W. Smellie, A Set of Anatomical Tables, with Explanations. London, 1754.—J. G. Roederer, Icones ut. hum. observ. ill. Götting., 1759.—W. Hunter, Anatomia ut. hum. grav. tab. illustr. Birmingham, 1774.—Luschka, Die Anatomie des menschl. Beckens. Tübingen, 1864, p. 346.—Hélie, Recherches sur la disp. des fibres musc. de l'uterus dev. par la gross. Avec Atlas. Paris, 1864.

CHAPTER III

DIAGNOSIS OF PREGNANCY

A.—METHODS OF EXAMINATION

An obstetrical examination can be made when the woman is in the erect, the recumbent, the lateral, or the knee-elbow position. By an examination in the erect posture the state of the vagina and that of the lower uterine segment can at once and easily be made out. For that purpose the woman must separate her thighs a little, and the accoucheur must kneel before her on the right knee if the left hand examines, and *vice versâ*, in order to be able to support the elbow of the examining arm upon the other knee, and put the other arm which is disengaged around her loins. But this method does not suffice for an accurate examination, since in this way only a very imperfect external examination can be made. By the internal examination in the erect posture we obtain no better result than if we had examined the woman in the dorsal posture. For towards the end of gestation the lower uterine segment does not descend much lower down, and the portio vaginalis deviates more backwards in consequence of the greater anteversion of the uterus, and is therefore less accessible to the exploring finger.

The knee-elbow and lateral positions offer exceptionally only any advantages. The dorsal position is therefore always to be recommended, because in it external as well as internal examination, and both combined, can be easily and accurately made.

1. *External Examination*

To make an external examination the abdomen of the woman must be uncovered.

By inspection the following points are made out: the size, form, and shape of the abdomen; the colour and other changes (cicatrices) of the abdominal walls, and the shape of the umbilicus.

By palpation the most important information is obtained at a later period of pregnancy. For this part of the examination the accoucheur must be seated upon the edge of the bed, with his back directed to the feet of the woman, and the hands placed flat upon her abdomen. If now, not only the tips of the fingers, but their whole length, are used for the examination, very accurate information will be obtained with regard to the position, shape, and consistency of the uterus, and also the position of the child. The uterus will give a peculiarly hard and elastic sensation, and only in extremely rare cases actual fluctuation is perceived in various places; according to the position of the fœtus, the uterus feels uneven. Usually a large part is felt in the fundus, on one side of it smaller parts more or less pointed, and at the other a considerable and uniform resistance. To find out what portion of the fœtus is placed, if any, on the brim, the accoucheur must rise and stand at the side of the bed. He must place both his hands upon the abdominal walls above the symphysis, in such a way that the tips of the fingers are directed towards it. If now sudden pressure is made on the abdominal walls with both hands simultaneously, or alternately, and if a large fœtal part is there movably placed, a distinct sensation of ballottement is obtained, that is, a sensation of a hard body being pushed away by the tips of the finger, and again striking against them. If the larger part is fixed in the pelvis, it is in this way also felt with equal distinctness; if this has already entered the pelvis, it can only be determined whether the tumour is continued into the true pelvis also or not. The breech especially of a very small fœtus and in an anomalous position may, by an inexperienced observer, be taken for a small fœtal part. But it always gives rise to the sensation of ballottement, and may therefore with certainty be distinguished from one of the smaller parts.

Percussion is rendered quite superfluous by the results of palpation in advanced pregnancy, and in the earlier months by the combined internal and external examination.

On auscultation of the abdomen of pregnant women different sounds are heard. They are the sounds of the heart of the fœtus, and sometimes a peculiar whizzing noise synchronous with the fœtal heart, and sounds originating in the umbilical cord. On the part of the maternal organism there is heard, besides the pulse of the aorta and gurgling in the intestines, the so-called uterine souffle which corresponds with the pulse of the mother.

Auscultation is best made by using the stethoscope. Immediate auscultation does not give any better results; it is disagreeable to the woman and to the accoucheur, and not applicable to every region of the abdomen.

The sounds of the fœtal heart are heard as frequent double beats. Their peculiarity is best studied by listening to the sounds of the heart of a newly born infant. Normally they are heard from the eighteenth to the twentieth week, but in exceptional cases much earlier. If the woman is healthy and the fœtus alive the heart's sounds can always be heard by a careful, and if necessary repeated, examination.

They vary from about 120 to 160 beats or more in a minute; their frequency is increased by the movements of the fœtus. Their intensity is also very variable. Sometimes they are so loud as to be heard even without applying the ear to the abdomen, at other times so feeble that they can only just be made out as such. But they must always be heard so distinctly as to be able to count each beat. If this is not possible, there is no certainty that they have really been heard. The sound in the umbilical cord is characterised by a whizzing murmur synchronous with and accompanying the sound of the fœtal heart, it is also heard on certain spots in the abdomen. This is by no means rare, and occurs, according to Hecker's and our own observations, in 14 to 15 per cent. of all cases. It, doubtless, originates in the umbilical cord, although the intimate conditions of its occurrence are not quite clear. It is not probable that the murmur is due to the coiling of the cord around the fœtus, for both Hecker and ourselves have found it less often in that condition than when there has been no twisting of the cord. Hecker's view seems the most probable—that it arises in the umbilical region from a flexure of the first piece of the cord at the umbilical ring. One circumstance, at least, is in favour of that opinion—that, as a rule, it is heard at places where the heart's sounds are loudest, whilst at more distant places it disappears and the heart's sounds are heard clear though loud. However, it is not of any practical importance during pregnancy. It is heard for a time and disappears without assignable cause, and then, after some time, may again be heard. If during parturition it be very loud and persisting it is well to watch the sounds of the heart carefully. Yet, even under those circumstances it does not in any way indicate an unfavorable prognosis. With the exception of peculiar and short noises dependent upon the movements of the fœtus, all the other sounds heard in the abdomen of a pregnant woman proceed from the maternal organism. Thus, often gurgling of the intestines is heard, the beat of the aorta, and the sound of the maternal heart conducted thither. If the latter, as often happens, are very frequent, they may be easily mistaken for those of the fœtal heart.

The so-called uterine souffle, formerly named placental bruit, is quite regularly heard. It is a blowing whizzing murmur of varying intensity, somewhat similar to the sound in the umbilical cord, but clearly distinguished from it by its different site. Sometimes it is very low, at others so loud as to completely hide the sounds of the fœtal heart at that spot. This sound appears early, not very often in the third month, but frequently in the fourth. It is usually audible on both sides, although louder on one; at times it may be entirely absent on one side. It is rare not to hear it on a careful examination. It very frequently changes its place, and is commonly heard low down on one side, rarely in the centre or over the fundus, sometimes at a very limited spot, at others over almost the whole uterus.

This sound is doubtless produced within the large vessels of the uterus. No safe conclusion as to the seat of the placenta can be drawn from it.

The mammæ may be examined in any posture of the woman. Attention is to be paid to the fulness of the gland, the colour of the areola, and to the nipple. With a little practice a few drops of the secretion can be pressed out from the gland as early as the third month of gestation.

2. *Internal Examination*

This may be made by the hand or by the aid of instruments. The former is made in the following way:—The woman being in the dorsal posture, one hand (the left is preferable, because it is always somewhat smaller and softer, and the right is then disengaged) is brought under the bedclothes, and the index, or index and middle finger, having previously been oiled, are passed along the perinæum into the vagina, so that the thumb is directed forwards and the other fingers come either to lie upon the sacrum or are flexed in the palm. This latter is especially to be preferred in the examination of the anterior wall of the vagina, and the more so in pluriparæ, whose perinæum is so yielding that it can be pressed upwards into the pelvic outlet. In introducing the fingers care must be taken not to take along with them any hairs or the labia pudendi, and if the growth of the hairs be abundant it is better, in order to avoid this, to free the vulva of them with the other hand.

In most cases the index finger is sufficient, but the introduction of two fingers is safer, and if it be done slowly and gently produces so little discomfort to a woman at term that it ought to be a principle always to introduce two fingers *if there be anything still to be made out.* This will rarely prove insufficient for a satisfactory examination, and there will seldom be an occasion in pregnancy to introduce the half or the whole hand. If this is necessary chloroform must be given, unless the genital organs are uncommonly wide.

It is well to make the examination in a certain order—to begin with the state of the vulva and to continue with the vaginal walls, the distension of the rectum, the vaginal portion, the fundus of the vagina, and the conditions of the pelvis.

An impervious vagina may render it necessary to make the internal examination by the rectum. But in all other cases this method is rendered superfluous by the vaginal examination, although in the first months of pregnancy exploration by the rectum may usefully supplement the latter in making the diagnosis.

Instrumental examination of pregnant women is made by means of the speculum. For that purpose a common milk-glass speculum is best. Such an examination is necessary if information is required upon the colour of the vaginal mucous membrane and the changes of the mucous membrane of the vaginal portion of the uterus. The vagina is at times very wide in pregnant women, so that to thoroughly examine its walls it is necessary to employ a Sims' instead of a milk-glass speculum.

If pregnancy is suspected the uterine sound is not to be used at all, or only by very practised operators.

3. *The Combined Examination*

The weight of the gravid uterus at term renders it superfluous to make a simultaneous examination with one hand in the vagina or rectum and the other on the abdominal walls. If the fœtus is very movable it has this great advantage, that by it the presenting part can be fixed. In the first months of gestation the combined examination is indispensable; for until the uterus has enlarged so as to be felt through the abdominal walls we can by this method alone ascertain its size, shape, and consistency, and these are the three conditions of the uterus from which the practised observer can very easily form a diagnosis.

The combined examination is so made that the hand placed on the abdomen always corresponds to the finger within the vagina (or rectum), and thus the pelvic organs are felt between the two hands. In this way the most accurate possible information is obtained concerning the uterus and the neighbouring organs.

B.—THE DIAGNOSTIC SIGNS OF PREGNANCY

1. *The Individual Signs of Pregnancy*

All the changes which gestation produces in the maternal organism are available as signs of pregnancy. But it is evident that some, occurring also under other circumstances, will be of slight diagnostic value, whilst others, accompanying pregnancy exclusively, or almost so, will enable a diagnosis to be arrived at with an absolute or approximate certainty. To the former class especially belong the subjective symptoms which, as a rule, accompany pregnancy; they are—a feeling of faintness and of general malaise, mental depression, vertigo, headache, toothache, and, above all, nausea and vomiting in the morning. All these may be due to another cause; but in pluriparæ, especially, in the absence of more reliable signs of pregnancy, they are of great value as showing that conception has taken place. As pregnancy advances these are set on one side by other and more valuable signs.

1. *The cessation of menstruation.*—If this sign is met with in a healthy woman who has, till now, always been regular, and in whom there is a possibility of pregnancy occurring, it is highly probable that conception has taken place; yet it must be borne in mind that menstruation may have ceased from other causes, and also that after conception the catamenia may once reappear. The regular persistence of menstruation during pregnancy, if it occurs at all, is one of the greatest rarities. The physician will not err if he immediately denies pregnancy in a woman who, though regular at the periods, yet supposes she is pregnant.

2. *The changes in the genital organs.*—These are of very great importance, and at times may afford the most positive signs of pregnancy. Certainly an enlargement of the abdomen does not prove pregnancy, and the latter is not clearly established even if the uterus be found to have enlarged. But after some practice in the combined examination, and with sufficient practical experience, there will rarely be any doubt on the significance of a uterine enlargement at about the third month of a supposed pregnancy, and sometimes much earlier. The softness of the vaginal portion, the round opening of the os uteri, the œdematous swelling, the velvety conditions and increased secretion of the vaginal mucous membrane, as well as the peculiar wine-lees-like colour, are valuable signs of the probability of pregnancy. They do not, however, possess the decisive significance which Holst ascribes to them. The changes in the mammæ also are very important. There are pathological conditions in which the glands enlarge and secrete, and where the pigmentation of the areola and the development of its nodules have greatly increased. But these signs—especially the pigmentation of the areola as well as of the linea alba—may reach such a high degree as is never seen in any other condition, and from such a pigmentation alone pregnancy may with certainty be concluded (at least, if the examination is not made soon after delivery, which can be easily determined). But there are three absolute signs of pregnancy namely—

(1) *The presence of fœtal parts.*—There is absolute certainty of pregnancy if the internal examination reveals the presenting head or actively moving smaller parts. But although the fœtal parts can almost always be felt towards the end of pregnancy by external examination, it must not be forgotten that there are pathological conditions which in external examination of the abdomen closely simulate small fœtal parts. This is the case with small, subserous, uterine fibroid tumours, and still more so with carcinoma of the peritoneum or of the omentum. If the tumours are surrounded by ascitic fluid, or this be incapsulated, the gravest mistakes can be made by attending to this sign exclusively.

(2) *The movements of the fœtus.*—If they are felt by a practised observer gestation is, of course, certain, since it is not likely that he could mistake the peristaltic movements of the bowels for them. The woman's own statement that she felt the fœtus moving is a very uncertain sign for diagnosis, on account of the many mistakes to which she is subject.

(3) *The sounds of the fœtal heart.*—This is, without question, the most certain sign of pregnancy. It is, however, possible that the very frequent sound of the maternal heart conducted to the abdomen, or the pulse of the aorta, may be mistaken for the sounds of the fœtal heart. They begin to be audible at the end of the first half of pregnancy. The uterine souffle can be heard much earlier, but this is by no means a sure sign of pregnancy. It is present even in chronic inflammation of the uterus, and still more frequently associated with fibroid tumours of the uterus.

2. *The Differential Diagnosis of Pregnancy*

It would be impossible to mention in the present chapter all those various conditions which have been mistaken for pregnancy. We presume that there is at least a circumscribed tumour in the abdomen, which rouses the suspicion of pregnancy.

The first thing is to find out whether the tumour is the enlarged uterus or not. For this purpose the combined internal and external examination is highly important. If by it there is found a small tumour with the cervix attached to it, and separate from a larger one, the small tumour is the uterus and the larger is extra-uterine. To confirm this diagnosis the uterine sound may be introduced, but for the experienced observer this is superfluous, and for the inexperienced too dangerous to be used in such a case. If the tumour is very large it may be very difficult indeed to recognise the normal uterus. But even the existence or non-existence of pregnancy can be easily determined in some other way.

Although the tumour be known to represent the uterus, it may, nevertheless, be impossible in the first two months of pregnancy to explain its enlargement with absolute certainty, whilst in the third month the diagnosis is, as a rule, not attended by any notable difficulty. For the shape and position of the uterus in various pathological states may resemble those of the gravid organ, especially in chronic inflammation, in interstitial and submucous fibroid tumours, in the former only if they do not form knobby projections, and in hæmatometra. The consistency of the uterus, however, is materially different in these conditions. The gravid uterus from the second to the fourth month is peculiarly soft and almost doughy, whilst if fibroid tumours are present it is harder; in hæmatometra it is firmly elastic, or even shows fluctuation. In chronic inflammation the uterus may be of a similar consistency, although it is usually much harder. Moreover, as a rule, in chronic inflammation the uterus is somewhat painful, and by means of the previous history and subjective sensations this affection can be excluded. If attention is paid to the history and the subjective symptoms there will scarcely be a question of hæmatometra, for it is distinguished from the gravid uterus by its greater firmness and by the early disappearance of the cervical canal. Under certain circumstances this last symptom may be important also for the diagnosis of submucous fibroid tumours. In spite of all the above-mentioned peculiarities it may be exceedingly difficult to arrive at a diagnosis between pregnancy and fibroids or infarctus after one examination. It is, therefore, necessary to repeat the examination after some weeks, and, since under pathological conditions the uterus does not enlarge as rapidly as in the first months of pregnancy, any doubtful point will thus be cleared up.

It must be readily admitted that even in the third month the exact diagnosis of pregnancy may not be made without difficulty, yet, as a rule, it is not the case. If by the combined examination the uterus is found slightly anteflexed and corresponding in size to the time of gestation elapsed, not painful on pressure, and of a peculiar softness, and the woman moreover is healthy, and her otherwise regular menstruation has ceased since a certain time, then the diagnosis of pregnancy is established with an absolute certainty.

At a later period the differential diagnosis of pregnancy becomes more and more easy, and in the fifth month there is scarcely a case where the experienced accoucheur can be in doubt after a repeated examination. At that time, and commonly also much earlier, everything points so distinctly to pregnancy that in cases in which the so-called absolute signs are wanting—such as a dead ovum or a mole—it can with certainty be diagnosed, and at the same time also the pathological product of gestation.

No doubt diseases may complicate pregnancy so as to mask it entirely, but it is in such cases far more frequent that it is overlooked, because the disease is easily recognised and pregnancy not thought of. If there be a suspicion of pregnancy repeated careful examinations will either confirm or remove it.

3. *Diagnosis between a first and a repeated Pregnancy*

It may be highly useful to be able to decide by an examination whether a woman is pregnant for the first time or whether she has been so before. The distinctive marks are characteristic to some extent, because preceding deliveries always leave distinct traces.

Now, in primiparæ the changes produced by pregnancy are in general those described above. To facilitate the object in view we shall here again group them as they appear at term.

The skin of the abdomen is firm and tense, the abdominal walls are difficult to compress, so that the uterus can at times hardly be felt through them. And unless the uterus were also harder this would be the case in a still higher degree, and this is also the reason why it is often so difficult to feel the fœtal parts. The great tension of the abdominal muscles in the last months of pregnancy causes solutions of continuity of the rete Malpighii, which are seen as reddish-brown or slate-coloured stripes. Commonly they are seen only on the abdominal walls, often also on the thighs and nates, and quite commonly also on the mammæ. The breasts also appear firm, roundish, and not pendulous.

The vulva does not gape much, if at all. The fourchette is intact; in the vaginal outlet the hymen is distinctly recognisable as a seam continuous along the base, but torn at one or more spots. The thickening of the urethra contained in the upper wall of the vagina often projects into the vaginal outlet like a bluish-red wrinkled *wedge*. The vagina is narrow and feels rough, which is due partly to the folds of the mucous membrane, partly to its swollen papillæ. The vaginal portion of the uterus is loose and soft, the os closed, or towards the termination of pregnancy often pervious for one finger. Its margin, consisting of the vaginal mucous membrane which covers the boundary between the cervical canal and the vaginal portion of the uterus, can be felt as a sharp edge. The os uteri forms a continuous circle without any interruption, and the smooth feel which it presents all around is broken only by one or more swollen follicles. Towards the end of pregnancy, when the head descends, the anterior lip of the vaginal portion is distended. If in such cases the cervix is pervious, it may be seen that the cervical canal has retained its normal length. In the last month, and sometimes much earlier, the head is, as a rule, already in the small pelvis, and pushes the anterior vaginal vault downwards, or it at least is fixed on the brim.

In pluriparæ there are always traces of the previous deliveries, and the signs just enumerated will therefore vary to some extent.

The skin of the abdomen is soft and wrinkled, and folds are formed if the hand be drawn over it. The usually flaccid uterus can nevertheless be felt through the thin and flaccid abdominal walls. Some fœtal parts are so very distinctly to be felt as to appear as though they were situated immediately beneath the integuments. The abdomen having previously been distended, the fundus uteri can almost always be easily defined, because the epigastrium can be easily and deeply compressed. This state is met with in primiparæ only in the last months of pregnancy. The cicatrices are the same as in primiparæ, only besides the cicatrices of a more recent appearance there are almost always old and white ones covered with transverse wrinkles, such as are only exceptionally seen in primiparæ during parturition or rarely in the last months of pregnancy. The breasts are less firm, pendulous, and in the integument there are always to be seen older cicatrices.

The vulva gapes more or less, and is frequently of a bluish colour, due to transparent normal or varicose veins. The fourchette is rarely in the same state as in primiparæ, but at its place there are often cicatrices from former ruptures. Instead of a distinct hymen, torn in primiparæ, but still continuous at its base, only traces of it are here found, as warty prominences, the so-called carunculæ myrtiformes. The hypertrophied vaginal walls, usually somewhat projecting into the vaginal outlet, are smooth and devoid of wrinkles, and the vagina is spacious and soft, and rarely swollen papillæ are felt. The vaginal portion of the uterus does not project as a conical wedge, but hangs down into the vagina as a soft and swollen flap. The external os uteri is open, and the canal gradually narrows to the internal os. The sharp edge of the external os, as seen in primiparæ, is wanting, and its continuity on the right and left side interrupted by distinct gaps, and frequently even by deep fissures. Though these solutions of continuity are sometimes very inconsiderable, there is, nevertheless, at each side a retraction by which an anterior and a posterior lip can be recognised, distinctly separated from each other. At the termination of pregnancy, and sometimes also at the end of the ninth month, the internal os opens, but the cervix remains in the state above mentioned, representing a funnel with the pointed end looking upwards, until parturition begins. Through the open os the presenting head can frequently be felt at the last period of gestation. The head, however, rarely descends into the true pelvis at that time, but remains movable on the brim. Often it is deviated somewhat to the right or to the left, and the commencing pains only fix it in the inlet.

4. *Diagnosis of the Time of Pregnancy*

It is very important to be able to determine from an examination the time during which pregnancy has gone on. The nature of the case, however, renders accuracy impossible, because certain changes do not accurately correspond to definite periods. By long experience it becomes less difficult in primiparæ to determine the time of pregnancy from an examination, because its changes are more characteristic and its course more typical than in pluriparæ. With this object in view, we shall enumerate the changes as they take place at different periods of pregnancy.

First month.—The uterus is already increased in size. The vaginal portion is somewhat loosened, and the vaginal mucous membrane secretes more. The changes are approximately the same as at menstruation, and the uterus only is larger and thicker. The enlargement may be taken as due to gestation if there has been an opportunity of examining the uterus by palpation before conception, but even then it is only a probability. At times the enlargement may be very apparent.

Second month.—Combined examination easily shows the uterus to be enlarged. The enlargement is that of a middle-sized orange, and the thickness also has greatly increased. The uterus is pretty hard, its axis rather more curved, so that the slight flexion over the anterior surface normally present is augmented. At the same time the fundus sinks forwards, and the anteversion is more distinct. The os uteri remains soft, loosened, and is somewhat roundish. The mammæ become fuller, areola and linea alba begin to assume a brown colour.

Third month.—The fundus uteri is by the combined examination to be felt in its whole extent as a soft and almost doughy body, situated in the anterior vaginal wall. It is almost as large as a child's head, and as it sinks somewhat more forwards the portio vaginalis lies more backwards, and is therefore much less easily accessible.

Fourth month.—The uterus is now of the size of a man's head. Its fundus can be felt extending above the symphysis, and by combined examination it is found to fill up the whole anterior part of the pelvis, and to lie somewhat upon the symphysis. The uterus is soft and uneven, especially in pluriparæ, and harder at some places (due to the fœtal body). By the combined examination also "ballottement" may often be produced. In this month, and often during the previous one, the uterine souffle is heard on one or both sides.

Fifth month.—The uterus can be distinctly felt externally in the middle line between the symphysis and umbilicus, usually somewhat more to the right. The vaginal portion is looser, and the external os in pluriparæ is already pervious for one finger. Towards the end of this month the mother feels the movements of the fœtus, and by auscultation the sounds of the fœtal heart can be heard.

Sixth month.—The fundus uteri reaches upwards to the umbilicus. Fœtal parts can often, though only indistinctly, be felt in primiparæ, whilst in pluriparæ they can, as a rule, be distinguished without difficulty. Pigment is now thickly deposited, and the mammæ are full and firm.

Seventh month.—The uterus is from two to three fingers' breadth above the umbilicus. The circumference of the abdomen at the umbilicus is about 91 cm., or 3 feet; at the middle point between the umbilicus and the symphysis, 94 cm., or 37 in., and the distance between the xiphoid process of the sternum and the symphysis is on an average 42 cm., or 16·5 in. The umbilicus is obliterated. The fœtal parts can be felt more distinctly. Through the hypertrophy and œdematous swelling of the mucous membrane and submucous connective tissue around the portio vaginalis and the vaginal vault, that portion of the cervix uteri projecting into the vagina becomes somewhat shortened. Whilst in primiparæ the cervix uteri is still perfectly closed, in pluriparæ often the whole cervix is pervious up to the internal os for the examining finger. In the former even now the head is felt presenting as a hard body, which disappears on being pushed by the finger, and immediately after falls back upon it (ballottement). The mammæ are much fuller, and commonly a thin milky fluid can be pressed out from them.

Eighth month.—The upper border of the uterus is between the umbilicus and the scrobiculus cordis. The circumference of the abdomen at the umbilicus is on an average 95 cm., or 37·5 in.; at the middle point, between the umbilicus and the symphysis, 97 cm., or 38 in.; and the distance between the xiphoid process and the symphysis, 43½ cm., or 17 in. The abdominal walls of primiparæ are so very tense that the epigastrium can only slightly be pressed in, whilst in pluriparæ this can easily be done. The umbilicus is perfectly smooth. The position of the fœtus can almost always be determined with ease and certainty by an external examination. In primiparæ the head is, as a rule, easily movable on the brim; in pluriparæ it is more frequently somewhat deviated to one side.

Ninth month.—The uterus reaches close to the scrobiculus cordis, and therefore acquires its highest position. The abdomen measures at the umbilicus 97½ cm., or 38·4 in., beneath it 99 cm., or 39 in., and the distance between the xiphoid process and the symphysis is more than 44 cm., or 17·3 in. The umbilical ring is a little convex. In primiparæ the external os frequently opens so much as to admit the tip of the finger, but the cervix is rarely pervious at this time. In pluriparæ it is easy to pass the finger up to the internal os, and this, also, is sometimes opened, so that the bag of waters or the presenting fœtal part can be felt. In primiparæ the head lies, as a rule, on the brim, and can only be moved with difficulty, and in the second half of this month it begins to descend into the true pelvis. In pluriparæ the head is often freely movable, and is more frequently deviated to one side, so that only by pressure from without it is accessible to the fingers examining within the vagina. Rarely small parts are felt presenting. From the mammæ a bluish liquid, mixed with thick whitish-yellow stripes can be pressed out.

Tenth month.—The uterus has again descended and is about on the same height as in the eighth month. The circumference of the abdomen is not diminished; at the umbilicus it is 99 cm., or 39 in., and beneath it 100 cm., or 39·4 in.; the distance between the xiphoid process and the symphysis measures 45½ cm., or 17·9 in. The epigastrium is, at present, easily compressible, even in primiparæ, since the fundus has descended, but the outlines of the uterus cannot be clearly made out. In pluriparæ this sign between the eighth and the tenth month is, for the reasons already stated, not so distinct. The fundus uteri sinks far forwards, and the umbilical region is prominent like a vesicle. The signs on examination of the vagina are quite different in primiparæ and pluriparæ. In primiparæ the whole anterior vault of the vagina is pushed downwards by the head, which has already entered the true pelvis, and sometimes even presents at the pelvic outlet. Thus, the fold of the vagina, which forms the anterior vaginal wall, disappears, the mucous membrane is continuous with the anterior lip on a level with the external os, and the anterior part of the portio vaginalis disappears. This, of course, does not indicate a distension of the cervical canal. The presenting head so alters the position of the cervix and the lower uterine segment that the external os looks to the symphysis, the internal to the sacrum. From this, also, it appears as if the head were situated closely above the internal os. If, as it frequently happens, the cervix is already pervious at this time, or if it is opened by the first contractions, it will be seen that in order to pass the finger into the cavity of the uterus it has to traverse the whole cervical canal, the length of which measures about 3 cm. (1 inch). When the finger has entered the uterine cavity the whole lower segment of the uterus may be drawn forwards by the finger curved like a hook, and thus the direction of the cervix will be altered. But even then it is seen that the cervix has lost nothing of its length. It, however, occurs exceptionally that contractions during pregnancy have forced asunder the internal os, and thus at the actual commencement of labour the head is found to be immediately upon the external os. In pluriparæ the external os is considerably wider than the internal, which is almost always pervious. During pregnancy the latter may even be pervious for two or three fingers. The head is frequently movable on the brim, often more or less deviated to one side. The mucous membrane of the vagina and vulva is softer and loosened and secretes abundantly a whitish mucus.

With regard to time, all those changes occur more regularly in primiparæ, and therefore the period of pregnancy can pretty accurately be determined under normal conditions. Abnormal conditions (*e. g.* twins or contracted pelvis) may here, also, give rise to great difficulties in determining the period of gestation. This is also chiefly and most frequently the case when the fœtus is not in the head position, or when the presenting head has not entered the pelvis. The vaginal portion is then not distended, and the appearances of the tenth month differ from those described above.

5. *Diagnosis of Multiple Pregnancy*

Various signs have been given as indicating twin pregnancy, but they are very uncertain. Those signs are—the large and broad abdomen, a distinct longitudinal furrow of the uterus, fœtal movements felt in both sides, which are, moreover, very active and uncomfortable, very great derangements during pregnancy, œdema of the legs, higher position of the lower segment of the uterus, and absence of a presenting part. The uncommonly great size of the uterus is a sign of value, and its presence always demands an accurate examination in this respect. The longitudinal furrow is, as a rule, absent in twin cases, and is sometimes very distinct in simple pregnancy (this furrow is a remnant of the fœtal origin of the uterus from two halves—uterus arcuatus); the other signs mentioned are still more uncertain. A sure diagnosis can only be arrived at by careful palpation and auscultation.

By the former very important information is obtained. Under favorable conditions it can be shown by it that the parts felt in the uterus belong to only one child, or that three large fœtal parts are felt, which cannot possibly belong to one (normally developed) child. This is always to be compared with the result obtained by an internal examination, and thus points may be cleared up which had remained doubtful by palpation alone; if one head, for instance, is low down in the pelvis, and by palpation two large fœtal parts are found in the abdomen. Auscultation also in connection with palpation and internal examination is highly useful, especially if the position of the presenting part can be recognised, which is frequently the case already during pregnancy. If the face or the head is felt presenting in the first position, small parts being felt in the left of the fundus, and heart's sounds audible to the right, it may be concluded that the head presenting in the vagina does not belong to the child found by the external examination. The diagnosis in such a way succeeds more frequently during parturition.

By careful attention to auscultation it is possible in most cases to hear the heart's sounds at two separate places. That those sounds do not come from the same heart may be concluded if they are equally loud at various places of the abdomen, whilst in the intermediate places they are low or disappear entirely; or if the sounds heard (at the same time by two observers) show a varying frequency. One of these places can always be found; the other will be more difficult to detect, and limited only to a very small spot, frequently immediately above a pubic ramus. If the heart's sounds vary greatly in their frequency this probably indicates different sexes, and the more frequent heart sounds correspond to the female.

To diagnose a triple pregnancy is extremely difficult, and even the most practised observers may fail in it under extraordinarily favorable circumstances.

6. *The Diagnosis of a Living or a Dead Fœtus*

It is always desirable, and may also be practically useful, to know even during pregnancy whether the fœtus is alive or dead.

Every healthy pregnant woman, unless there be special reasons to the contrary, may be supposed to bear a living fœtus. Movements perceived by the hand placed upon the abdomen, and distinctly audible fœtal heart sounds, indicate with certainty that the fœtus is alive.

The following circumstances may excite the suspicion that the fœtus is dead, viz.—

Diseases of the mother which experience has shown to be frequently followed by the death of the fœtus, as syphilis; cessation of the fœtal movements, which have previously been distinctly felt by the mother (this sign has no absolute value, and is devoid of any significance when the uterus has begun to contract, since the woman only occasionally perceives a fœtal movement); diminished size and greater softness of the uterus; flaccidity and shriveling of the mammæ; sensation of the mother as if a heavy body were tumbling to and fro in her abdomen; debility, bad taste in the mouth, and other anomalous sensations which did not exist before.

Certainty of death having taken place is obtained only when the os is open and allows the loose cranial bones to be felt distinctly; also when the sounds of the fœtal heart, which, in the absence of other pathological conditions can always be distinguished by a repeated careful examination, cannot be heard.

Literature.—W. J. Schmitt, Sammlung zweifelhafter Schwangerschaftsfälle u. s. w. Wien, 1818.—Hohl, Die geburtshülfliche Exploration. Halle, 1833.—W. F. Montgomery, An Exposition of the Signs and Symptoms of Pregnancy, &c. London, 1837.—F. H. G. Birnbaum, Zeichenlehre der Geburtshülfe. Bonn, 1844.—H. Deventer, Neues Hebammenlicht. Jena, 1717. Cap. XIII-XXII.—M. Puzos, Traité des accouch., publié par Morisot Deslandes, 1759. Chap. V, p. 55.—M. A. Levret, L'art des accouch. Paris, 1761, 2e éd., § 448.—J. G. Roederer, Elem. art. obst. Gottingæ, 1753.—J. L. Baudelocque, L'art des acc. 8e éd., Paris, 1844, § 371 seq.—Jörg, Taschenbuch für ger. Aerzte u. Geburtsh. Leipzig, 1814, p. 65.—Kiwisch, Klinische Vortr. über Krankh. d. weibl. Geschl. 4 Aufl., Prag, 1854, B. I, p. 26.—Holst, Beiträge zur Gyn. u. Geb. 2 Heft, Tübingen, 1867, p. 63.—Veit, Krankh. d. weibl. Geschlechts. 2 Aufl., Erlangen, 1867, p. 252.—Mayor, Bibl. univ. des sciences, &c. T. IX, Genève, Nov., 1818.—Lejumean de Kergaradec, Mémoire sur l'auscult. appl. à l'étude de la gross., &c. Paris, 1822.—Ulsamer, Rhein. Jahrb. für Med. u. Ch. 1823, B. VII, p. 50.—Haus, Die Auscultation in Bezug auf Schwangerschaft. Würzburg, 1823.—Ritgen, Mende's Beob. u. Bem. aus d. Geb. Göttingen, 1825, B. II, p. 38.—D'Outrepont, Gemeins. deutsche Zeitschr. f. Geb. 1832, B. VII, p. 21.—Kennedy, Observations on Obstetric Auscultation. Dublin, 1833.—H. F. Naegele, Die geburtsh. Auscultation. Mainz, 1838.—Depaul, Traité theor. et prat. de l'ausc. obst. Paris, 1847.—Martin, M. f. G., 1856, B. 7, p. 161.—Frankenhäuser, M. f. G., B. 14, p. 161.—Hüter, M. f. G., B. 18, Suppl., p. 23.—F. A. Kiwisch, Ritter v. Rotterau, Klinische Vorträge über sp. Path. und Th. d. Krankh. des weibl. Geschlechts, II Abth., 2 Aufl. Prag, 1852, p. 298.—Birnbaum, Ueber die Veränderungen des Scheidentheiles. Bonn, 1841.—Holst, Beiträge zur Gyn. und Geb., I, p. 130 u. 150, und II, p. 164.—Hecker, M. f. G., B. 12, p. 401, und Klinik d. Geb., I, p. 32.—Hüter, M. f. G., B. 14, p. 33.—Schroeder, Schw., Geb. u. W., p. 9.—Duncan, Edinburgh Medical Journal, March and April, 1859, and Sept., 1863, and Researches in Obstetrics, p. 243.—Taylor, American Medical Times, June, 1862 (s. Schmidt's Jahrb., B. 117, p. 178).—Spiegelberg, M. f. G., B. 24, p. 435, und De cerv. ut in gravid. mutat., &c. Regimonti, 1865.—P. Müller, Unters. über die Verkürzung der Vaginalportion, &c. Würzburg, 1868 (Scanzoni's Beiträge, B. V, Heft 2).

CHAPTER IV

DIETETICS OF PREGNANCY

It is urgently necessary to regulate on rational principles the mode of life of a pregnant woman. For although pregnancy is a physiological process, yet the state it produces deviates so greatly from what is commonly considered a healthy condition that even a slight error in diet, which would otherwise be of no significance, may now be attended by injurious consequences to both mother and child.

The first principle is to allow the woman to adhere to her usual habits of life, to warn her against extraordinary exertions or other grave hygienic errors, and to comply with the chief demands of hygiene, especially cleanliness and fresh air.

The genital organs are to be washed frequently, and general baths of a temperature not more than 88° F., as well as exercise in the open air, are to be recommended. Anomalies in the composition of the blood, deranged digestion, and sleeplessness, too easily follow a persistent sitting or recumbent mode of life. Extraordinary and great exertions, such as raising heavy weights, dancing, riding in jolting carriages or on horseback, are to be avoided.

Also, with regard to food and drink, the woman must adhere to what she has been accustomed, and avoid everything difficult of digestion, and especially highly spiced meats or heating drinks. The stomach, especially in the evening, must not be overloaded. The often fanciful appetites of pregnant women may be satisfied provided no harm can be done by it. A regular action of the bowels is very important, but by no means must laxative or drastic medicines be continually used throughout the whole period of gestation. It is different, however, at the end of pregnancy, and if constipation has been persistent purgatives may now be more freely used. Very mild laxatives, such as light magnesia, which not only regulates the bowels but diminishes the very disagreeable acid eructations, may from time to time within the last months be very useful. Local or general bloodletting is never demanded by pregnancy itself, and ought only under very exceptional circumstances to be had recourse to in pregnant women.

The clothing of the woman is to be so arranged as not to be too tight, and to keep her feet and abdomen warm.

Special attention with regard to the subsequent nursing is to be paid to the mammæ, and particularly to the nipples. The breasts must be kept moderately warm and protected from pressure. If the nipples are very tender and sensitive, they are best hardened during pregnancy by washing with cold water or with spirituous substances. If they are greatly depressed attempts may be made to draw them out; this, however, is to be done cautiously, in order not to excite prematurely uterine contractions.

Tranquillity of mind is of great value to pregnant women, especially to those who have no fixed employment that occupies their time. We must endeavour to keep them in a cheerful and contented disposition, to guard them against emotions, and to dissipate as much as possible by rational persuasion the often exaggerated fear of parturition.

Vomiting, pains in the back and in the side, toothache, fainting, &c., which so often occur in the last months, may be treated symptomatically. Treatment, however, is often unsuccessful, and if these affections have not reached an unusually high degree, and if delivery is not far distant, the woman may be told that they are the attendants upon her condition, which she has to support, but that they will spontaneously disappear with her delivery.

Literature.—L. J. Boër, Natürliche Geburtshülfe. 3 Aufl., B. 1. Wien, 1817 p. 48.—F. A. v. Ammon, Die ersten Mutterpflichten und die erste Kindespflege. 13 Aufl. Leipzig, 1868.

PART III

PHYSIOLOGY OF PARTURITION

CHAPTER I

THE CAUSES OF NORMAL LABOUR

NORMAL parturition is the expulsion of the fœtus at term from the uterus through the maternal passages by natural forces.

Two factors, therefore, are chiefly concerned—(1) the expelling powers, and (2) the obstacles which they meet with.

The latter is caused by the ovum, the object to be expelled, as well as by the state of the parturient canal which the ovum has to traverse.

The normal conditions of the parturient passages and of the ovum, which, as far as they themselves are concerned, ensure a normal labour, have already been spoken of. We have now only to consider the expelling powers, their mode of origin, and the way in which they act.

I.—CAUSES OF THE SETTING IN OF LABOUR

The natural forces which expel the ovum chiefly consist in the contractions of the unstriped muscular fibres of the uterus.

It certainly is one of the most interesting subjects of midwifery to study how those contractions are excited at a certain time, and how they are continued, gradually increasing in power until the ovum is expelled.

Anatomical preparations and physiological experiments have conclusively shown that the sympathetic is the motor nerve of the uterus. Whether the sacral nerves also have any influence in causing uterine contractions is very doubtful; more probably they are inhibitory nerves.

The influence of the sympathetic in the expulsion of the ovum is called forth by the irritation which the ovum exerts upon the terminations of the nerves on the inner surface of the uterus; by reflex action that irritation is transformed into motor activity. Only towards the end of the tenth month this irritation begins to be exerted, because only then, through fatty degeneration of the decidua, the ovum becomes a foreign body in the uterus. The fatty degeneration of the peripheral layers of the ovum causes a separation between the ovum and the uterus, which the resulting contractions complete. The process is essentially the following:

Towards the end of pregnancy a gradually increasing fatty degeneration of the decidua takes place. The organic connection which until then existed between ovum and uterus is now solved by the degeneration of the cells between those structures. At all the places where the degeneration has reached a certain stage the terminations of the nerves are irritated. But to obtain a reflex action, namely, the contraction of the uterine muscles, a certain amount of continuous irritation is necessary. This sum once obtained, a reflex action takes place in the form of a contraction, which, however, is slight at the beginning. Then a pause follows, until the sum of the irritation is again sufficient to cause a contraction. By the increase of the intensity of the contractions the uterine wall is removed from the ovum, and this separation becomes a new source of irritation to the uterine nerve-fibres. The reflex action, in the form of labour-pains, becomes more and more powerful, until they follow at last in rapid succession and complete the expulsion of the ovum.

The concurrent observations of Brown-Séquard, Obernier, and Kehrer, that the irritability of the uterine nerves progressively increases with the advance of pregnancy, would tend to explain the regular setting in of labour, as well as the not unfrequent retardation of the pains in cases where the separation of the ovum has been premature.

Literature.—C. C. Th. Litzmann, Artikel "Schwangerschaft" in R. Wagner's Handwörterb. d. Phys., III, 1, p. 107.—G. Veit, Verh. der Gesellsch. für Geb. in Berlin. 7 Heft, 1853, p. 122.—F. A. Kehrer, Vergl. Phys. d. Geb. d. Mensch. u. d. Säugethiere. Giessen, 1867, p. 8.

CHAPTER II

THE EXPELLING POWERS

THE powers which expel the ovum from the uterus are chiefly vested in the contractions of the unstriped muscular fibres of the uterus. The contractions and the elasticity of the vagina and the abdominal muscles are to be considered only as auxiliary forces.

A.—CONTRACTIONS OF THE UTERUS

With each contraction the uterus alters its shape. Whilst towards the end of pregnancy it has a flattened ovoid form, it becomes more globular during a pain. The longitudinal and transverse diameters are somewhat diminished, whilst the antero-posterior diameter enlarges. By this and the contractions of the muscular fibres of the broad and round ligaments of the uterus the fundus is somewhat tilted forwards, and consequently the circumference of the abdomen at the umbilicus becomes proportionately greater than that between the umbilicus and the symphysis. The size of the entire abdomen becomes greatly diminished in the course of parturition, and this is due partly to the contractions of the uterus and to the escape of the liquor amnii, but chiefly to the descent of the presenting fœtal part, so that its measurements become those of the eighth month. The circumference at the umbilicus is on an average 96½ cm., or 39·2 in.; at the middle point, between the umbilicus and symphysis, 96 cm., or 39 in.; and the distance between the xiphoid process and the symphysis 40¾ cm., or 16 in. These measurements are considerably smaller than in the seventh month.

As in almost all unstriped muscular fibres, the reflex action following upon an irritation is slow and gradual, and, according to the degree of irritation, of varying intensity and duration. As already explained above, a certain sum of irritation is necessary to cause a contraction. At the commencement, corresponding to the slight irritation, the contraction is feeble and short, and the time which is required to obtain the necessary sum of the persistent irritation for a new reflex action as comparatively long. But with the increasing separation of the ovum from the uterine wall the irritation also becomes stronger, and the contractions, therefore, gain in intensity and duration, and follow each other in progressively shorter intervals. At the height of the expulsive stage the irritation is so considerable that the contractions are interrupted only by short pauses. The contractions slowly increase in strength, remain a short time at their acme, and with the exhaustion of the irritation they gradually remit.

The sympathetic ganglia being the motor centres of the uterus, the contractions, therefore, are involuntary. They cannot be produced by volition, or slackened or suppressed by it. Yet mental excitement is decidedly capable of altering the force of the contractions, and it remains to be settled whether the sacral nerves are inhibitory in that respect.

It is also doubtful as yet whether a normal pain has a peristaltic course, or whether it simultaneously attacks the whole organ. In experiments on bitches and rabbits it can easily be seen that the uterine contractions of those animals are clearly peristaltic, but Obernier remarks that in some cases the contractions are so rapid that the single stages of them can hardly be distinguished. The peristaltic is proved only as regards the uterus bicornis. To judge from analogy, it is probable that in woman also the pain has a peristaltic course (and then it certainly begins in the muscles of the ligaments, and proceeds from the fundus towards the cervix), but from the rapidity with which the wave of the contraction passes over the whole organ it will be practically preferable to consider it simultaneous.

To explain the expulsion of the ovum, the assumption of a peristaltic contraction is, on the whole, unnecessary. The muscular fibres of the fundus are more strongly developed than those of the body, and they again more so than those of the cervix. In the cervix, again, there is a spot which is entirely devoid of muscular structure. If, therefore, a simultaneous contraction of the whole organ takes place, the place devoid of muscular fibres must always be more and more enlarged, and this process of enlargement is only very slightly impeded by the circular fibres which surround that opening. The dilatation of the cervix is effected in the following way:

The presenting part of the fœtus, or the bag of waters, is pushed by the contractions of the uterus against the internal os uteri, so that it is forcibly separated. By its dilatation the cervix becomes a portion of the uterine cavity, and thus the presenting part gradually opens the external os uteri. The dilatation of the internal os uteri takes place exceptionally in primiparæ towards the end of pregnancy, but here also by contractions of the uterus. The more the cervix is dilated the thinner it becomes, until it is finally drawn back over the presenting part. The uterus and vagina then form one common cavity. At that time, usually somewhat earlier, the membranes rupture, and by the partial escape of the waters, and by the consequent diminution of the bulk of the ovum, the separation of the ovum from the uterus becomes more complete. The now more powerful contractions expel the fœtus, and immediately afterwards the membranes with the placenta. By the detachment of the placenta the large vessels which carry the maternal blood to it are opened up; but they are soon compressed by the firm contraction of the uterus, and the supply of blood to them is kept within narrow limits, so that the formation of thrombi is favoured in the former placental insertion.

As the name—pains—already indicates, the contractions of the uterus are attended by pain. It has its principal seat in the loins, and thence radiates into the abdomen and the thighs. It is chiefly caused by the pressure to which the terminations of the nerves contained within the contracting muscular fibres are exposed. The pain is of varying intensity, but most considerable when the fœtal head passes through the very sensitive vulva supplied by cerebro-spinal nerves. Moreover, the intensity of the pain varies in different individuals. Sometimes the woman is greatly excited and bathed in perspiration; or vomiting may set in, and at the moment of the birth of the head the woman may become quite unconscious.

The influence of a pain upon the whole organism is shown by the increase of pressure in the arterial system. The frequency of the pulse increases from the beginning of a pain to its acme, and slowly diminishes with its cessation.

The force which the uterus is capable of exerting also varies in different individuals. From his very exact researches Schatz has found that in the passive organ during pregnancy the pressure which the "tonus" of the uterine and abdominal muscles exerts upon the uterine contents amounts to 5 centimètres of mercury. That pressure remains the same during parturition in the intervals between the pains. The intra-uterine pressure rises in the intervals between the pains, only when the muscles of the contracting uterus begin to thicken after a part of the fœtus has been expelled from the uterus. Under normal conditions the increase of pressure scarcely amounts to one half of the uterine pressure which existed at the commencement of parturition. During the pains the uterine pressure rises considerably. In the cases observed by Schatz the pressure of the pains, together with that of the abdominal muscles necessary to terminate delivery, amounted to 80—250 mm.

B.—THE CONTRACTIONS OF THE VAGINA

The comparatively narrow vagina at first opposes by its elasticity the progress of the ovum, but the presenting part of the fœtus soon forcibly distends it. If the greatest circumference of the fœtus has passed, the expulsion of the succeeding fœtal portions, together with the afterbirth, is considerably facilitated by the elasticity and the contractions of the unstriped muscular fibres of the hypertrophied vaginal walls.

C.—ABDOMINAL PRESSURE

The abdominal pressure has a considerable influence upon the progress of the child. It can be voluntarily called into play. However, at the acme of the labour-pains the muscular action of the abdominal walls is involuntarily exerted by reflex action. Abdominal pressure acts in the following way:—The extremities are pressed against some firm support, and thus the trunk is fixed; by deep inspiration the diaphragm is pushed downwards; the abdominal muscles now contract, and the diaphragm, which descends still deeper, partly by its own contraction, but chiefly by the powerful action of the expiratory muscles (whilst the glottis is still closed), exerts an equal pressure upon the whole abdominal contents.

Literature.—Wigand, Die Geburt. d. Menschen. 2 Bd. Berlin, 1820, p. 197, u. a. a. St.—G. Veit, Verh. der Berliner geb. Ges. 7 Heft. 1855, p. 131.—Hohl, Lehrb. d. Geb. 2 Aufl. Leipzig, 1862, p. 385.—Kehrer, Vergl. Phys. u. s. w., p. 41.—v. Scanzoni, Lehrb. der Geb. 4 Aufl. Wien, 1867, 1 Bd., p. 227.—Schatz, Beiträge zur physiologischen Geburtskunde.—Kehrer, Vergl. Phys., &c., p. 51.—Schatz, Der Geburtsmech. der Kopfendlagen. Leipzig, 1868, p. 23.

CHAPTER III

THE COURSE OF PARTURITION

It is best to distinguish three stages in the regular course of parturition:

1. The stage of dilatation of the os uteri.
2. " " expulsion.
3. " " the afterbirth.

The stage of dilatation commences with the end of pregnancy, and terminates with the perfect dilatation of the os uteri. Sometimes, and especially in pluriparæ, the pains begin rather suddenly, so that there is scarcely a clear boundary between pregnancy and parturition; usually, however, and especially in primiparæ, the transition is gradual. The contractions, which had already been taking place for some time, and which were less observable to the pregnant woman herself than to the examining hand, now increase in intensity and frequency. The pregnant woman becomes restless and seeks support for her back, by pressing her arm against the loins or by leaning against some firm object. If, as frequently is the case with primiparæ, the os is not yet open, the first pains now render it pervious for the finger. As a rule, the cervix is found to exist in its whole length, and the dilatation of the internal os just commences. Only exceptionally have the contractions during the last period of pregnancy been so strong (*travail insensible*) that the cervix has already disappeared as such—that is to say, the canal of the cervix has become an integral part of the uterine cavity, the latter extending down to the external os. As a rule, also, in primiparæ the head is already entirely within the pelvis, but in pluriparæ it is fixed in the pelvic inlet, at least during a pain. If the internal os is dilated the tension of the external os produced by its circular fibres can be felt during a pain. The external os gradually becomes thinner, its edges quite sharp, and within it the membranes, tensely filled with liquor amnii, can be felt. If only a small quantity of water is present between the membranes and the head, the head can also be felt through them. In the interval of a pain the os and the membranes again become flaccid, and the head can be clearly distinguished. When the contractions increase in force the external os also becomes dilated, and the vaginal mucus is tinged with blood (the show), due to slight lacerations of the edge of the os. The bag of waters protrudes through the os (sometimes into the vagina in the shape of a sausage) and remains tense during a pain. When the dilatation of the os has reached a diameter of three to four inches the membranes rupture, and the liquor amnii in front of the bladder (the forewaters) runs away. The rest of the waters is retained by the head acting as a *tampon*, but usually at the beginning of a pain a portion of the waters passes by the side of the head, and is pressed out from the uterus.

Not infrequently the membranes rupture when the os is only slightly opened. The dilatation then is to be effected by the head itself; and now, since a portion of the fœtal head freely projects through the os, whilst the rest of the body is still exposed to the equal pressure of the contracting uterus, a sero-gelatinous exudation takes place under the aponeurosis of that portion of the head which, lying in the os uteri, is not subjected to compression. Thus the caput succedaneum is formed. As a rule, a small amount of blood is found under the aponeurosis, and most frequently between the bone and the epicranium.

Sometimes the membranes rupture a long time after the os has been dilated and whilst the child is passing through it, or the child is born pushing the membranes entire in front of it. It often occurs that water escapes during a pain whilst the bag of waters is still presenting. This is rarely due to an accumulation of water between the ovum and uterus; more frequently it occurs when the membranes are ruptured within the os uteri, but higher up, so that the waters can escape, although the bag protrudes. But in that case the membranes gradually become flaccid and insensibly retract, and the rupture is not so evident. A short pause in the activity of the pains usually sets in after the rupture of the membranes. Soon, however, the pains begin again more powerfully, until the os becomes so dilated that the head can pass through it; the first stage of labour is then completed.

The stage of expulsion begins with the retraction of the os over the head, and terminates with the complete expulsion of the child. Here the pains become more and more powerful, and follow each other with short intervals. Abdominal pressure becomes involuntarily engaged in the act of expulsion. The progress of the child during a pain is clearly observable, whilst in the interval of a pain the elasticity of the floor of the pelvis again presses the head somewhat backwards. The retrograde movement of the head is to some extent also caused by the cessation of the abdominal pressure, which depresses the uterus and with it the head. The way in which the head passes through the pelvis is of great interest, and an accurate description of that mechanism will be given at a later period. When the head is in the pelvic outlet the perinæum is arched forward during a pain, the labia gape, and a portion of the head becomes apparent between them. Just as within the pelvis, so here also the head somewhat recedes at the end of a pain, but the next pain again pushes the head further forward. The perinæum becomes more and more distended, the anus protrudes, and involuntary defecation at times takes place. Frenulum and labia surround with a thin ring the head, which advances with every pain, until by the next pain, or even during an interval by the abdominal pressure alone, the head is expelled through that ring. Generally the same pain or the succeeding one expels the trunk. After the birth of the head, and still more so after the expulsion of the whole child, the liquor amnii which had been retained flows off. It is usually mixed with blood from the placenta, which is already partially or entirely detached.

The stage of the afterbirth begins after the expulsion of the child, and terminates with that of the membranes and placenta. By the very considerable contraction which takes place in the uterus after the expulsion of the child the diameter of the area of the placental insertion in the uterus also becomes diminished, and, since the placenta cannot follow that contraction, it is consequently detached from the uterine wall. The separation takes place at the expense of the mother; for the villi of the fœtal chorion remain intact, and a portion of the decidua serotina belonging to the maternal uterine mucous membrane adheres to the placenta. Thus, the maternal sinuses are widely opened, and unless they be closed by the contractions of the uterus may give rise to a considerable loss of blood. For a while placenta and membranes remain within the uterus. After a varying interval the uterus again begins to contract, and pushes the afterbirth into the vagina, and usually also out of the vulva, aided by the contractions of the vagina itself.

Delivery is attended by considerable general excitement, and almost always by an elevation of the temperature. The latter, however, is not quite constant, and often very slight. It may rise to 39° C., or 102·2° F., and more, without there being any pathological changes to account for it. In general the temperature of parturient women follows the daily fluctuations. At the time approaching delivery it rises somewhat, but very frequently it falls immediately before delivery. This may, perhaps, be due to an increased radiation of heat.

The duration of parturition is very varying; at times in pluriparæ it is over in a few hours, at others it lasts many days. In primiparæ it is, on an average, much longer than in pluriparæ; in the former about twenty, in the latter about twelve hours, according to Veit.

The greater part of that time is spent in the first stage; the second stage lasts in primiparæ, on an average, an hour and three quarters, in pluriparæ one hour; but the natural duration of the third stage is very varying, and is seldom observed, because it is usually shortened by the intervention of art. Parturition may begin and terminate at any time of the day. In the majority of cases it begins in the evening between 9 and 12 o'clock, and terminates in the morning between 12 and 3 o'clock.

CHAPTER IV

THE VARIOUS POSITIONS OF THE FŒTUS

It has already been mentioned that there is during pregnancy great variety in the position and attitude of the fœtus within the uterus. Shortly before the commencement of labour in primiparæ the head of the fœtus has, as a rule, already entered the pelvis, but in pluriparæ the head is movable above the pelvic inlet, or somewhat deviated to one of the two sides. Exceptionally that deviation is so great that the commencing uterine contractions are unable to bring the longitudinal axis of the fœtus into that of the uterus, so that during parturition the child is in a more or less transverse position. In other cases it happens that under the conditions mentioned above the breech is above the pelvic inlet, or at least nearer to it than the head. A breech presentation is, then, produced when the commencing pains try to fix the fœtus in its longitudinal axis.

It follows from what has just been said that transverse presentations during parturition are very rare in primiparæ, and that they are always caused by considerable abnormalities (mostly pelvic deformities), whilst in pluriparæ it is more frequently the case.

If the head is directed downwards, and the fœtus has its normal attitude, the skull will be presenting; but if the chin is removed from the chest, and the head is flexed on the nape, the face will be the presenting part.

If the pelvic end is directed downwards, and the fœtus has its normal attitude, the breech will be the lowest part; perhaps the feet may be felt by the side of it, but they will remain behind in the further course of parturition. In rare cases the thighs may be removed from the abdomen, and the feet will then descend into the os uteri before the breech. Or, as very seldom occurs, the legs may remain in apposition to the thighs, and the knees can be felt within the os. If only one foot has descended whilst the other remains fixed against the abdomen, this presentation is called imperfect. The positions of the fœtus are, therefore, to be distinguished as—

I. The longitudinal.
- 1. Cephalic end — *a.* Head, *b.* Face — presentations.
- 2. Pelvic end — *a.* Breech, *b.* Footling — presentations.

II. The transverse.

All agree in this—that the transverse positions belong to the pathology of parturition, because as such they are usually insuperable by the forces of nature. It is far more difficult, however, to classify precisely other positions. First of all, face and footling presentations cannot be considered normal, because the fœtus has lost its normal attitude. And this is the less advisable because footling presentations, if left solely to the forces of nature, are very unfavorable for the child, and face presentations are not only tedious both to mother and child, but delivery is slow, and this alone clouds the prognosis. Breech presentations proper may, indeed, be placed in a mid position between normal and abnormal. For, on the one hand, they are relatively frequent (about 3 per cent.), the mechanical passage through the maternal parts is here rather more easy, and certainly not more difficult than in head presentations, and the prognosis is here also at least as favorable for the mother as in the latter. On the other hand, the prognosis for the child is far more unfavorable. According to Ch. Bell ('Monthly Journ. of Med. Sc.,' September, 1853, p. 225), of 2367 breech presentations, 519 children were stillborn, *i.e.* 22 per cent., and not infrequently they require in the last stage the assistance of art for the delivery of the head, so that they can certainly not be called quite normal labours. A normal parturition proper is exclusively represented by a head presentation, and even in that deviations in the position and attitude of the head occur which cannot be considered normal.

If from this place, which is devoted to the physiology of parturition, we exclude the transverse positions and consider all the longitudinal, it is not because all the latter are normal, as has been explained above, but it is done from a practical point of view. Apart from the diagnosis, the transverse positions are excluded, because in them there cannot usually be the question of a mechanism of parturition proper. Therefore, to obtain uniformity in the description of the way in which the fœtus is expelled through the pelvis, all the other positions will be considered together.

In head and breech presentations the position of the fœtal parts is of great importance for the consideration of the mechanism of parturition. A distinction is therefore made according to whether the back of the child is placed opposite the left or the right side of the uterus. The former, twice as frequent, is called the first, the latter the second head or breech position. By a vaginal examination the position of the child can be made out from the presenting part. The small fontanelle, the forehead in face presentations, or the sacrum correspond to the side towards which the back of the child is situated. It is of importance also to ascertain whether the back of the child accurately lies towards the side, or whether it is directed more forwards or backwards. Most frequently, but by no means exclusively, the back lies more forwards in the first and more backwards in the second head presentation. This observation can easily be explained by what has been stated with regard to the causes of the position and presentation of the fœtus. The circumstance also that the first head presentation is twice as frequent as the second cannot appear strange if we consider that out of twenty-four hours the woman passes sixteen in the erect and eight in the recumbent posture.

The most frequent of all presentations is that of the head; it is seen in 95 per cent. of all cases of labour (once in 1·05 cases). Face presentations occur in 0·6 per cent. (one in 166 cases); pelvic presentations in 3·11 per cent. (once in 32·1 cases); and transverse positions in 0·56 per cent. (once in 178 cases).

The frequency of the first head presentation to the second is as 2·26 to 1 (Hecker).

That of the first face presentation to the second is as 1·4 to 1 (Winkel).

The first pelvic presentation is met with three times as often as the second. Breech presentations are twice as frequent as footling.

Knee presentations are exceedingly rare; 1 in 185 breech presentations.

CHAPTER V

DIAGNOSIS OF THE FŒTAL POSITIONS

It is highly important to know, as early as possible after labour has set in, the position of the fœtus. If the presenting part is low down, and the membranes are ruptured, the position can usually be recognised without difficulty by an internal examination. But if the membranes are very tense, and the fœtal parts too high for the examining finger, the position must be determined from an external examination. By it excellent results are obtained if the uterus is flaccid in the intervals of the pains. For a less practised observer, under not too unfavorable conditions, the fœtal position is less liable to be mistaken from an external examination than from an internal made alone. The former ought never to be omitted, because it gives an excellent control over the results obtained by the internal examination.

A.—THE EXTERNAL EXAMINATION

The most important points for the diagnosis of the fœtal position by an external examination will be obtained by palpation. First of all we must ascertain if the position be longitudinal. For that purpose the accoucheur stands at the side of the bed, and places his hands upon the abdomen of the pregnant woman, so that the tips of the fingers are directed to the symphysis, the wrist to the navel of the woman. By short tapping movements it can be ascertained if a large part lies movably above the pelvic inlet, for if that is the case a distinct sensation of ballottement is obtained. If that part is slightly deviated to one side, one of the two hands feels it still more easily. If that large part is fixed in the pelvic inlet, it can also be easily felt through the lower segment of the uterus. It is, however, more difficult to feel if it has entered the small pelvis to a great extent or entirely. Nothing more can be made out then but that the perceptible tumour is continued into the small pelvis, since the hand cannot pass deeply between the symphysis and the lower segment of the uterus. It is evident that in such cases the internal examination will afford very material assistance. But to obtain information as to the position of the other large part it is best to sit on the edge of the bed, and to place both hands in such a way upon the abdomen of the parturient woman that the tips of the fingers are directed towards the sternum, and now to endeavour by short tappings to produce ballottement also in the fundus uteri. In that way the large part lying in the fundus can easily be made out.

If the position be vertical, we have to determine whether the head or the breech presents. The chief distinction between these two is that the head feels harder, and therefore ballottement is more distinct and more easily produced. Again, the head is larger and less convex. Then the discovery of the limbs, which are closely attached to the breech, may aid in the diagnosis, for in the head presentation there is first the interval of the neck, and then only the attachment of smaller parts. Fassbender draws attention to the fact that frequently the parchment-like crepitation of the cranial bones may be felt under favorable conditions through the abdominal walls, and thus the head can be distinguished with certainty from the breech. Sometimes, especially in a small fœtus, the head feels remarkably small and pointed, so that it could be easily mistaken for some other fœtal part. From it, however, it is distinguished by ballottement, which is never obtained with small parts.

The question whether the presenting part is in the first or second position is usually not difficult to decide. First of all the large part situated above must be made out by palpation; a hasty decision must not be made as to its absolute position in the abdomen; careful examination must show on which side of the large part the small ones can be felt. For it is found in great lateral deviation of the uterus that a large part can be decidedly felt on one side, say the right, of the abdomen, so that the assumption would be that the position is the second. Yet by careful palpation no other fœtal parts can be felt on the left, whilst they can be distinguished on the right of the large part; the position, therefore, is the first. It may also be difficult to feel the small parts if the back is almost entirely situated in front, and in such cases, if the head is not yet fixed in the pelvis, the back may not infrequently be brought now to the left, and at another time to the right.

Auscultation materially assists in determining whether the position is the first or the second. However, it is not sufficient to hear the sounds of the fœtal heart, but the place must be found where the sounds are most distinctly heard. In the first position the heart sounds are heard in the left lower region of the abdomen, somewhat to the outer side, and rarely only near the middle line. In the second position the sounds are very distinctly heard (corresponding to the situation of the heart in the left thoracic cavity) usually to the right, close to the linea alba, rarely far externally, but sometimes a little to the left of the linea alba. Considerable deviation of the uterus to the right or left may easily give rise to mistakes if the linea alba is taken as a limiting line, and not the middle line of the uterus. As an aid to diagnosis between head and breech presentations, auscultation is only to be used with great caution. Frequently, indeed, the sounds of the fœtal heart are heard very high up in the region of the umbilicus in breech presentations, but this is by no means constant, and occurs also in head presentations.

On the other hand, auscultation is highly useful to distinguish between face and head presentations, for in the former the thorax lies so much in that half of the uterus which does not correspond to the back that there the heart sounds are best heard. Thus, even in the first face presentation the breech is felt high up to the left, to the right of it are the small parts, and the heart's sounds are heard to the right of the linea alba. If, moreover, the occiput is felt projecting over the left pubic ramus, a face presentation is certain.

It is most difficult of all to decide from an external examination whether the breech or the feet present. The latter may be diagnosed if the breech is somewhat deviated to the side where the back lies, *i. e.* to the left in the first position.

Transverse positions are recognised by a large movable part in each side of the uterus whilst the fundus is empty, and the examining hand can more or less easily pass between the symphysis and the lower uterine segment. The distinction between head and breech is often easy, but sometimes it is, on the contrary, very difficult. The heart's sounds are usually heard more to the side in which the head lies. If through the anterior abdominal walls small parts are distinctly felt, the back lies behind and somewhat below.

B.—THE INTERNAL EXAMINATION

By the internal examination, especially when the presenting part is sufficiently low down, and when the membranes are flaccid or already ruptured, a practised observer is able to obtain precise information, not only as to the presentation, but also as to the position of the child. But to a less practised one this internal examination gives rise to a variety of mistakes, and it is therefore advisable always to control the results of an internal by an external examination.

The head is known to present from the large round body of uniform bony hardness, which easily produces ballottement if it is still high up and movable. If fixed it is usually easy to feel on it the sutures, and at least one fontanelle. For further information the examining finger must pass along the head towards the maternal sacrum. In this way the finger comes to a suture, which, in the great majority of instances, is the sagittal suture. It must be followed to both sides as far as possible until a place is reached where several sutures meet. The shape of the two fontanelles serves to distinguish them from one another. Yet they can be mistaken, especially if the bones are so tightly compressed that the space of the large fontanelle entirely disappears; the small fontanelle also may be strikingly like a deficiency in the bone if the apex of the occipital bone is pushed under the parietal bones. Exceptionally, a fissure with diverging margins is found at the superior angle of the occipital bone, or a Wormian bone is inserted, and then the shape of the small fontanelle is just like that of the larger. Not infrequently a rhomboidal opening is met with along the sagittal suture, about one centimètre in front of the small fontanelle, which may also lead to mistakes.

The small fontanelle is recognised by three sutures proceeding from it. Four sutures proceed from the larger. It is, of course, easy to distinguish between the fontanelles if both can be felt.

Not infrequently in internal examination the sagittal is mistaken for one of the lambdoidal sutures. But this would be less frequent if, from a previous external examination, information has been obtained as to the position of the child. The following may guide in the diagnosis:—The sagittal suture pursues a straight, the lambdoidal a convex, course. Along the sagittal suture the parietal bones gradually become very thin and yielding, whilst the margins of the bones forming the lambdoidal sutures are far more acute and ragged. In the lambdoidal sutures irregularities in their downward direction can often be felt with great distinctness and precision, which is not the case with the sagittal suture. In great compression of the head the parietal bones may be placed so as to meet one another at an acute angle, but the sagittal suture always remains a straight line, and the edges of the bones are always much softer than those of the lambdoidal sutures. It is well to remember that the upper angle of the occipital bone is almost constantly pushed under the parietal bones, and thus there is scarcely any difficulty in making out the various sutures. The sagittal is rarely mistaken for the coronal suture. It is seldom so low down (in the os), and the larger fontanelle can usually be felt so well that from the shape of the fontanelle, the acute angle of which enters between the frontal bones, the sutures can be recognised. For the same reasons the frontal suture is known, and, since the finger cannot pass along it very far without meeting the glabella and the bridge of the nose, it is not possible to mistake it for the sagittal suture.

A sure diagnosis of the presentation of the head or of its position may be rendered impossible by the bag of waters and by the formation of the caput succedaneum. The latter may be so enormous that the whole presenting portion of the head appears soft to the touch; but then the cranial bones can easily be felt if the finger is passed up immediately behind the symphysis.

The presentation of the face is easily made out if it is not too high up after the os is dilated. The parts composing the face—viz. the nose, the eyes (especially the supra-orbital margins of the frontal bones), the frontal suture, and below the mouth and chin—are so characteristic that they cannot be easily mistaken. If the finger is put into the mouth of the fœtus at times a distinct suction is perceived, yet the diagnosis may become difficult on account of a considerable swelling of the face. Those parts are sometimes so much deformed that even after delivery they cannot be recognised.

A breech presentation is, in the great majority of instances, equally easily recognised when the os is dilated; as a rule, the soft part (the gluteal region) is felt in front, and behind it the opening of the anus. On both sides are the tuberosities of the ischium: towards the back the tip of the coccyx, and higher up the spinous processes of the sacral vertebræ can be felt, whilst towards the abdomen the vulva or the scrotum is reached. The latter may be so much swollen that it feels like a large bladder. However, it is not always easy to reach the genital organs, for not infrequently the sacrum presents. At the lower end of the sacrum, a few centimétres above the anus, the skin is sometimes so strongly depressed that it feels very much like the anus, but is distinguished from it by resting upon bone. It is well to be aware of this, or it may lead to mistakes. If the breech is low down, the groin can often be felt in front.

The presentation of the feet can be known when the os uteri is still closed by their characteristic and sudden jerks and their quick disappearance. When the os is dilated they are distinguished from the hands without difficulty.

Transverse positions should always be thought of as soon as no presenting fœtal parts can be felt. Frequently a hand is found lying in the os, which is known by the greater length and mobility of its fingers, and distinguished from a foot by the absence of a heel. Whilst the foot makes short jerking movements, the hand usually remains quiet, or grasps the examining finger, or is slowly drawn back. In a later stage of a transverse position the shoulder is felt, more or less, tightly presenting over the pelvic inlet. Sometimes an arm projects into the vagina, and is easily recognised as such. The shoulder is known from the armpit, in which the ribs are felt. From the direction of both it can be concluded on which side the head is situated. The clavicle in front and the scapula behind, together with the projecting series of the spinous processes of the vertebral column, indicate whether the back is situated in front (towards the symphysis) or behind (towards the sacrum).

C.—THE COMBINED EXAMINATION

A combined internal and external examination is superfluous when the presenting part is firmly impacted or movable only with difficulty. If this is not the case, it is well to fix it externally in order to learn its position. This mode of examination becomes still more important if the presenting part is deviated to the side, or if the child is entirely in a transverse position. In the first case the simultaneous external pressure makes the parts easily accessible to the finger examining within the os, and in transverse positions it not infrequently in the commencement of parturition brings one large part within the reach of the finger, so that it can be distinctly recognised.

CHAPTER VI

THE MECHANISM OF PARTURITION

A.—HEAD PRESENTATIONS

UNDER normal conditions no obstacle is given by the bony pelvis to the entrance of the head. It is the soft parts alone which, as a rule, prevent its entrance during pregnancy; and this is usually the case when, as in pluriparæ, the yielding abdominal muscles do not check the forward growth of the uterus. But if the abdominal walls have not yet been stretched, they exert such a pressure upon the growing uterus from above that its lower segment containing the head is pushed into the small pelvis towards the end of pregnancy. In primiparæ it is therefore normal to find in the beginning of labour that the head has already entered the small pelvis. This in pluriparæ is effected during labour by the pressure of the abdominal muscles, which push the head, still contained in the lower segment of the uterus, into the small pelvis, or it is brought about by the first pains, which force the head out of the uterus and into the small pelvis.

The position in which the head enters depends upon how it was situated above the pelvic inlet. Since the antero-posterior diameter of the head is too large for the antero-posterior diameter of the pelvic inlet, the head cannot enter in that position. In all others its entrance is possible, and actually occurs. It may assume any one of the following six positions:—

1. Occiput directly to the left (*l.*).
2. „ behind and to the left (*h. l.*).
3. „ in front and to the left (*v. l.*).
4. „ directly to the right (*r.*).
5. „ behind and to the right (*h. r.*).
6. „ in front and to the right (*v. r.*).

Fig. 8.

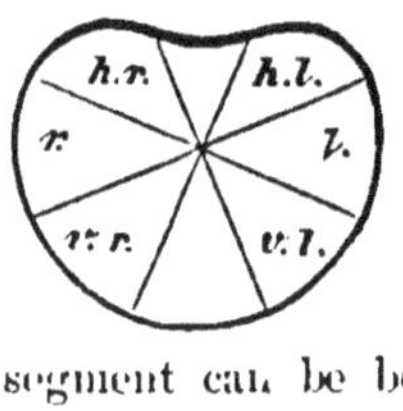

We do not consider it superfluous to draw attention to the fact that usually too narrow limits are allowed for the entrance of the head in the transverse direction. The head enters in the transverse diameter if it is placed, not entirely transversely, but approaching more to the transverse than to the oblique diameter. The extent of each segment can be best represented if (Fig. 8) around the point of section between the antero-posterior and the transverse diameter a circle is constructed, and so divided into eight equal segments that the pelvic diameters bisect those segments. Each segment corresponds to an angle of 45°; the head never enters the anterior and posterior segments with its long diameter, so that the six only which have been marked will have to be considered.

Since, as mentioned above, the back of the child, if allowed to follow its own gravity, falls to the left and forwards in the erect, and to the right and behind in the recumbent, posture of the mother, the head most frequently enters the pelvic inlet in the first oblique diameter with the occiput to the left and in front, or to the right and behind. But there are numerous exceptions, and not infrequently the head enters transversely, and often in the left oblique diameter.

In the beginning of labour, therefore, the head contained in the lower uterine segment lies in primiparæ within the small pelvis, in pluriparæ above the pelvic inlet. The attitude and position of the head are quite normal. The chin is flexed upon the thorax, so that the large and small fontanelles are on about the same level; the occiput is directed to one side, usually somewhat forwards or backwards; and the sagittal suture runs in the midst between promontory and symphysis, since the axis of the uterus approximately coincides with that of the pelvis, thus forming a right angle with the pelvic inlet.

The mechanism by which the uterus effects the expulsion of the fœtus is a double one:

First, by the contractions of the unstriped muscular fibres the contents of the uterus—viz. the ovum, considered as a whole—are exposed to a uniform pressure, which Schatz calls the "internal uterine pressure" (*inneren Uterusdruck*). Now, since the muscles of the fundus are more strongly developed than those of the lower portion of the uterus, and since the internal os is devoid of a muscular structure, the ovum is pressed against that latter point, and consequently dilates it.

Secondly, the uterus tends to assume a round form by its contractions. This is obtained at the expense of the longitudinal and the transverse diameters, the antero-posterior being relatively very small in the uncontracted condition. Thus, the uterus shortens from above downwards and transversely. Schatz calls that tendency of the uterus to assume the round form the "form-restitution power" (*die Formrestitution-kraft*).

Whilst the internal uterine pressure uniformly acts upon the contents of the ovum, viz. the fœtus and the liquor amnii, the innate power in the uterus to reassume its previous form tends, by the shortening of its transverse diameter, to place longitudinally what may perhaps have been situated very obliquely, straightening at the same time the fœtus, which is strongly bent over its anterior surface. And the breech also, which by the shortening of the transverse diameter is forced more towards the fundus, is subject to the same pressure, and this, acting through the vertebral column, pushes the head into the pelvic inlet.

Where the fœtus is so small in proportion to the other contents of the ovum that its longitudinal axis finds sufficient room in the diameter of the globular uterus, the effect of the form-restitution power is, of course, lost. Here the head sinks only downwards to the os, because its specific gravity is greater than that of the liquor amnii, but it is not forced down by a pain. This also explains why in hydroamnion, when the head presents at the os uteri, ballottement can as easily be produced during a powerful pain as in an interval.

But where the fœtus is large, so that the fundus uteri acts upon the breech during a pain, whilst the fœtus itself is straightened by the shortening of the transverse diameter of the uterus, the internal uterine pressure forces the ovum, as a whole, towards the dilating os, but the form-restitution power, acting upon the breech, brings the head of the fœtus into the lower portion of the ovum.

If the os uteri is so far dilated as to closely surround the head when forced into it during a pain, that portion of the liquor amnii which is contained in the membranes in front of the head is more or less completely separated from the rest of the liquor amnii, and is called the "forewaters."

Let us assume the separation to be absolute.

In that case the ovum, upon which the internal uterine pressure acts, only consists of the parts situated above that line where the os closely surrounds the head. For, since the head completely separates the lower portion of the ovum, after the retraction of the lower uterine segment from the last-mentioned part, the internal uterine pressure can only act upon the remaining portion of the ovum situated above the line of separation. The internal uterine pressure, therefore, forces the whole ovum, with the fœtal head foremost, out of the lower uterine segment. The effect of the form-restitution power, however, is different. That force acts upon the progress of the fœtus within the ovum by pushing it towards the bag of waters. Now, if the membranes, which contain the waters, are so formed as not to burst under that force, the waters cannot give way in any possible direction, and since water is almost entirely incompressible, the form-restitution power cannot have effect under such circumstances.

In point of fact, it is rare for the bag of waters to be so completely separated. Usually, as soon as by the internal uterine pressure the tension of the contents of the upper part of the ovum becomes greater than that of the bag projecting, water rushes past the head, and now the bag is more filled and becomes very prominent and tense. But not even now would the form-restitution power have effect, because water is only then forced into the bag of membranes when the strength of the form-restitution power is surpassed by that of the internal uterine pressure. If the reverse is the case the waters go back from the bladder into the fœtal covering, and the head is closely surrounded by the membranes forming the bladder.

The case is different when the membranes are ruptured. Whilst the internal uterine pressure in the same way tends to expel the whole ovum as far as it still exists, the form-restitution power now forces the head out of the fœtal membranes. Therefore, both powers can exert their full strength only when the membranes are ruptured.

In proportion as the uterus diminishes in size so the fundus descends with the downward progress of the child. The uterine contractions continue to act in the same way upon the breech during the whole duration of labour, and their effect becomes the more powerful the more the liquor amnii escapes from above the head.

The contractions of the abdominal muscles never act directly either upon the ovum or upon the child. They only force the uterus, and with it its contents, deeper into the pelvis. At the same time they have some indirect influence on the dilatation of the os. When the uterus has been pushed into the pelvis its ligaments assist in dilating the os.

The vagina at first opposes the exit of the child by its elasticity and its contractions, and only after the greatest circumference of the child has passed through it does it expel the succeeding portion.

The head makes various regular movements in its passage through the pelvis. These are produced by the form-restitution power entirely if the membranes are not ruptured, and the fore-waters not separated from the rest of the amniotic fluid. Under other circumstances they are also in part produced by the internal uterine pressure. But the first of these movements can be only effected by the form-restitution power.

The first change in the position of the head consists in rotation around the transverse diameter, by which the small fontanelle comes lower down. This takes place in the following way:

The expulsive power acts upon the head through the vertebral column. Were this attached to the middle of the head, the latter, meeting everywhere the same resistance, would evidently be moved onwards in the same direction in which it presented in or above the pelvis, that is, occiput and forehead in the same plane. But the vertebral column is attached to the head nearer to the occiput, and by this attachment the antero-posterior diameter of the head is divided (so to speak) into two arms of a lever, the longer of which is towards the forehead. Since the resistance, equal on both sides, becomes more powerful by acting upon the longer arm of the lever, the forehead must remain behind the occiput Under normal conditions the head, therefore, in pluriparæ enters the pelvis at the commencement of labour with the sagittal suture directed more or less obliquely, and the small fontanelle the lower; in primiparæ at this time the head will be in the pelvis, having the same inclination. The small fontanelle will naturally descend the more the stronger the expulsive power and the more considerable the resistance is, that is, the more powerful the uterine contractions are and the more unyielding and firm the soft parts. This obliquity of the sagittal suture and depression of the posterior part of the head can be seen in a marked degree in cases of pelvic deformity, where the progress of the head is opposed by a uniform resistance on the part of the bony pelvis. Were the relations of the pelvis to remain the same through its entire course, so that it somewhat resembled a cylinder, the head would then pass out through the whole pelvis in the same position. But, whilst on the sides of the pelvis the resistance remains the same, it becomes totally different on the anterior and posterior walls. For the anterior wall is formed only by the short symphysis, the posterior by the sacrum and coccyx. The resistance from the posterior wall soon preponderates over that from the anterior. This is still further increased by the attachment to the posterior wall of the powerful, unyielding, pelvic floor, whilst the anterior wall is not strengthened by any soft parts. By the resistance of the posterior wall the head must be pressed forwards. Since the head passes slowly through the pelvis, that portion of the head which presents will have first to assume the forward direction. If, therefore, the small fontanelle is the lowest down, it will be pushed forwards by the resistance of the posterior wall. The head, whatever may be the position of the small fontanelle, when it enters the pelvis, to the side, backwards, or forwards, always continues to descend in the same direction until it meets the resistance of the floor of the pelvis. It is this which presses the small fontanelle forwards, and the lower the head descends the more it turns forwards, and at the exit of the head the sagittal suture is not quite but approximately in the antero-posterior diameter of the pelvis. The increased resistance of the posterior wall also presses the whole head forwards, that is, out of the vulva. The expulsive power acting in the direction of the axis of the pelvis, and the resistance which the pelvic floor presents in an opposite direction, do not quite counterbalance each other. It therefore results that the perinæum is slowly distended, and that during this time the upper portion of the pelvic floor pushes the occiput forwards from beneath the symphysis. A further progress in that direction becomes impossible as soon as the nape of the neck is pressed against the posterior surface of the symphysis. At this moment a short pause takes place. The expulsive power of the uterus pushes the head still further down, and protrudes the perinæum until this is so much distended that the upper portion of the outlet can act upon the face from behind and above. By it the chin is removed from the thorax, and by the pressure from behind the head is gradually forced along over the perinæum. At this stage the following parts appear after each other, viz. the large fontanelle, the frontal suture, the glabella, the nose, mouth, and chin. When the chin is born the head again becomes flexed. The shoulders enter the pelvis in the opposite oblique diameter to which the sagittal suture entered, and in the outlet they are also placed approximately in the conjugate diameter. The one shoulder, lying in front, is fixed against the pubic arch, and the other sweeps over the perinæum. The trunk then follows both by its own weight and by the elasticity and the contractions of the vagina.

To recapitulate, therefore, briefly, the mechanism of labour in the first head presentation, with the back of the child to the left, is as follows:—As soon as the pains have begun to advance the head of the child, the small fontanelle is the lower down, and the sagittal suture is directed almost transversely or more or less in the oblique diameter. If the suture is followed to the left the small fontanelle can be reached without difficulty, and it may be found directed somewhat backwards, but more frequently forwards. The large fontanelle is so high up that it can only be reached with difficulty or not at all. But even if the small fontanelle is, in the beginning of labour, directed somewhat backwards, as soon as the head presses against the outlet it turns forwards, and then stands opposite the left thyroid foramen. As the head now stands lower down, usually the large fontanelle can also be felt, and it is to be found somewhat higher up and opposite the right sacro-iliac synchondrosis. In passing through the vulva the posterior upper fourth of the right parietal bone is first seen, then the occiput appears under the symphysis, and head and face sweep over the perinæum. The face, when born, is directed towards the inner surface of the right thigh of the mother. The shoulders enter the brim in the second or left oblique diameter; within the cavity of the pelvis they are placed almost in the congugate diameter, and whilst the right shoulder is fixed against the symphysis the left sweeps over the perinæum.

In the second head presentation the occiput is to the right side of the pelvis, but from the reasons above stated it is at the commencement of labour most frequently directed somewhat backwards, but in the further progress of labour it regularly turns forwards. On the whole, the relations are just the same as above described if right is substituted for left and *vice versâ.*

The caput succedaneum, which always forms on the portion situated in the os, is in the first head presentation on the posterior upper angle of the right parietal bone, and very frequently it extends to the sagittal suture of the small fontanelle. If the latter is very low down, the caput succedaneum may also be found upon the squamous portion of the occipital bone. In the second head presentation the caput succedaneum is upon the same place on the left side of the skull.

By the greater pressure to which the posterior parietal bone (*i.e.* the left in the first head presentation) is subjected from the posterior wall of the pelvis, it is, as a rule, even in the normal pelvis, somewhat pushed under its fellow along the sagittal suture. For the same reasons it is also found much more flattened. These changes are found in a greater degree in the flattened pelvis, and may, therefore, give rise to a very considerable degree of asymmetry of the cranium.

The mechanism just described is the normal; it is, however, not infrequently subject to deviations. Thus, it occurs in pelves whose conjugate diameters have not quite the normal length (because the resistance which the head meets with in the conjugate lies nearer to the occiput than to the forehead), or in abnormally shaped heads (because the posterior arm of the lever has nearly the same length as the anterior), that the large and small fontanelles remain on the same level even after their entrance into the pelvis. If these anomalous conditions do not change in the cavity of the pelvis, the descent of the small fontanelle does not take place, the sagittal suture remains nearly in the transverse diameter, and the head may be born in this transverse direction. Yet this very rarely takes place, since usually the occiput descends after all.

In still greater deviations from the normal, where the conjugate is considerably shortened and the head unusually deformed (either too large, with very strongly developed occiput, or too small and round, so that the less prominent forehead does not meet with the usual resistance), the large fontanelle is lower at the entrance into the pelvis. It then continues the lower, and in marked cases may be felt in the pelvic axis. As soon as the greater resistance of the posterior pelvic wall preponderates it also turns forwards like the small fontanelle when that is lower down, and for the same reasons. In the great majority of instances these relations change within the pelvic cavity—sometimes only at the outlet, and, finally, the small fontanelle descends after all. After its descent it turns forwards, and the head is delivered in the regular way. Rarely the large fontanelle remains lower down up to the last; if so, it then turns more to the front, and the head is delivered with the occiput directed backwards—anterior parietal presentation, Wigand and Hecker. In the vulva the angles of the frontal and parietal bones forming the large fontanelle first become visible. Then the corresponding frontal bone (the right in the first and the left in the second position) is fixed against the pubic arch, the occiput sweeps over the perinæum, and at last the face appears from beneath the pubic arch. The shoulders pass through the pelvis in a direction opposite to that of the sagital suture. Corresponding to the position of the large fontanelle, the caput succedaneum is also on the anterior corner of the parietal bone close to the large fontanelle. Sometimes it is on the large fontanelle itself, or on the adjacent angle of the frontal bone.

In the passage of the head through the outlet it not infrequently occurs that the small fontanelle is turned to the opposite side, so that, therefore, in the first head position it appears in the last moments of labour beneath the right pubic arch. The face then looks to the left and behind, and the shoulders pass as in the second head position. This superrotation (ueberdrehung) is caused by the abnormal position of an arm with the head or by the shortness of the umbilical cord.

B.—FACE PRESENTATIONS

In face presentations, also, there is nothing definite in the way in which the face is placed at the brim. The length of the face may be almost in the transverse or in one of the oblique diameters, and the chin in front or behind. The direction in which the force conveyed through the vertebral column acts upon the progress of the child in face presentations divides the length of the face into two unequal arms of a lever, of which the longer is situated towards the forehead. Therefore the equal resistances of both sides will more strongly act upon the longer arm of the lever, and consequently the forehead will remain behind whilst the chin descends. For the reasons already stated in the mechanism of head presentations, the chin when lower down must always turn forwards in the progress of labour, even if it had originally been directed backwards.

In the first face presentation the face is so situated at the brim that the forehead is directed to the left. The length of the face runs almost in the transverse or in one of the two oblique diameters. Chin and forehead are approximately on the same level, or the forehead is somewhat lower down. As soon as the face has entered the pelvis the chin descends and turns forwards within the pelvic cavity. The rotation is naturally the greater the more backwards the chin has been originally directed. In passing through the vulva the right cheek and the angle of the mouth are first visible, the chin then appears from beneath the descending ramus of the right pubic bone, and whilst the neck is fixed against the pubic bone the nose, forehead, and the whole head sweep over the perinæum. When the head is born the face looks forwards and to the right. The shoulders enter the pelvis in the first oblique diameter, the right shoulder is fixed against the left descending ramus of the pubic bone, and the left shoulder sweeps over the perinæum.

In the second face presentation the delivery takes place in the same way, *mutatis mutandis.*

A swelling is formed on the presenting face, for the same reasons that in head presentations a caput succedaneum is met with. It is situated close to the mouth at that angle which is directed forwards —in first face presentations the right angle of the mouth. If the swelling is more considerable it extends over the whole cheek, the nasal region, and sometimes also to the other side, so that the face is frightfully disfigured, and bluish-black from suffused blood. The disfiguration disappears in a few days after delivery.

In face presentation the shape of the fœtal head is usually altered by passing through the pelvis. The change consists in a compression of the head in the vertical diameter. The upper surface of the head is flattened; in the region of the large fontanelle there is sometimes a kind of saddle-shaped depression; the occiput is drawn out and the longitudinal diameter of the head is lengthened. This change of shape also disappears a few days after birth. In treating of the etiology, it will be mentioned that a hereditary development of the occipital bone predisposes to face presentations, so that a greatly prominent occiput may be still more lengthened by labour.

According to the statistics of Winkel, face presentations occur as frequently in primiparæ as in pluriparæ; they are more frequent in a narrow pelvis than in the normal.

The causes of face presentation are essentially those which result from the contractions of the uterus. These, by diminishing its transverse diameter, tend to change oblique positions of the fœtus into longitudinal. The longitudinal positions are obtained by the part which is situated high up being pushed into the fundus, *i. e.* the breech in most cases, whilst the part situated below—the head— is pushed into the brim. In large children this is often attended by great difficulties. If the child is so situated that its back is directed downwards to the pelvis, the breech having been forced into the fundus by the uterine contractions, the pressure of the other lateral wall of the uterus must chiefly act upon the forehead, whereby the curvature of the child is increased and the chin is closely pressed against the thorax. But if the child is situated with the back directed to the fundus and the abdominal surface to the pelvic inlet, the breech having again been pushed into the fundus, the pressure of the outer lateral wall of the uterus chiefly acts upon the occiput and forces it tightly against the nape of the neck. At the same time the contracting uterus forces the head downwards. The occiput is pressed against the uterine wall, whilst the forehead does not meet with any resistance. Under these conditions, although the expelling force is nearer to the occiput than to the forehead, yet the forehead descends lower, because the greater resistance, acting through the short arm of the lever (the occiput), preponderates over the slight resistance acting on the long arm of the lever (the forehead). When the head has entered the brim under favorable conditions, the chin may be so far removed from the thorax that the diameter of the head (represented by a line from the chin to the large fontanelle) may be divided into two unequal arms of a lever by the direction of the expulsive force; the shorter of the arms is situated towards the chin. That point once reached, the head must necessarily remain behind, whilst a face presentation fully develops. That change in the attitude of the fœtal head is very easily brought about when the occiput is more strongly developed, as the preponderance of the resistance to the occiput may then be much less. The direction also of an existing obliquity of the uterus is not without influence. If the uterus lies to the right, and the child with its back upwards and its head to the right, the uterine wall, running from the left below to the right above, must rather tend to approach the occiput to the nape, and to produce a face presentation, than when the head lies to the left. Since the obliquity of the uterus is normally to the right side, it can be easily explained in that way that the relation of the first face position is to the second as 1·4 : 1, whilst that of the first head position to the second is as 2·26 : 1. In a very similar way, Duncan ('Edinburgh Medical Journal,' May, 1870) explains the relative preponderance of the second over the first face presentation. The narrow pelvis also favours the occurrence of face presentation. It prevents the entrance of the head into the pelvis, which in primiparæ takes place towards the end of gestation and tends to obliquities. According to the etiology given, face presentations would, as a rule, be produced at the commencement of labour—secondary face presentation; but in primiparæ, in whom, during pregnancy, already distinct uterine contractions have taken place, they may be produced during that period—primary face presentation.

Deviations from the usual mechanism of labour are very rare in face presentations. What most frequently occurs is that the forehead and chin remain on the same level in the pelvic cavity, and the face passes through it in the transverse diameter. A passage through the pelvis with the chin backwards and lower down is therefore impossible, because the occiput and chest would have to pass together through the conjugate.

However, in all these cases the chin turns forwards if it is the lower down. If the forehead is lower down, the chin remains directed backwards for reasons already stated, and the head, under especially favorable conditions only, is thus born—the forehead and large fontanelle first appear, then the cranium, at a spot close to the large fontanelle, is pressed against the pubic arch, and the face sweeps over the perinæum.

It occurs much more rarely—indeed, can only occur under the most unusual conditions—that the forehead is fixed against the anterior pubic wall, and the nose and eyes first appear, then the mouth and chin sweep over the perinæum, and, last of all, the forehead and cranium clear the pubic arch.

These deviations are, however, extremely rare and only possible under such exceptional circumstances that the rule is constant and ought to be adhered to in practice, that in face presentations, with the chin directed backwards, delivery can only naturally take place if the chin turns forwards. If the chin does not turn forwards delivery is only possible and actually only occurs when the forehead is lower down than the chin.

If the descent of the forehead is still more considerable it passes into the pelvic axis and the head assumes a position intermediate between the normal head and face presentation.

In abnormal conditions, especially in the narrow pelvis, the head not infrequently enters the brim in that position. The vertebral column then divides the cephalic diameter—represented by a line from the chin to a point in the sagittal suture situated twice as far from the large as from the small fontanelle—into two unequal arms of a lever, so that if the direction of the expulsive power and the resistance were to remain unchanged the head would have to pass through the whole pelvis in the same direction, but this, however, rarely occurs in the further progress.

Almost always the direction of the expulsive force deviates gradually to one side. As soon as this occurs one arm of the lever becomes longer than the other, and if the now longer lever arm is situated towards the chin the cranium descends, the frontal presentation becomes a head presentation; and if the longer lever arm is situated towards the cranium the chin descends, the frontal presentation becomes a face presentation.

Only in very rare cases the head enters the pelvis with the forehead presenting.

According to Von Helly, if after the escape of the waters the head is found entering the pelvis with the forehead presenting, the coronal suture will be in the transverse diameter. In its descent the forehead turns forwards, the rest of the head backwards. The forehead first appears in the vulva, then the eyes, whilst the superior maxilla is fixed against the pubic arch and the cranium sweeps over the perinæum. The superior maxilla, the mouth, and the chin, do not appear from beneath the pubic arch until after the whole of the cranium has been delivered.

Sometimes the head maintains its transverse position; then the face, with the exception of the inferior maxilla, which remains behind, is born, appearing beneath one pubic bone; the occiput appears from beneath the other, and, finally, the chin passes through the vulva.

The caput succedaneum is then situated on the forehead, extending from the root of the nose to the upper angle of the large fontanelle. This gives the head quite a peculiar appearance, which is still more increased by its altered shape. This is very high in front, so that the distance from the chin to the forehead is very long. The parietal bones recede backwards and rather abruptly from the larger fontanelle, and the back of the occiput is applied to the nape of the neck. The skull is compressed in a direction from the chin towards the small fontanelle.

Labour is very tedious in face presentations, because the entrance of the head in this abnormal position is not infrequently due to slight pelvic deformity. In considerable deformities the head could not enter in this way. The passage through the outlet is the most difficult of all, since the greatest diameter of the head must pass through the antero-posterior diameter of the outlet. As face, and still more so forehead, presentations are abnormalities and belong to the pathology of parturition, it is just as well to say a few words on the prognosis and treatment of those presentations.

The prognosis in face presentations, even with the usual mechanism, is far more unfavorable than in head presentations. Whilst in the latter nearly 5 per cent. of the children are stillborn, that proportion rises to 13 per cent. in face presentations. For the mother, also, face presentations are more unfavorable than head presentations, because, on an average, the labour lasts much longer.

Of course, the prognosis is still more unfavorable if the mechanism is abnormal, that is, when the chin does not take its forward turn. Then scarcely a child will be born alive.

In forehead presentations the prognosis is very bad for the child. In Massmann's cases, out of forty-one children, twenty-one were stillborn. The disproportion between the head and the pelvis and the long duration of labour render the prognosis equally unfavorable for the mother.

The treatment of face presentations differs essentially from that of forehead presentations. In the former, when the face is in the pelvis, a more expectant treatment is the only proper one, provided it can safely be done. The forceps should only be used if there are urgent symptoms. If the rotation of the chin forwards is delayed very long attempts may be made to hasten its descent, upon which the rotation depends, by exerting counter-pressure upon the forehead during a pain. In forehead presentations, when the os is sufficiently dilated and the forehead is still movable, it is better to turn. If this is impossible we may try, as advised by Hildebrandt, to change the forehead presentation into a head or face position.

In the first case pressure is exerted against the face, and in the second against the cranium. Very properly, Hildebrandt calls attention to the fact that this must be tried during a pain. The posture of the woman must also be considered. If, for instance, in a forehead presentation, the woman is placed on the right side, the trunk of the child falls over to the right, and as by this the direction of the vertebral column through which the expelling force acts is changed, the arm of the lever towards the head is shortened and that towards the face lengthened. Thus, the conversion into a head position is favoured. But if this change does not succeed, it would be necessary to perforate the child, if it be dead, as soon as mechanical difficulties arise. If the child is living we may wait as long as possible, or, if not, the forceps must be used. The direction of the traction is, at first, downwards until the superior maxilla emerges from beneath the pubic arch, and then the occiput is raised over the perinæum. When delivery by the forceps fails, the question of perforation of the living child will have to be considered.

C.—BREECH PRESENTATIONS

In breech presentations the back of the child at the commencement of labour may be in various positions, as in head presentations, *i. e.* either looking to the left or to the right, and either more to one side or more forwards or backwards.

The positions are distinguished in the same way as in head presentations, and the first breech presentation is the one where the back looks to the left, the second where the back looks to the right. The expulsive power acts upon the breech through the vertebral column, the sacrum must descend and, being the presenting part, must turn forwards for the reasons previously mentioned.

In the first breech presentation the sacrum is found accordingly at its entrance into the brim looking almost directly to the left or somewhat forwards or backwards. Often the heels can be felt on the other side of the uterus. When the breech has passed into the pelvis the feet remain behind, the sacrum descends and regularly turns forward when the pelvic floor has been reached.

The transverse diameter of the hips is, therefore, in the further progress of the first breech presentation, in the left oblique diameter of the pelvis. The left buttock is first seen at the vulva, and after it has cleared the symphysis it is pressed against it until the right buttock has swept over the perinæum.

The abdomen of the fœtus then looks backwards and to the right. The shoulders also pass through the brim in the left oblique diameter, and the arms are born at the same time as the chest, in their natural position.

The head enters the brim with its longitudinal axis in that of the right oblique diameter of the pelvis, the chin resting in its natural position on the chest; whilst the occiput is fixed against the left ramus of the pubic bone, chin, face, and vertex sweep over the perinæum.

The description last given applies also to the second breech presentation, if left is substituted for right and *vice versâ*.

The same causes which in the presenting head had determined the formation of the caput succedaneum also produces a tumour in breech presentations upon the presenting buttock, which may be very large and of a dark blue colour from suffused blood. Not infrequently, also, the scrotum participates in it and appears as a large and very tense bladder.

In footling presentation the mechanism is just the same, only, when at the commencement the back is situated behind, the rotation forwards usually takes place somewhat later, because the breech without the thighs is so small that the resistance which it meets with on the floor of the pelvis does not suffice to cause the rotation. In this case, however, the rotation takes place when the back comes to be pressed against the floor of the pelvis, so that rotation of the back forwards always takes place except in the unusual cases where the child is immature, the pelvic outlet very wide, or the floor of the pelvis very flabby. The position of the feet is regular. They are placed sometimes close to each other, sometimes they are crossed. Imperfect footling presentations resemble breech presentations. These are only rare exceptions to the ordinary mechanism. Still more frequently than in head presentations a so-called super-rotation takes place. So that, for instance, in the first breech position, when the trunk appears at the vulva the back is turned from the left and forwards to the right and forwards.

It has already been stated that breech. and especially footling, presentations are far less favorable to the child than when the head presents. There is danger to the life of the child when the delivery of the head is somewhat delayed, for as soon as the trunk is delivered beyond the umbilicus the cord is liable to be compressed, and thus the placental circulation is stopped. And the uterus also, in order to expel the after-coming head, must contract so powerfully that the placenta is detached from its inner surface.

At any rate, even if the umbilical cord is not compressed when the expulsion of the head is delayed, the communication of the fœtal blood with the maternal is impeded, consequently the fœtus makes inspiratory movements and draws foreign bodies into the air-passages. The child becomes asphyxiated and dies unless the head is immediately delivered naturally or by the intervention of art.

Under normal conditions the prognosis for the mother is the same in breech as in head presentations.

From what has just been observed it can easily be seen that presentations of the pelvic extremity not infrequently require operative assistance. It is, of course, desirable not to interfere with the natural course of parturition until after the delivery of the breech, provided there be no urgent symptoms, since by premature traction on the feet or on the breech the arms may be slipped above the head and the chin removed from the thorax and chest, producing a further impediment to the expulsion of the child. Even at that time it is necessary to attend to the sounds of the fœtal heart.

As soon as half of the child is born, unless the delivery will soon terminate spontaneously, it is the duty of the accoucheur to extract the head. Unless there be other anomalies, the extraction is usually easy and successful. But it must be done as quickly as possible, since we cannot foresee the moment when the first inspiratory movement will take place, and the inspiration which then blocks up the lungs with foreign bodies places the life of the child in jeopardy.

Literature.—Fielding Ould, A Treatise of Midwifery. Dublin, 1742.—W. Smellie, A Treatise on the Theory and Practice of Midwifery. London, 1752.—R. W. Johnson, A New System of Midwifery, &c. London, 1769.—M. Saxtorph, De div. partu ob div. cap. ad pelv. rel. mut. Havn., 1771, and Ges. Schritten, herausg. von Scheel. Kopenh., 1803.—J. Bang, Tent. med. de mech. part. perf. Havniæ, 1774.—F. L. J. Solayrés de Renhac, Diss. de partu virib. mat. abs., &c. Paris, 1771, and Comment. de p. v. m. a. herausgegeben von E. C. J. v. Siebold. Berlin, 1831.—J. L. Baudelocque, L'art des acc. Paris, 1871.—L. J. Boër, Abh. und Vers. geburtsh. Inhalts. 1791—1807.—W. J. Schmitt, Geburtsh. Fragmente. Wien, 1804.—J. H. Wigand, Die Geburt des Menschen. Herausgegeben von F. C. Naegele. Berlin, 1820, II.—F. C. Naegele, Ueber den Mechanismus der Geburt. Meckel's Archiv für die Physiol. 1819, 5 Band. 4 Heft, p. 483, and Ueber der Frau Lachapelle Pratique des acc. Heidelberger Jahrb. d. Liter. 1823, 5 Heft.—C. F. Mampe, Bemerk. über den Herg. d. menschl. Geb. Meckel's Archiv, 1819, 5, 4, p. 532, and De partus hum. mech. D. i. Halis, 1821.—Mme. Lachapelle, Prat. des acc., publ. par Ant. Dugés. Paris, 1821.—H. F. Naegele, Die Lehre vom Mechanismus der Geburt. Mainz, 1838.—M. Duncan, Edinburgh Medical Journal, 1861; Obstetric Researches, p. 344; Edinburgh Medical Journal, June, 1870, and Ritchie, c. l., June, 1871.—H. L. Hodge, Principles and Practice of Obstetrics. Philadelphia, 1864, and American Journal of the Medical Sciences, Oct., 1870, p. 325.—W. Leishman, An Essay, Historical and Critical, on the Mechanism of Parturition. London, 1864.—Ritchie, Medical Times and Gazette, 1865, vol. i, p. 381 and 408.—H. Hildebrandt, De mech. partus cap. pr. norm. et enormi. Reg., 1866.—O. Spiegelberg, M. f. G., B. 29, p. 89.—Schroeder, Schw., Geb. u. Wochenbett. Bonn, 1867, p. 43.—F. Schatz, Der Geburtsmech. der Kopfendlagen. Leipzig, 1868, and Wiener med. Presse, 1868, Nos. 30, 32, 42, and 43, and 1869, No. 29.—Kueneke, Die vier Factoren der Geburt. Berlin, 1869.—H. Lahs, Zur Mechanik der Geburt. Marburg, 1869, and Archiv f. Gynaek., B. I, H. 3, p. 430.—De Soyre, Etude hist. et crit. sur le mech. de l'acc. sp. Paris, 1869.—L. Bourgeois, dite Boursier, Observ. div., sur le sterilité, perte de fruict, &c. Paris, 1609. —Paul Portal, La pratique des accouch., &c. Paris, 1685.—Johann van Hoorn, Die zwo um ihrer Gottesfurcht und Treue willen von Gott wohl belohnten Weh-Mütter, Siphra u. pua, &c. Stockholm u. Leipzig, 1726.—M. F. A. Deleurye, Traité des accouch., &c. Paris, 1770.—Simon Zeller's Bem. über einige Gegenstände aus d. prakt. Entbindungskunst. Wien, 1789.—L. J. Boër, Abh. und Vers. geburtsh. Inhalts, 1791-1807.—Wigand, Die Geb. d. Menschen, B. 2.—F. C. Naegele, Meckel's Archiv für die Phys. 1819, 5, 4, p. 513.—Mme. Lachapelle, Pratique des accouch. Paris, 1821.—Winkel, M. f. G., B. 30, p. 8, and Klin. Beob. zur Path. d. Geburt. Rostock, 1869, p. 47-131.—Hecker, Ueb. d. Schädelform bei Gesichtslagen. Berlin, 1869 (s. das Referat von Schultze, Archiv f. Gynaek., I, p. 355), and Arch. f. Gyn., B. II, p. 429.—Fasbender, Berliner B. z. Geb. u. Gyn., B. I, p. 100.—Historisches s. H. F. Naegele, Die Lehre vom Mechanismus d. Geburt., p. 146, and W. A. Freund, Klin. Beiträge zur Gyn. Breslau, 1864, 2 Heft, p. 179.—Mauriceau, Traité des mal. des femme grosses. Six. éd. Paris, 1721.—De la Motte, Tr. compl. des acc., &c. Paris, 1722.—P. Portal, La prat. des acc., &c. Paris, 1685.—J. van Hoorn, Die zwo Weh-Mütter Siphra u. Pua, &c. Stockholm u. Leipzig, 1726.—H. van Deventer, Neues Hebammenlicht. Jena, 1717.—Solayres de Renhac, Diss. de partu vir. mat. abs. Paris, 1771.—Baudelocque, L'art des acc. Paris, 1781.—L. Boër, Natürliche Geburtshülfe. Wien, 1817, B. I, 3 Aufl., 3 Buch.—E. v. Siebold, Neue Zeitschr. für Geburtsk., B. 26, p. 175.—H. F. Naegele, Die Lehre vom Mech. d. Geb., &c., p. 222.—Hodge, American Journal of Medical Sciences, July, 1871, p. 17.

CHAPTER VII

TWIN LABOURS

Twin labours, as a rule, only present deviations of a minor importance. The first child is easily born in the same way as in simple labour, the second frequently follows in the course of ten minutes to half an hour. As a rule, both children present with the head, but exceptions are very frequent, so that the first child may present with the vertex, the second with the breech, or *vice versâ*, or both may be breech presentations. Transverse positions also are by no means rare with the second child. The placentæ always, whether connected or not, are expelled after the delivery of the second child.

The diagnosis of twins is often attended by far greater difficulties during labour than in pregnancy. Through the contracted firm uterine walls the individual parts of the child can less easily and less quickly be felt. A careful auscultation, also, is much more difficult in the parturient than in the pregnant woman. One circumstance, however, facilitates the diagnosis. It is the accuracy with which the presenting fœtal part and its position can be determined. If this does not correspond with the result of external palpation there is reason to suppose that the presenting parts belong to another child, and, consequently, that there are two present.

All the other signs of twin gestation which have been mentioned, though uncertain, are, of course, equally to be considered. It will rarely be possible to diagnose twins, from the fact that movable cranial bones can be felt on internal examination (a dead child), and that fœtal heart sounds are heard on external examination (a living child). It is equally rare to find two bags of waters. Under favorable conditions the presenting parts of one child may be pushed on one side and the other parts artificially made to present, and thus the second child may be diagnosed.

If one child is born the diagnosis of the probable presence of a second is easy. The examination of the abdomen of a woman just delivered must suffice to show whether another child is still within the uterus, or whether it contains only soft parts, such as the placenta and blood. Tumours, and especially fibroids, can easily be distinguished. If the hand is passed high up into the uterus the presence of a second child is indicated either by another bag of waters or by fœtal parts. In the second child it is of great importance to attend to the heart sounds. The sudden and considerable diminution of the uterus following the delivery of the first child often causes the complete or partial detachment of the placenta of the second, so that on account of the danger which arises therefrom the second child is to be immediately extracted. For the same reason the second bag of waters must not be ruptured, since by it the uterine contents would be still more lessened.

If, after the delivery of the first child, the heart sounds of the second cannot be heard with certainty, extraction must be performed immediately. This can safely be done because under such circumstances the operation is easy and always successful.

Literature.—Ed. v. Siebold, M. f. G., B. 14, p. 401.—Chiari, Braun u. Spaeth, Kl. d. Geb. u. Gyn. Erlangen, 1852, p. 5.—Hecker u. Buhl, Kl. d. Geb. Leipzig, 1861, p. 72, and Hecker, B. II. Leipzig, 1864, p. 63.—Winkel, Zur Path. d. Geburt. Rostock, 1869, p. 132.—Kleinwächter, Lehre von den Zwillingen. Prag, 1871.—S. auch § 44, *sequ.*

CHAPTER VIII

THE DIETETICS OF PARTURITION

Labour is a physiological process, and medical assistance, properly so called, is unnecessary in normal parturition. Since each labour is not only attended by pain and great excitement in the parturient woman, which may be lessened by skilful assistance, but also accidents or slight deviations from the normal may arise, which are attended by great danger to both mother and child, it appears urgently necessary always to have proper assistance for a woman in childbed.

It is the duty of that attendant to assist the woman by word and deed, and to watch carefully the course of the labour, in order that anything which interferes with its normal course may be removed, and that the actual occurrence of accident or danger may be immediately recognised and proper means at once taken.

As a rule, these duties are performed by trained midwives, but it cannot be denied that they are only very improperly qualified for their task. It is just the prophylactic treatment of the lying-in woman that requires, not only a skilled obstetrician, but a physician thoroughly acquainted with all the branches of his science. Whilst by the most simple means, as, for instance, change of posture, dangers, if early recognised, may be averted which otherwise threaten the life of mother and child, a midwife is only able to recognise pathological conditions fully developed, and after she has recognised them is forced to summon a medical man, so that the loss of time is certainly not conducive to the safety of the parturient woman.

If a medical man is summoned to a case of labour, and the place is close to his residence, he should provide himself with a stethoscope and a silver male and a thin elastic catheter. We often have to use the catheter, and in difficult cases it is the male catheter alone which will succeed. The elastic catheter may probably be wanted for the treatment of asphyxia in the child. It is useful also, to have the forceps at hand. But if the case of labour is far distant from the residence of the medical attendant he must not neglect always to take his obstetric bag with him.

The usual posture of the parturient woman is either on the back or on the side, but it must be admitted that neither of them realises what may be reasonably expected from a rational position. The posture of the parturient woman is to be so arranged that first the direction of the expulsive powers act as perpendicularly as possible to the plane of the pelvis through which the fœtal head is to be pressed, for thus only the amount of friction will be the least, and, therefore, very little force will be lost, and, secondly, the gravity of the child itself will not be prevented from aiding. This last demand is complied with when the plane of the pelvis, through which the head has to be forced, lies horizontally.

If in each stage of labour these two points are complied with the posture then is the most perfect.

The first demand—that the expulsive force should act perpendicularly to the plane of the pelvis through which the head passes—can only be satisfied at the commencement of the pelvic canal, and, as a rule, it is so here as soon as the axis of the uterus and that of the brim have approximately the same direction. But when the head has passed through the excavation, that is, when the pelvic axis is curved forwards, the axis of the uterus and that of the corresponding plane of the pelvis form an angle. This will be the smallest when the uterus lies as far backwards as possible.

By accurate measurement Schultze has shown that the mobility between the lumbar vertebræ and the pelvis is not inconsiderable. The more obtuse the angle is which the lumbar vertebræ form with the brim the more backwards the uterus comes to be placed and the more the passage will realise our expectations.

Secondly, it is required that the gravity of the child shall not be impeded in its action. This is obtained if the plane of the pelvis through which the head passes lies horizontally. The woman, therefore, until the head is in the excavation, must be in a half-sitting and half-recumbent position, because then the inlet is horizontal. When the head descends, the back of the woman must be raised still more, so that in the further progress of the delivery she has a sitting posture, and when the head passes through the outlet she must be bent forwards. Since the entrance of the head into the inlet is favoured by the half-reclining half-sitting posture, both these demands are perfectly satisfied by it and that position is, therefore, undoubtedly the most rational at the commencement of labour. The passage of the head through the outlet and the vulva is, most of all, favoured by the weight of the child when the woman is bending forwards, and since in that position, through the bending of the sacrum, the lumbar vertebræ form a very obtuse angle with the inlet, that position perfectly corresponds with all rational demands for the last period of the expulsive stage.

As regards the customary postures, the dorsal position with the trunk somewhat raised is very suitable for the entrance of the head, and the more so the higher the back lies.

In the expulsive stage, however, the dorsal position is irrational, since in it the head must be forced opposed to its own gravity over the ascending inclined plane of the pelvic floor. In the dorsal position, therefore, the expulsion is not only not favoured by the weight of the child, but the latter is directly opposed to the expulsive force—the head would have to be pushed over an eminence, so to speak.

If it is preferred to deliver the woman in the dorsal position parturition may be facilitated as far as it is possible in that position if a pillow is placed under the sacrum, so that the angle between the lumbar vertebræ and the inlet becomes as large as possible.

Lying on the side is unsuitable as long as the head has not yet entered the brim, provided always that there are no definite therapeutic plans connected with this position. For here the head can easily deviate to the opposite side.

For the delivery of the head the lateral position does not offer the same inconvenience as the dorsal, but it also has no advantages, since the gravity of the child cannot be made use of. To conclude, therefore, it is most suitable to place the woman at the commencement of labour in the dorsal position with her back raised as much as possible. The expulsion of the child is, however, most facilitated if the woman is bending forwards.

Similar positions can easily be assumed in bed. If the woman is lying on her side she must turn a little more and kneel on the bed and grasp with her hands the head of the bed, or if she is in the dorsal position she must raise herself and support the upper part of the body by resting on the foot of the bed. By this the sacrum is strongly drawn inwards and thus the position corresponds with all that may be justly demanded, provided that the abdominal walls are tense.

If this is not the case, and there be a somewhat pendulous belly, the uterus must be pressed against the lumbar vertebræ by means of a binder. The pressure of the abdominal muscles can in that position also more powerfully come into play than in a dorsal or lateral. Its advantages for the preservation of the perinæum will be mentioned later on.

When summoned to a woman whom he has not seen before, the accoucheur, before commencing his examination, should ascertain whether his patient is a primipara or a pluripara, the duration of the pregnancy, and the time when labour-pains set in. He will do well before examining internally to commence with palpation and auscultation. Before examining internally no long discourse is necessary in order to prove to the woman the necessity of such an examination; let him show that it is necessary to do it by simply asking for some oil or cold cream, some water and a towel. If he examines her without many preliminaries he will not meet with any resistance or refusal on her part, since a woman who sends for the accoucheur must be prepared to be examined. He must, of course, proceed with all possible decency, but on no account whatever must he omit any method of examination if he considers it necessary.

After the examination he will invariably be asked how the case will go on and how long it may probably last. If nothing pathological has been found he may assure those around the woman that at present everything is in perfect order.

As regards the second question he must tell the woman that the duration cannot be determined beforehand, and that it entirely depends upon the force of the pains; if she have slight pains it will last longer, but if the pains become very powerful she will soon be relieved from pain. It is always great consolation to a parturient woman to know that the stronger the pains the nearer the end of labour approaches. A time should never be fixed as regards the duration of labour, as experience has taught that this is subject to considerable variations.

As long as the pains are feeble and do not sufficiently dilate the os the woman may walk about in the room. To empty the rectum in time it may be advisable to give an enema. The bladder also should be emptied, since a full bladder impedes the entrance into the pelvis of the presenting part and makes the pains irregular.

If the head of the fœtus has already entered the pelvis the woman may walk about until the os is dilated. But if the head has not entered the brim she must be careful with her walking and should rather be put to bed. The same also when the os is fully dilated and the rupture of the membranes may be expected. When the membranes are ruptured and the os dilated the woman may be allowed to bear down during a pain. If during the stage of expulsion she feels the desire to empty her rectum she must not be allowed to go to the night-stool, but a bed-pan should be at her disposal.

When the head is in the vulva the most important task is to avoid lacerations of the perinæum. In the dorsal position it is certainly very difficult, since in it the head is by its own weight not pressed against the pubic arch, but directly against the perinæum. By suitably applied pressure with the ball of the thumb or the fingers against the head, still covered by the perinæum, it may be gradually directed more towards the symphysis, and if the head, whilst passing from beneath the symphysis during a pain is somewhat kept back and during the interval of the pains the edge of the vulva is pushed backwards over the head, we have the best chance for preserving the perinæum which is possible in the dorsal position. But experience has shown that even under these conditions in primiparæ slight lacerations, at least, are always unavoidable in the dorsal position. The lateral position is more useful, indeed, since in it the head of the child, at least, is not pressed directly against the perinæum. The most favorable of all is the position where the body is bent forwards, as has already been described. Here the head is forced by the gravity of the child against the anterior wall of the pelvis and the perinæum is only slightly distended.

We have found that in primiparæ delivered in the dorsal position the frænulum remains intact in only 39 per cent., whilst in 37·6 per cent. there were actual lacerations of the perinæum. In the position recommended the frænulum remained intact in 57 per cent., and lacerations of the perinæum occurred only in 24·4 per cent. of all the cases.

When the head is large and the rima vulvæ very narrow the latter is usually torn. In such cases it is well not to wait until the perinæum is lacerated, the extent of which cannot be estimated beforehand, but to make two lateral incisions in the direction of the tuberosities of the ischium, which would sufficiently dilate the vulva for the passage of the head.

When the head is born the nose and mouth of the child are to be freed from mucus that air may easily enter. If the trunk is not expelled soon afterwards its delivery is to be hastened, especially if the cord should be twisted round the neck of the child. For that purpose Kristeller's method of "expression" is very suitable. Or the woman may be told to assist by bearing down whilst the head is pressed somewhat backwards in order to make the shoulder pass from under the symphysis, and then the head is carried forwards in order to draw the other shoulder over the perinæum. If this does not succeed moderate traction on the chin and occiput may be made, or two fingers are put into the axilla lying posteriorly and thus the child is extracted. The cord twisted round the neck must be somewhat loosened and the trunk made to pass through that loop, but if the cord cannot be loosened it must be divided; the fœtal end is then to be compressed by the fingers and the child quickly extracted. When the shoulders are born the rest of the child soon follows. The child is placed between the thighs of the mother, so that the cord is not stretched, and the mouth and nose are free.

A few minutes after the expulsion of the child, when the umbilical arteries cease to pulsate strongly, a ligature is placed about three centimètres distant from the navel and another at the same distance further on towards the placenta; these should be tied very tightly, especially if the cord is very gelatinous. Afterwards the cord is divided between the two ligatures. The ligature situated towards the child is decidedly necessary, since dangerous hæmorrhage occurs, not only when it is not ligatured, but even when it has been badly done. The other is superfluous and useless when the placenta is already delivered. If it is not entirely detached the double ligature is advisable, because the placenta filled with blood is more easily detached than an empty one.

The delivery of the placenta may be left to the contraction of the uterus. Should it, however, be delayed for some time, since it is desirable that the woman should be at repose as soon as possible, Credé's method for the removal of the afterbirth is now almost universally adopted. His proceeding deserves to be strongly recommended, since it does no harm and permits an easy and safe removal of the placenta.

After the birth of the child the uterus should be examined to see if it is sufficiently contracted. If it is flaccid slight friction will cause it to contract, and as soon as that takes place the fundus is grasped with one or both hands and in this way the afterbirth is pressed out, so to speak. Should this not succeed we may wait a short time and then repeat the process. It almost always succeeds in bringing away the placenta in a few minutes or a quarter of an hour after the delivery of the child.

Goschler's remarks also deserve to be noticed, viz. that the placenta is not infrequently retained by antiflexion of the uterus. In Credé's method such an impediment is obviated by the elevation of the fundus. Sometimes the membranes, or at least a part of them, remain in the vagina. They may then be removed by slight traction, but care must be taken that they are not lacerated, for which purpose the placenta may be twisted so as to form a string.

After the removal of the afterbirth the external genital organs must be carefully cleansed with a sponge and lacerations of the vulva looked for. As a rule, they are found in primiparæ, but in the great majority of instances they are also met with in pluriparæ. They are chiefly more or less slight rents through the frænulum. Frequently lacerations are found to be behind it in the fossa navicularis, most frequently they are found in the mucous membrane at the side of the labia minora, and still more frequently at the side of the urethra or between the urethra and the clitoris. The last may give rise to profuse hæmorrhage, but if these lacerations are slight and do not bleed much it is best to leave them alone.

After the uterus has been repeatedly examined to see that it is sufficiently contracted the woman may be left to enjoy the necessary rest.

CHAPTER IX

ON THE USE OF CHLOROFORM IN PARTURITION

In general, chloroform is used with advantage in midwifery in the same class of cases for which it is employed in surgery. In all difficult, and especially in painful operations, it is of immeasurable benefit to the patient; it also materially facilitates operations, and therefore deserves to be always used in such cases. But in midwifery a new question arises. Parturition is solely a physiological process, which is attended by pains often very intense. Is it, therefore, allowable to mitigate or remove these pains which depend upon a physiological process? According to our views there is never for a moment a doubt on the advisability of chloroform in such cases, for it is one of the most pleasant duties of the physician to alleviate suffering.

To effect parturition the contraction of the unstriped fibres of the uterus—which, indeed, cause the pain—are necessary, but by no means the perception of the pain, and there is, therefore, no reasonable ground why the chloroform should not be used. Unless it can be proved that it is attended by danger and materially interferes with the process of parturition it should always be used.

No doubt, as is shown in many cases, the use of chloroform is not entirely without danger; but the amount of anæsthesia which we want for the present purpose is quite free from danger.

Generally, parturient women are easily put under the influence of chloroform. This occurs frequently after a few inspirations and without any uncomfortable symptoms. There is no need to produce absolute unconsciousness if nothing else is desired but the mitigation of the pain. A few whiffs of chloroform at the commencement of a pain easily suffice to suppress the loud expressions of pain; the woman is still conscious, she replies in a drowsy way to loud questions, the abdominal muscles act powerfully, and yet the pain is suppressed. Anæsthesia not continued any further than this is never dangerous either to the mother or to the child.

Although there is no doubt that profound anæsthesia continued for many hours (as is sometimes necessary, for instance, in eclampsia), may be transferred to the child also and prove fatal to it, yet past experience has shown that complete anæsthesia lasting for a short time has no influence whatever upon the child, and still less, therefore, can the use of the chloroform as described above have any influence upon it.

But there is another question—whether chloroform does not materially retard parturition or give rise to dangerous accidents, such as hæmorrhage? Now, Winkel has shown by accurate researches that parturition is, indeed, somewhat delayed by the use of chloroform, as the acme of a pain is somewhat shortened and the intervals somewhat lengthened. This delay, however, is, even in complete anæsthesia, as induced by Winkel himself, only very inconsiderable, and, as at first the anæsthesia is only employed to lessen the pain, there is no delay whatever. Moreover, since intense pain prevents the assistance of the abdominal muscles, after the use of chloroform not infrequently the pains are seen to be far more efficacious, since the woman is able to exert greater pressure.

Experience also has shown that by careful supervision of the stage of the afterbirth hæmorrhages do not occur more frequently than otherwise. If we add to this that chloroform causes a decrease of temperature, since, according to the experiments of Scheinesson, the production of heat is diminished, we have as the result of our considerations in slight anæsthesia no unfavorable effects whatever, but only favorable ones.

It must, however, be mentioned that not seldom after chloroform vomiting sets in. This is frequent in parturition without the use of chloroform, and has no unfavorable influence.

To the other dangerous consequences of chloroform women in labour are little subject, and they always feel very comfortable if they are allowed to sleep quietly and are not disturbed when the narcosis is over. It cannot, therefore, be questioned that the use of chloroform is advisable in normal parturition to suppress the sufferings in connection with it. Of course, we do not mean to indicate by this that it ought to be used in every labour, since there are many which are attended by slight pain only, and, as a rule, its use is subject to various considerations, such as the high price of the drug, the time which is required to give it, &c. But if these circumstances do not prevent it, and the suffering is very intense, there is no reason to abstain from its use. It is in all cases so beneficial to the woman who is suffering severely from pain that even women who at first could with difficulty be persuaded to take it urgently demand its continuance, and always require it in any subsequent labour.

It appears that chloral has until now been used only in a limited number of normal labours. Lambert recommends it strongly, but he prefers it in small doses of fifteen grains every quarter of an hour to a larger dose given at once. It by no means interferes with the contractions of the uterus, but rather favours them. We can bear testimony to this from our own experience. For even if by its use the interval between the pains becomes somewhat lengthened the pains themselves increase in efficiency. The internal and subcutaneous use of morphia is also in many cases of great benefit.

Literature.—Simpson, Edinburgh Monthly Journal, March, 1847, and Lancet, 11 Dec., 1847.—Kaufmann, Die neue in London gebr. Art der Anw. d. Chloroform. Hannover, 1853.—Houzelot, De l'emploi du chlorof. d. l'acc. nat. Meaux, 1854.—Krieger, Verh. d. Ges. f. Geb. in Berlin, 3 and 8 Heft.—Scanzoni, Beitr. z. Geb. u. Gyn. 1855, 2 Bd., p. 62.—Spiegelberg, Deutsche Klinik. 1856, No. 11, *sequ.*—Chapman, Chloroform and other Anæsthetics; their History and Use during Childbirth. London, 1859.—Kidd, Transactions of the Obstetrical Society, II, p. 340, and V, p. 135.—Lambert, Edinburgh Medical Journal, August, 1870, p. 113.—Gerson da Cunha, Lancet, 1870, vol. ii, p. 432.—Du Hamel, American Journal of Medical Sciences; s. Berl. Klin. W., 1871, No. 8.

PART IV

THE PHYSIOLOGY OF THE LYING-IN STATE

CHAPTER I

THE CONDITION OF THE MOTHER

Although the puerperal state is a physiological one, yet it must be admitted that it differs greatly from other physiological conditions, and that we find changes occurring there which in other circumstances would be called pathological. For the acute degeneration of the uterine substance is a process which, under other conditions, where not the mode but the extent and rapidity of its progress is thought of, would be considered undoubtedly pathological. This is still more the case in the change which is going on upon the inner surface of the uterus. After labour a part of the mucous membrane remains within the uterus. This exfoliates and a new one is formed from some of the inferior layers by the abundant proliferation of new cells and the pouring out of serous exudation, a state which is usually designated as catarrhal inflammation.

Then, again, the occlusion by thrombosis of the patulous openings of the internal vessels is a process standing alone, and quite exceptional in physiology.

The puerperal state also predisposes to a variety of diseases; thus, the lacerated vessels may give rise to hæmorrhage, and the considerable changes which the genital organs undergo in this state easily lead to inflammations and displacements. Nevertheless, although the puerperal state remains within physiological limits, it is proper, nay, absolutely necessary, to consider the lying-in woman a patient, and to watch over her carefully.

From what has already been stated it will easily be seen that it is very difficult to distinguish between the physiological and pathological puerperal state. It may be considered normal when the changes in the individual organs proceed favorably, and no disturbance can be found in them or in the general health.

Let us consider, therefore, the principal features of the normal puerperal state. It begins with the expulsion of the placenta and lasts from four to six weeks, after which time the retrogressive metamorphosis of the genital organs is almost completed.

When labour is over, the woman, though exhausted, feels easy and comfortable. Sometimes a slight shivering occurs, which is without significance, and only indicates the rise of the temperature, which is normal within the first twelve hours of the puerperal state. The sudden loss of blood and the unavoidable uncovering during labour favours its occurrence. Immediately after delivery the temperature begins to rise, and in the second twelve hours it completely falls again.

Independent of the processes which immediately after labour cause a temporary rise of temperature to above 39° C. (102·2° F.), it is chiefly the time of the day at which the delivery has taken place, which determines the height of the temperature. Its greatest rise takes place when the woman has been delivered in the course of the forenoon, because the normal and daily evening exacerbation then falls within the first twelve hours of the puerperal state. The succeeding decrease is very considerable when the delivery occurs in the first hours of the morning. Under such conditions the rise of temperature takes place within four to six hours, and the greatest fall between twenty and twenty-two hours after delivery. The rise amounts in pluriparæ, on an average, to 0·5° C., in primiparæ above 0·3° C., and decreases, on the contrary, in the former 1° C. and in the latter $1\frac{1}{2}$° C.

The highest temperature reached is, on the average, 38° C.; it may be somewhat more, even over 39° C., without any actual disease being present; the lowest is 37° C. (98·6°).

On the next day the temperature is at the highest towards 5 o'clock p.m., the lowest between 11 and 1 a.m. In exceptional cases only the exacerbation occurs in the morning and the remission in the evening.

Compared with other physiological conditions the temperature is somewhat elevated. Not infrequently it rises above 38° C., whilst the pulse is very slightly increased in frequency and the general condition remains excellent. The elevation of temperature is caused by the gradually increased production of heat through rapid metamorphosis of the uterus.

It would, however, be still greater if a considerable amount of heat were not rendered latent through the evaporation of the profuse perspiration, and if a quantity of organic matter only partially oxidized did not leave the system in the lochia and the milk.

By the enormous excretions from the lungs, skin, genital organs, and mammæ, lying-in women lose, on an average, 4500 grammes in weight in the first week of the normal puerperal state.

The pulse is, independent of individual variations, very slow, usually about 60, frequently less, even below 50, and sometimes even below 40. This decrease in the frequency of the pulse is of very favorable prognostic significance.

As a rule, very soon after delivery and lasting throughout the first week women show a great tendency to profuse perspiration, which chiefly occurs during sleep. Also after the first week the activity of the skin is increased.

The capacity of the lungs increases in most cases in proportion to that during pregnancy. Respiration is moderately frequent, between 12 and 25 in a minute. The appetite is less; thirst is increased on account of the high temperature and the profuse excretions. The bowels are torpid and frequently do not act for many days. The excretion of urine is increased, the desire to micturate, however, lessened, and the absolute excretion of urea decreased. Very frequently there is a retention of urine lasting for twelve or fourteen hours or even longer.

With regard to the gradual return of the genital organs to their non-puerperal state, the change in the uterus begins even during labour. The powerful contractions quickly following each other consume the contents of the cells of the unstriped muscular fibres, and by simultaneous compression of the afferent vessels the restitution of the oxidized protoplasm is prevented.

After the expulsion of the fœtus and its secundines the oxidization of the unstriped fibres still proceeds. At first the remainder of the cell contents is still able to perform its function, and is shown to do so by its contraction during the puerperal state. But gradually the albuminous substances of the protoplasm are converted into fat, which is easily reabsorbed, and by that gradual reabsorption the enormously enlarged cells are destroyed.

At the height of that process the new formation of cells begins in the external layer of the organ, and they are designed to build up the new uterus. In from six to eight weeks that process is terminated and then the lochia usually cease. In women who do not nurse menstruation returns, whilst in those who do it usually returns much later.

The uterus, which immediately after delivery weighs at least two pounds, a week after only weighs one pound, and a fortnight after only three quarters of a pound, and in about six weeks it has returned to its normal size. A variable amount of mucous membrane remains upon the inner surface of the uterus after the expulsion of the ovum. Sometimes the ovum is expelled only covered by the decidua reflexa, but usually upon the decidua reflexa are shreds of the decidua vera varying in size and rather intimately connected with and adherent to it.

The principal mass of the decidua vera, *i. e.* the whole glandular layer and the bulk of the cellular layer of Friedländer, remains behind on the whole inner surface of the uterus. According to that writer the formation of the new mucous membrane takes place in the following way: All that has remained behind of the cellular layer richly infiltrated with blood, as well as the upper portions of the glandular layer, gradually exfoliates and is discharged in the lochia. The flatly compressed glandular tubes situated close to the muscular coat are opened up and their cylindrical epithelium forms the new mucous epithelium of the internal surface of the uterus. The connective tissue situated between the tubular glands accordingly proliferates and becomes reorganized. In consequence of the increase in the thickness of the mucous membrane the previous shallow depressions of the epithelium are deepened, and in that way the uterine glands are also reformed in the new mucous membrane.

After delivery the same changes take place at the placental insertion as in the other parts of the uterus, only the sinuses penetrating it are opened up. A portion, however, of these sinuses have already begun to be closed by thrombosis at the eighth month of pregnancy, partly by the immigration of the larger cells of the serotina. The open sinuses are now closed by thrombi, and they become organized into young connective tissue, which starts either from the endothelium or from the immigrated white blood-corpuscles. The shrinking of that gelatinous tissue proceeds only very slowly, and sometimes four or five months after birth the placental insertion can be distinctly recognised. The gradual involution of the uterus gives rise to signs which can be found on examining the lying-in woman. After the delivery of the placenta the body of the uterus is pressed forwards by the pressure of the abdominal muscles acting upon its posterior surface, and can be felt between symphysis and umbilicus as a round and hard body. A few hours afterwards it becomes somewhat softer and is pressed somewhat higher up, since a full bladder mechanically forces the fundus backwards and upwards. It is then again to be felt on the level of the umbilicus. The position of the uterus is corresponding to that during pregnancy, most frequently somewhat to the right. It then begins to become smaller, and until the tenth or twelfth day, frequently somewhat longer, it can be felt from the outside. By internal examination immediately after delivery the soft and flaccid cervix can be felt like a loose sail hanging down into the vagina, and in that position the cervix is formed for the anteflexed uterus. The internal os first contracts; it frequently remains patulous, however, up to the tenth or eleventh day. The external os, in which, as a rule, rents can be felt, remains longer patulous for one finger. Not until after five or six weeks has the vaginal portion returned to its former condition as far as that takes place. As the uterus is inclined forwards and the cervix is formed again, corresponding to the length of the vagina, it naturally follows that in a puerperal state there will be normally an anteflexion. It is regularly found even in the first days, but on account of the progressive metamorphosis it disappears and is only apparent as a slight curve of its anterior surface.

The contractions of the uterus in the puerperal state are felt as after-pains. In very slow labours they do not occur at all, whilst they are very severe when the uterus has been much distended and labour is accelerated. Consequently in exceptional cases only are they found in primiparæ, whilst in pluriparæ they occur as a rule. They are distinguished from other pains by their periodical occurrence, by their peculiar child-bearing pains (dragging from the loins towards the lower part of the abdomen and into the thighs), by the contractions of the uterus felt by the hand applied to the abdomen, and by the absence of sensitiveness on pressure. This placing of the hand on the abdomen, however, may produce contraction of the uterus and give rise to pain.

The women themselves frequently state that what they experience is like labour-pains. By the application of the child to the mammæ the pains are increased. Frequently they appear only on the first day, sometimes they last up to the third or fourth, rarely to the sixth, day or a longer period. The transformation in the vagina is more slow and imperfect. After delivery the anterior wall hangs loosely down into the cavity. There is no distinct diminution in the size until the third or fourth week. Then folds begin to form in it, but it never regains its former narrowness and wrinkled condition. In exceptional cases the vagina, and especially the vulva, are even a few days after delivery so markedly narrowed that they scarcely differ from what they were before.

The above-described small lacerations of the mucous membrane are, as a rule, found in primiparæ and very frequently also in multiparæ. In primiparæ portions of the torn hymen are suffused with blood and destroyed by gangrene, so that in the vulva some warty or tongue-like projections remain (carunculæ myrtiformes). The frænulum, also, is frequently destroyed; the external genital organs are gaping and are reformed only imperfectly. The skin of the abdomen also remains flaccid and wrinkled for some weeks, so that puerperal women are subject to meteorismus in consequence of the great extensibility of the anterior abdominal walls. The discharge from the genitals—the lochia—consists, according to Werthheimer, for some hours, of pure blood with fibrinous coagula, and then the exudation of a serous fluid of alkaline reaction mixed with vaginal mucus commences. For the first two or three days the blood is so abundant that the lochia have a dark red appearance.

On the third and fourth, and sometimes even on the fifth day there is less blood, and the lochia then have a pale red appearance. Examined under the microscope blood-corpuscles, squamous epithelium, mucus-corpuscles, and sometimes the remains of the decidua, are seen. Organic compounds, as albumen, mucine, and fat, and also various salts, are found. From the fifth to the seventh or eighth day the lochia are still serous. The blood-corpuscles decrease in quantity, and now numerous pus-corpuscles are found. From the eighth to the ninth day the secretion is of a greyish-white or greyish-yellow appearance, of creamy consistence, and of a neutral or acid reaction. It chiefly contains pus and young, unshaped, round epithelium, also young spindle-shaped connective tissue cells with fat-globules, free fat and cholesterin-crystals. An infusorium, the trichomonas vaginalis, is found in the vagina of puerperal women.

It must also be stated that very frequently fresh blood is found in the lochia, even after the fifth day, especially if the woman leaves her bed. As regards the quantity of the lochia Gassner has estimated it to be—

About	1 kilog.	up to the	5th day,	when it is	red and bloody,
,,	0·28	,,	6th	,,	serous,
,,	0·205	,,	7th	,,	white,

so that within the first eight days 1·485 kilog. are lost by the lochia. The quantity is as much again in women who do not nurse.

The duration of the secretion of the lochia varies very much. After a fortnight or three weeks it is usually very slight, especially in women who nurse, whilst in those who do not nurse it lasts much longer. But this is by no means the rule. Not infrequently the secretion lasts in strong healthy women who do not nurse a much shorter time than in those who do so, but are of feeble constitution, and who for a long time subsequently suffer from fluor albus.

The changes in the breasts commence during pregnancy and are continued during the lying-in state. The glandular epithelium undergoes fatty degeneration, and a secretion flows from the mammary glands. For the first few days it constitutes the so-called colostrum, a thick fluid of a lemon-yellow colour, which coagulates on boiling. Under the microscope it is found to contain a number of large granular corpuscles (larger than white blood-corpuscles) with a nucleus, contractile protoplasm, and numerous small fat-particles, the so-called colostrum-corpuscles. There are also, even now, milk-corpuscles, round fat-globules strongly refracting light, which, according to Kehrer, have a membrane which consists of casein and not of albumen. They are at first of very variable size. After the third or fourth day milk is secreted, a thinner liquid of white colour, which does not coagulate on boiling. In it colostrum-corpuscles are only sparingly found, but the milk-globules are in large masses and of equal size about that of a red blood-corpuscle. There are also the free nuclei of colostrum-corpuscles, and globules, quite similar to milk-corpuscles, which on addition of carmine show a nucleus. The fats of the milk doubtless arise from albuminous substances, and the albumen contained in colostrum is transformed into casein. Kemmerich has shown that in fresh colostrum, after its evacuation from the mammary gland, casein is increased, whilst the albumen proportionately decreases; but according to Zahn there is still some albumen in the fully formed milk. When milk is boiled it forms small coagula which, by careful decanting of the supernatant fluid, remain at the bottom of the vessel. Casein is contained, according to Kehrer, in the débris of the glandular cells which forms a thin mucus-like fluid with the serum of the milk and becomes the vehicle for the fat-globules.

The commencement of the secretion of milk is usually attended by a slight elevation of temperature. It takes place on the third or fourth day, usually amounting only to a few tenths of a degree; yet not infrequently, especially in those who do not nurse or when lactation is begun, on the third or fourth day it rises to a fever heat, the breasts become tumefied and painful, and the superjacent skin reddened. From that time the normal secretion of milk continues and lasts for a varying time. If the mother does not nurse it soon stops. In women who nurse the quantity of milk increases up to the sixth or seventh month, after the eighth month it usually decreases, and since at that time the child requires somewhat more solid food it is well to wean it gradually, so that from the ninth to the tenth month it is no longer fed on the mother's milk.

Literature.—F. Winkel, Die Path. u. Th. d. Wochenbettes. Berlin, 1866, pp. 1–11.—Hecker, Charitéannalen, V, 2. 1854.—Winkel, M. f. G., B. 22, p. 321.—V. Gruenewald, Petersb. med. Z. 1863, Heft 7, p. 1.—Lehmann, Nederl. Tijdschr. voor Geneesk. 1865 (s. M. f. G., B. 27, p. 229).—Schroeder, M. f. G., B. 27, p. 108, and Schwang., Geb. u. Wochenbett., p. 177.—Wolf, M. f. G., B. 27, p. 241.—Baumfelder, Beitr. zu der Beob. d. Körperwärme, &c., D. i. Leipzig, 1868.—Lefort, Études cliniques, &c., Strassbourg. Thèse, 1869.—Scherer, Artikel Milch in Wagner's Handwörterbuch der Phys. 1845, B. II, p. 449.—Becquerel and Vernois, Comptes rendus, T. XXXVI, p. 188, and L'Union, 1857, 26.—Moleschott, Phys. der Nährungsmittel. Giessen, 1859.—Hoppe, Virchow's Archiv. 1859, B. XVII, p. 417.—V. Gorup-Besanez, Lehrbuch der physiol. Chemie. Braunschweig, 1862, p. 385.—Beigel, Virchow's Archiv. 1868, B. 42, p. 442.—Langer, Stricker's Handb. d. L. v. d. Geweb. Leipzig, 1870, IV, p. 627.—Kehrer, Arch. f. Gyn., B. II, p. 1.

CHAPTER II

THE DIAGNOSIS OF THE LYING-IN STATE

THE diagnostic signs of childbed consist partly in changes produced during pregnancy, which are visible for some time after it, partly in the traces which the process of parturition has left behind, and partly in the other peculiar changes which take place in the genital organs and the mammæ during childbed. They are the following:

The skin of the abdomen is flaccid and wrinkled. In it are seen small white cicatrices with transverse wrinkles; the linea alba is greatly pigmented, the vulva somewhat swollen, the labia gaping, and in the vaginal outlet small ulcers or very vascular cicatrices are always found; soon after delivery, of course, recent lacerations of the mucous membrane. The vagina is wide and smooth, and in it is a copious secretion of a peculiar stale and sometimes fetid odour, and with the characteristic condition of the lochia. The uterus is more or less enlarged and anteflexed, and this enlargement can be easily made out, without the sound, by a combined internal and external examination. If the external os is still pervious the wide uterine cavity is felt, and in it also is a copious secretion, and a somewhat prominent and rough place covered with small plugs indicating the former seat of the placenta. The mammæ are large and tense, the areolæ pigmented, and the known secretion flows or can be pressed from the gland, which if soon after delivery is colostrum, but if later milk.

The value of these signs is, of course, different, yet a great number of them are characteristic. The flaccid, shrivelled abdominal walls, covered with wrinkles and cicatrices, are a sure sign, since conditions in which enlargement of the abdomen had existed, as from ascites, ovarian tumours, &c., can easily be excluded. The pigmentary deposits also are frequently more intense than are met with in any other condition except pregnancy and parturition.

The slight lacerations in the vulva, also, are characteristic; the lochia, at least within the first few days, cannot be mistaken for anything else. The size and shape of the empty uterus, as found in the puerperal state, do not occur in any other condition, and the placental insertion covered by thrombi places the diagnosis beyond a doubt. The changes also of the mammæ suffice to ensure the diagnosis. Though a slight pigmentation of the nipple and a secretion from it occur in some pathological conditions, yet the deposit of pigment and the quantity of the secreted milk in the puerperal state are almost always much greater than is met with under any other condition Thus, on the whole, it is easy to establish the diagnosis of the lying-in state in the first few weeks by objective examination alone. It is, nevertheless, sometimes difficult accurately to determine the time which has elapsed since delivery, and the difficulty increases the longer the interval. If there are still lacerations at the vaginal entrance, and they are quite recent, the woman has only lately been delivered; if, on the contrary, there are already distinct cicatrices, the first stage of childbed is past.

From the changes of the lochia above described valuable assistance can be derived, only it must be kept in mind that it is just in those cases which are of forensic importance that the sanguineous lochia usually last much longer. The most important point for a practised accoucheur is the size of the uterus. After a great many puerperal women have been examined internally and externally, a sound and pretty accurate judgment is acquired of the size of the uterus corresponding to the different periods of childbed, yet it also is subject to individual variations. The internal os, which is rarely pervious after the tenth day, gives very valuable indications with regard to the time. We have also to take into consideration that the uterus will be smaller, and the cervix more contracted, after premature labour than after delivery at term. The secretion of the mammary gland is also important as indicating the early period, the colostrum would point to the first days after delivery.

By a careful consideration of all these circumstances it is possible in the first fortnight of childbed to fix the time of delivery within a few days with accuracy, but later on it is frequently necessary to be satisfied with more uncertain data.

CHAPTER III

THE CONDITION OF THE CHILD THE FIRST FEW DAYS AFTER ITS DELIVERY

THE changes which take place in the fœtal circulation at birth, as well as the obliteration of the purely fœtal vessels, have already been mentioned. The umbilical cord dries up when the circulation ceases in it. Between the skin of the abdomen and the sheath of the umbilical cord a line of demarcation is formed, and the umbilical cord is thrown off with slight suppuration, at times on the third day, but, as a rule, on the fourth or fifth day, or even much later.

The umbilical ring is for some time moist, and gradually cicatrices. On the head of the child we almost always see traces of hyperæmia. The mucous membranes are swollen and secrete copiously, the conjunctivæ are injected, and the sclerotic is not infrequently the seat of half-moon shaped hæmorrhages which surround the cornea. The whole skin of the head is somewhat infiltrated, and the spot which during labour was situated in the os is frequently the seat of a circumscribed tumour, which consists of a sero-gelatinous infiltration of the connective tissue, and usually also of slight hæmorrhage into it. Under the epicranium also, at times, small effusions of blood are found. The infiltration, however, as a rule, almost always entirely disappears within twenty-four hours. Soon after birth the meconium, the greenish-black contents of the intestines, is voided. After a few days the excreta of the child have a feculent appearance; the secretion of the kidneys becomes very copious. The skin, which was covered by the vernix caseosa becomes denuded of its epithelium; the red colour of the skin passes away, and in consequence of the previous hyperæmia, even on the third day at times, a very intense icteric colour is seen, which later on gives way to the normal colour. The icterus may also arise in part from the liver, in which the greatly decreased blood pressure after birth must favour the entrance of the already formed bile into the capillaries.

The small mammary glands very frequently swell soon after birth, in boys and girls equally. They become sensitive, the skin which covers them is reddened, and by slight pressure a serous milky secretion can be evacuated from them. This inflammatory irritation soon disappears spontaneously.

In the first days after birth a loss of weight has frequently been observed; this, however, appears to be physiological, as it is probably caused by the loss of meconium and urine. It appears, at least, if the puerperal woman is healthy and has sufficient nourishment this loss may be almost entirely avoided. Later on, a healthy child must daily gain in weight; during the first four months twenty to twenty-five grammes, and from the fifth to the tenth month fifteen grammes. According to Odier and Blache, a child ought to weigh at the end of four months twice as much as at the time of birth, and at the end of sixteen months again twice that last weight.

According to Pollak, a child from eight days to two and a half months old passes, in twenty-four hours, from 250 to 410 cc. of urine. It is of a pale straw colour, slightly acid, and of a sp. gr. of 1005 to 1007. It contains very little urea, uric acid, and phosphates, some mucus, with a little albumen and very little sugar (somewhat more than adults). The urine of a new-born child, of which (on an average) 7·5 cc. are found in the bladder, is still paler, sometimes as clear as water, of a sp. gr. of 1001·8 to 1006, and usually contains no albumen.

CHAPTER IV

DIETETICS OF THE LYING-IN STATE

A.—ATTENTION TO THE MOTHER

It has already been pointed out that, although childbed as well as pregnancy and labour are physiological processes, yet during them changes occur in the organism such as are never seen except under pathological conditions. If the usual mode of living is continued during childbed serious dangers may result, either immediately or in the future. Therefore it is urgently necessary to consider the puerperal woman a patient and to watch over her carefully. It naturally follows from this that the woman should at first be kept in bed and allowed gradually to resume her old mode of living and her usual occupation. She should remain in bed for at least one week, and, if possible, still longer; otherwise, on account of the size and weight of the uterus and the softness and flaccidity of its ligaments and of the vagina, lasting deviations of the uterus may easily develop. The usual occupations should not be undertaken before six weeks, for not until after that time have the organs of generation (as far as such is possible) returned to their previous conditions. Immediately after delivery the uterus is to be examined, to ascertain if it is well contracted and whether there are any considerable lacerations of the soft maternal passages. If delivery has been normal the uterus can be felt contracted externally. If little blood escapes and there are no doubtful symptoms in the general state there is no reason to make an internal examination.

We should never neglect to examine the vulva, since considerable lacerations of the perinæum may require to be attended to. The genital organs of the woman should be cleansed with a proper linen towel, and directions given always to preserve scrupulous cleanliness.

The linen soiled during parturition and the wet sheets must be removed and replaced by proper ones. It is desirable, also, to have a freshly warmed bed close to where the woman has been delivered, in which she should be carefully placed.

Rest is now what the woman most urgently requires; under the supervision of an intelligent nurse she may be allowed to sleep.

The further treatment during the puerperal state is a merely expectant one, as a repeated internal examination of the puerperal woman is unnecessary and, under the circumstances, also not without danger. It is best to abstain from it if the temperature is regular and the frequency of the pulse noticed. The thermometer so exactly indicates the least disturbance in puerperal women, that so long as the temperature remains under 38° C., or only a little above it, no internal examination need be made. If the pulse is very slow there is every hope that the course of the puerperal state is normal. The due contraction of the uterus may be ascertained in the first few days by the application of the hand to the abdomen, but it will be desirable before the woman is dismissed from treatment to make a careful examination.

The external genitals may best be cleansed by the puerperal woman herself, either with a sponge or with a towel; a frequent change of sheets is decidedly necessary.

The lying-in room must be as large, as high, and as airy as possible. It is not necessary that the room should be kept darkened; in many respects it is uncomfortable. In the summer one window may be opened all day long; in winter the room is to be frequently aired. Special attention must be paid to the emptying of the bowels and the bladder. If the bowels have not acted on the third day an enema is to be given, or one or two tablespoonfuls of castor oil. If the latter cannot be taken on account of its taste, or if it causes vomiting, a weak solution of senna, or some salt of magnesia, or soda, may be administered. Slight diarrhœa in the puerperal state is not to be dreaded, and drastic purgatives should not be used. Should there be retention of urine the catheter should be passed at least once in twenty-four hours until the urine is spontaneously voided. The diet of the puerperal woman must be a light and natural one. Immediately after delivery there is little or no appetite, and therefore broths, soups, an egg and some bread, are sufficient. As a beverage milk is to be recommended. If the woman, however, should have an appetite she may have sufficient meat, and on the following days also some vegetables, until she gradually resumes her previous diet.

Every mother ought, unless there be special reasons against it, to nurse her child, since this depends in the first days of its extra-uterine life upon the mother for its maintenance. Only when the mother is ill and too feeble, or when experience has shown that the milk does not agree with the child, should nursing be prohibited. It may, however, be rendered impossible by the absence of milk, and by such a condition of the nipple that the child cannot grasp it.

After the mother has rested for a while the child is put to the breast, a few hours, not later than twelve, after delivery. It is quite unnecessary to give the child previously tea or sugar water.

It is, at times, a troublesome process to get the child to take the breast. Primiparæ must be taught how to do it. In the first few days the child may be put to the breast as often as it has appetite, but even from the commencement it is advisable to accustom the child to regularity. The child is soon accustomed to take the breast every third hour during the day, and during the night at an interval of six or seven hours. Later, six times in twenty-four hours is quite sufficient.

If the secretion of the milk is relatively or absolutely too abundant, so that the mammæ become sensitive, inflammation may be best prevented by full doses of salines, as derivatives to the intestinal canal. The same remedy also can be used if the mother cannot or does not wish to nurse, in order to stop the secretion of milk.

B.—ATTENTION TO THE CHILD

After the umbilical cord has been divided the child is to be put into a warm bath of about 82° F., and cleansed with a sponge from the blood, liquor amnii, and vernix caseosa. If the latter substance be very abundant a little oil rubbed into the skin will help to remove it easily. The child must then be dried in warm sheets, the stump of the umbilical cord covered with a piece of oiled lint, turned upwards, and gently fixed by a binder. The dressing of the stump must be renewed in a few days, until it has fallen off. As long as the umbilicus remains moist a piece of oiled lint may be applied to it and fastened by a binder. Later all dressings may be omitted.

The temperature of the bath must not be too high. Midwives and nurses, on account of their frequent occupation with warm water, are often unable to determine correctly the temperature of the bath. Keber states that one midwife lost, in two years, in 380 deliveries, 99 children from trismus, doubtless in consequence of using too hot baths.

The clothing of the new-born child must be so arranged as to keep the child warm without impeding the movements of the extremities or those required for ample respiration. It is best to leave the child without a binder; a chemise and a jacket are sufficient. The legs are to be loosely wrapped up in a napkin and covered by a long woollen frock, or a large cloth, which is to be loosely fastened by a short and broad binder. The head is to be covered with the lightest possible cap.

The child is then placed, with the head somewhat raised, in its bed, which should contain a mattress of finely divided straw. Upon this a waterproof and linen sheets are placed. The child is to be covered with a blanket or with a light plumeau.

The principal means of favouring the thriving of the child consist, besides suitable food, in attention to scrupulous cleanliness. The child must be bathed every day, and as soon as the napkins are soiled they must be replaced by proper and dry ones. The mouth also must frequently be washed with a piece of linen, especially each time after having taken food.

If the mother does not nurse the child it is doubtless best to provide a wet-nurse. She must be perfectly healthy, and there must not have been any hereditary disease in her family. The mammæ and nipples must be normal and well developed. Besides the necessary qualifications with regard to character and disposition, she must not be either too old or too young. It is desirable that the time of her delivery and that of the birth of the child be not too far distant from each other. It is difficult to decide whether a wet-nurse has good and sufficient milk, and whether she will retain it. When the child has not quite finished sucking the milk will easily flow out in a stream when the tumid mamma is pressed; the milk should be of a white colour and not too thin. If she has already nursed her own child for some time the appearance of the latter is a good criterion by which to judge of the value of her milk. If the child appears well-nourished and strong the nurse has for the present sufficient milk. Whether she will retain it is another question. In order not to prevent the secretion of the milk, the customary food and style of living of the wet-nurse is not to be too much changed. It often happens in the better classes that very strong country nurses soon lose their milk, because all day long they sit comfortably in an easy chair and are fed with roast meats and puddings.

If a child is to be reared without the mother's or a nurse's milk, cleanliness and regularity are the chief agents to make it thrive. Cow's milk, the usual substitute, contains more fat and is less sweet than the milk of woman; it must therefore be diluted and some sugar added. It is desirable to let the child have the milk of the same cow as fresh as possible. In the first days twice as much water is to be added, in the next fortnight water and milk in equal parts, and later on two thirds milk and one third water. To sweeten it a little sugar of milk is added, which, at the same time, gently excites the action of the bowels.

The food should always have the same temperature, about 82° F., and be given in a feeding bottle. The bottle as well as the stopper ought to be always in fresh water when not used. If the food is of acid reaction a little carbonate of lime is to be added and the fluid decanted from the undissolved powder which has fallen to the bottom. In the first month the milk suffices; later, chicken or veal broth may be added, or a pap of wheaten flour with sugar and milk may be given. Of the numerous preparations Scharlau's substitute for mother's milk and Liebig's nutritive powder are most to be recommended.

Literature.—V. Ammon, Die ersten Mutterpflichten und die erste Kindespflege. 13, Aufl. von Grenser. Leipzig, 1868.—C. Mayer, Verh. d. Ges. für Geb. in Berlin, I, 1816, p. 56.—Wegscheider, M. f. G., B. 10, p. 81.

PART V

V. THE PATHOLOGY OF PREGNANCY

We have already stated, whilst speaking of the physiology of pregnancy, that disturbances regularly attend this physiological process, and that they would be considered pathological if met with under other circumstances. As long as they do not give rise to consequences which are serious at the time of their occurrence, or threaten to become so at some future period, they are supposed to form an integral part of the normal course of pregnancy. From this it will be easily understood that pregnant women are predisposed to a variety of diseases. This is greatly increased by the changes which the genital organs undergo during pregnancy, and which may lead to inflammation, alteration in position, &c. Since the ovum also is liable to various kinds of disease, the pathology of pregnancy therefore includes both the affections of the maternal organism and those of the ovum.

CHAPTER I

1. DISEASES OF PREGNANT WOMEN

It is evident that the present chapter will not contain a full description of all the internal and surgical diseases which may possibly complicate pregnancy. Only the acute and chronic affections which essentially modify pregnancy, or which as complications of it are of a greater interest to the accoucheur, will therefore be mentioned. We shall treat here of the diseases which are due to pregnancy itself, and which are localised either in the whole organism or in the genital organs.

A.—PECULIARITIES OF SOME ACUTE AND CHRONIC DISEASES DURING PREGNANCY

Formerly the opinion was prevalent that the later period of gestation offered, at least, great immunity to enteric fever, but recent observations have shown that such is not the case.

It has been found that pregnancy is often prematurely interruped by an intercurrent attack of enteric fever, but that on the whole the disease does not take a more unfavorable course in pregnant women than in other individuals. Pregnancy may be primarily interrupted by the activity of the uterus prematurely roused, but commonly the interruption is due to the death of the fœtus. Deeply enveloped within the maternal soft parts, the fœtus will already have a temperature equal to about that of the maternal body, and as the fœtal body itself produces heat, it follows that its temperature must exceed that of the mother. Therefore, in acute febrile diseases of the mother the death of the fœtus is often due to an accumulation of heat. According to Kaminsky an elevation of the temperature of the mother above 104° F. is very dangerous to the child. At first its movements become more active, and the sounds of the heart more frequent; later on, however, they become slower, the movements decrease, and the fœtus dies.

From former experience it was supposed that relapsing fever most frequently caused abortion; yet Weber found that amongst sixty-three pregnant women premature labour set in only in twenty-three (36·5 per cent.). One was delivered in the hospital, after the termination of the fever, of a fully developed living child; the other women left the hospital still pregnant and cured. The tendency to abortion appears to be least in typhus.

In variola abortion may occur and the fœtus be born dead or alive, as a rule without, rarely with, smallpox. Often an eruption of smallpox is observed after birth. Not infrequently pregnancy goes on to term, and the child is born quite healthy or covered by cicatrices. It also occurs that children are born with variolous pustules whose mothers were not affected by smallpox, and that of twins one only is covered by the eruption. Vaccination does not, as a rule, succeed in children whose mothers suffered from variola during gestation.

In cholera the death of the fœtus usually occurs in the stage of asphyxia; this is easily explained by the cessation of the exchange of gases in the placenta. Pregnant women do not exhibit a special predisposition to cholera, and the prognosis is here hardly more unfavorable to the mother than when the disease is not complicated by pregnancy. When the mother has reached the stage of asphyxia the life of the fœtus is in danger, yet scarcely as much as Baginsky states he has experienced.

Of the other acute diseases, acute yellow atrophy of the liver is so far of greater interest as pregnant women are especially liable to it. It may occasionally be due to phosphoric poisoning, or may be the most prominent symptom of a septic infection, and as such it is frequent in puerperal women, or it may have primarily begun as a parenchymatous hepatitis. The explanation given by Davidsohn most likely applies to the great majority of cases. According to him, catarrhal icterus is the primary affection, and as the excretion of the deleterious biliary acids by the kidneys is impeded during pregnancy, the icterus easily becomes the source of a general blood-poisoning, with consecutive fatty degeneration of the larger abdominal glands and the muscles, especially those of the heart.

Amongst the chronic diseases syphilis is the most important. If by a fruitful coitus mother and fœtus are infected, there is, as a rule, premature death of the fœtus and abortion. If the fœtus remains alive, it is always prematurely born, is badly developed, and almost always dies a very short time after birth. If the syphilis of the father has been latent at the time of the coitus, the unfavorable influence upon the fœtus remains the same, and the mother may equally be affected, probably through the fœtus.

Syphilis in the mother apparently diminishes the power of conception, but by no means entirely abolishes it. If conception occurs the influence of the syphilitic mother on the fœtus is the same or even more unfavorable still than that described above. If the syphilis of the mother be latent at the time of conception, apparently healthy children are born, which sicken later.

If the father or the mother at the time of the conception have tertiary symptoms, the child is often not infected; children may, however, be born with secondary symptoms.

If the mother be healthy at the time of conception and be infected only during gestation, the child is healthy at birth and remains so, unless it be infected by the mother during her delivery or in the puerperal state.

The very important question in practice, whether syphilitic pregnant women may undergo treatment with mercury without danger to themselves or their offspring, must decidedly be answered in the affirmative. For the frequency of premature births and the number of children dying during gestation are considerably lessened by treatment with mercury.

Chorea is relatively a frequent complication of pregnancy. It occurs very easily in those pregnant women who have suffered from chorea in infancy, and it may then recur in successive pregnancies. Chorea complicating pregnancy not infrequently endangers the life of the mother (seventeen times in fifty-six cases, Barnes), and may leave psychical disturbances behind. In very favorable cases chorea ceases, and gestation reaches its normal end. In other cases there is spontaneous abortion, not always of a putrid child, after which the disease regularly abates. If chorea threatens the life of the mother, premature labour and under circumstances also artificial abortion is to be induced, whilst in less severe cases bromide of ammonium with liquor arsenicalis and bromide of potassium may render signal service.

B.—DISEASES CAUSED BY THE CHANGES IN THE WHOLE ORGANISM DURING PREGNANCY

In the present section of the pathology of pregnancy, all those changes of the whole organism will be described which, although previously mentioned as belonging properly to pregnancy, have yet reached such a degree that they must necessarily be considered pathological.

Whilst the blood even in normal pregnancy is constantly altered, so that the number of the red blood-corpuscles and the quantity of albumen are diminished, yet the alteration in the state of the blood may go so far that diseased conditions are the necessary consequence. If the number of the red blood-corpuscles is chiefly diminished, the pregnant woman is chlorotic; if a decrease of albumen coexists, the pregnant woman suffers from hydræmia. It is certainly exceptional to find such a state of the blood resulting from pregnancy alone. Usually that pathological condition is met with in those individuals who have previously suffered from some anomaly in the composition of the blood. The symptoms as well as treatment are the same as in the non-pregnant state. It has only to be pointed out that in anæmia of the mother hæmorrhage during pregnancy or delivery may become very serious.

The consequences of hydræmia, viz. œdema of the inferior extremities and of the abdominal walls, are very tedious. The œdema becomes very great if at the same time the distension of the uterus is excessive, and so by pressure upon the veins prevents the return of the blood. The œdema is sometimes so great that walking is almost impossible, the labia are swollen to the size of a child's head, or even more, and from the abdomen a large bag hangs down, which is formed by the abdominal integument and is filled with water. If the tension is so great that gangrene is threatening, the dropsical parts must be punctured. It is to be remembered that needle punctures of the swollen labia very easily excite uterine action. The varicose veins may burst and give rise to dangerous hæmorrhages, and the venous blood spouts forth in a curve for some yards, as if it came from a large artery. The bleeding may be easily stopped by compression. The discomfort produced by the swollen legs can be diminished by diuretics and diaphoretics; directions also must be given that in such cases pregnant women should walk as little as possible, and that in sitting they should keep the feet raised, or wear elastic stockings, which exert an equal pressure upon the legs. At the end of the gestation, or the commencement of the tenth month, when the uterus descends and falls somewhat more forwards, all these troubles are considerably lessened.

The pressure of the gravid uterus may also give rise to a considerable disturbance on the part of the rectum and of the bladder. It may cause an obstinate retention of fæces, so that the whole colon is filled with hard fæcal masses, and long-continued enemata, applied by means of a long elastic tube, will be necessary in order to empty the bowel. Tenesmus is the most frequent disturbance on the part of the bladder. This appears to be due less to the pressure upon than to the irritation of the neck of the bladder from being drawn upwards. Complete incontinence and retention of urine are rare in the normal position of the uterus. Retention of urine may become so obstinate in the last month of pregnancy, when the head presses the urethra against the symphysis, that a repeated application of the catheter is necessary.

Obstinate vomiting—hyperemesis—is a very serious disease of

pregnancy. Nausea and vomiting on an empty stomach or shortly after a meal are the most common symptoms of gestation in the first months. The women do not greatly suffer by it, and their nutrition often even increases. There are cases, however, where vomiting lasts throughout the whole period of gestation and becomes so excessive that the stomach can no longer retain anything, and extreme nutritive disturbances follow. In such grave cases symptomatic treatment is often of little avail; the best result is obtained by small pieces of ice, champagne, and subcutaneous injections of morphia. Pippinsköld has cured it by enemata containing broth, a little oil, and wine. If gestation be so far advanced that the fœtus is viable, the induction of premature labour is, as a rule, indicated. If at an earlier period vomiting is so severe that by its continuation death is inevitable, and if it cannot be subdued by any remedy, it may become necessary to induce artificial abortion. It must not be forgotten, however, that the obstinate vomiting may also be due to a round ulcer of the stomach.

Excessive salivation must be restricted by astringent mouth washes, and the very tedious toothache often evades every treatment. This insupportable pain can for a short time only be mitigated by the internal, or better by the local, application of narcotics.

Amongst the neuroses of pregnant women eclampsia is the most important. But as it is more frequent during parturition, it will be considered as a whole in the pathology of parturition. Epileptic and hysterical convulsions are not to be mistaken for eclampsia. They are not, however, very frequent in pregnancy, and do not show deviations from their usual course, nor disturb the normal progress of pregnancy.

C.—ANOMALIES OF THE GRAVID ORGANS OF GENERATION

1. *Anomalies of the Uterus*

(A) *Malformations.*—Malformations of the female genital organs which prevent conception do not interest us here. It may be remarked that conception is everywhere possible where normal ova are expelled from the ovary, and where their passage from the ovary to the vaginal outlet is not absolutely impossible at any one spot. Therefore in the uterus unicornis, as well as in all kinds of double uterus, pregnancy may take place provided the condition just mentioned is complied with.

The course of pregnancy is greatly modified if the ovum is situated in a rudimentary cornu (*Nebenhorn*). Even if its cervix were entirely imperforate, impregnation would be possible in two ways. The spermatozoa may pass either through the well-developed cornu and the appertaining Fallopian tube, traversing the abdominal cavity to the ovary of the other side, and there fecundate the ovum, which then enters the Fallopian tube of the rudimentary cornu (*external migration of the spermatozoa*); or the spermatozoa may fecundate the ovum of the ovary on the side of the fully developed cornu, and the ovum itself may then be received into the tube of the other side (*external migration of the ovum*). Gestation in the rudimentary cornu has the greatest similarity to tubal pregnancy, which will be described further on. The rupture of the sac with its fatal consequences occurs between the third and sixth month. Turner describes a case in which the end of pregnancy was reached. The very severe labour-pains ceased after a few days, and on post-mortem examination of the woman, who died six months later of phthisis, the fœtus was found in the occluded left cornu. The rupture usually is met with at the most feebly developed part of the cornu. Exceptionally a lithopædion may form in the atrophied cornu. Köberle extracted such a fœtus twenty-one months after its death by the Cæsarian section. The normally developed cornu is in the same state as in extra-uterine pregnancy, that is, it hypertrophies and forms a decidua.

In the living subject a diagnosis of pregnancy in a rudimentary cornu can only be made with some degree of probability. Though in the unimpregnated state that anomaly may be detected under favorable circumstances by means of palpation and sounding from the laterally bent and pointed shape of the pervious cornu, yet in pregnancy the shape of the cornu becomes so greatly modified that a distinction is scarcely possible. Even in the dead body this state can be very easily mistaken for tubal pregnancy. A diagnostic sign is, however, furnished by the course of the round ligament. In double uterus it passes outwards from the gravid organ, and in tubal pregnancy between the sac containing the ovum and the uterus. The circumstance also that generally a rudimentary horn ruptures later than a Fallopian tube may serve to establish a probable diagnosis.

If in uterus unicornis, with a rudimentary horn, the fully developed horn encloses the ovum, gestation, as a rule, terminates normally. This anomaly, however, is not quite inaccessible to diagnosis. The enlarged uterus is then felt bow-shaped and pointed above towards the iliac bone of one side, and by combined internal and external examination a small tumour can be felt on the other side connected with the uterus by a short cord. Eccentric insertion and abnormal shortness of the vaginal portion assist in the diagnosis.

If both halves of the genital canal are sufficiently developed, but separated by a more or less perfect septum—*uterus bicornis and uterus septus*—pregnancy often occurs in one or both horns simultaneously. In complete separation of both halves down the vagina (a condition normal in the marsupalia) simultaneous pregnancy of both horns, if seen at all, is very rare. Though, *à priori*, the possibility of such an occurrence cannot be disputed.

All kinds of double genital canal, provided that each half is sufficiently developed, may contain children up to the normal end of pregnancy, and there is no doubt that such regularly takes place. Whether abortion is here more frequent than in the normal uterus is as yet doubtful. The course of pregnancy is normal in all these forms of double genital canals. In order to detect such an anomaly a careful examination must be instituted in all cases of a double vagina. If in each vagina a cervix is felt, the presence of the uterus, divided also above the cervix, is almost absolutely certain. If a double cervix opens into a simple vagina the body may be simple also, but this is certainly very rare. In all cases where two vaginal portions are present the diagnosis will offer no great difficulty, for in pregnancy of one horn this lies so distinctly to one side that a sound may safely be introduced into the cervix of the other. If pregnancy is not too far advanced the unimpregnated uterus will be accessible also to palpation by combined internal and external examination, but if the two horns are simultaneously impregnated a deep furrow in the middle of the abdomen, from the fundus uteri to the symphysis, will serve for a correct interpretation. Those cases in which the vagina and cervix are simple, or where only one vagina is pervious, are very difficult to diagnose. Under the latter circumstances the imperforate vagina may lead to an accumulation of the menstrual discharge, and gestation and hæmatometra may coexist. Where the cervix is simple it is possible to recognise the actual state from a considerable lateral deviation of the gravid uterus and a small palpable uterus on the other side. If the septum be so small that the fœtus can lie in both halves, or if there be only a depression of the fundus indicative of a bipartite uterus, *uterus arcuatus*, the shape of the organ as determined by an external examination is the sole point of reliance for diagnosis.

(B) *Changes in the position.*—1. *Flexions and versions.*—Although sterility is a frequent consequence of anteflexion of the uterus, pregnancy, nevertheless, often occurs without or after suitable treatment, and in the absence of any other complication it may terminate normally. The metritis often causes abortion. The fundus uteri lies, in the first months, more forwards than usual, but at a later period it is in no way distinguished from the normal, so that the anteflexion which had existed previous to gestation can no longer be seen.

A peculiar form of anteversion sometimes occurs towards the end of pregnancy. This, the so-called pendulous belly, is produced by various causes in which abnormal relaxation of the abdominal walls and a separation of the recti muscles are chiefly concerned. A distorted pelvis, great pelvic inclination, and lordosis of the lumbar vertebræ, also have a share in its production.

Slight degrees of pendulous belly are of daily occurrence; the forward inclination of the fundus may, however, be so great that it forms a ventral hernia and reaches to the knees. The discomfort arising from it can be remedied by the wearing of a suitable bandage.

Retroversion is not a very frequent change of position; it is sometimes a cause of sterility. The fundus does not, in the majority of cases, lie beneath the promontory. If pregnancy occurs, the body of the gravid uterus rises spontaneously, as there is nothing to prevent it. If, exceptionally, prevented by some impediment from doing so, it may be retained beneath the promontory to the end of the third or fourth month so as to give rise to symptoms of incarceration. Its structure, as a rule, has become so flaccid in the mean time that the cervix is not directed upwards as it ought to be if a version were only present, but at the place of the internal os a flexure exists and the vaginal portion either looks towards the symphysis or more frequently downwards. A retroflexion is then closely connected with the retroversion of the gravid uterus. A peculiar shape of a gravid uterus has been described as partial retroversion. The cervix lies here closely behind or above the symphysis, and the posterior and lower uterine segment is gradually depressed into the small pelvis between the cervix and promontory by a fœtal part or by the bag of waters.

The greater portion of the uterus and of the fœtus lies above the promontory, although Oldham in one case felt the head through the posterior wall of the vagina, and the pelvic end in the os, which was situated about three inches above the symphysis. This condition may easily be recognised if an examination with the whole or part of the hand can be made, and it ought to be recognised even if the vaginal portion cannot be reached, from a careful consideration of all the circumstances relating thereto. The course of pregnancy is, as a rule, not modified by the condition.

Retroflexion of the uterus is the most common alteration of position in women who have borne children, and is rarely attended by sterility even if it be of a very high degree. It is seldom met with in women who have not been pregnant, but if it occurs it frequently causes sterility.

After conception the uterus at first enlarges, retaining its actual position, but having acquired such a bulk that the true pelvis can no longer contain it it ascends above the promontory. The fundus usually falls forwards, and the retroflexion which had previously existed is now converted into anteflexion and anteversion.

The ascent and the falling forwards of the gravid organ, as a rule, take place so gradually that they are not noticed by the woman. In other cases, and more frequently in complications with chronic metritis, the uterine activity is prematurely excited and abortion follows. Far more rarely the uterus is not able to ascend above the promontory; it continues to enlarge within the small pelvis, and generally at the commencement of the fourth month gives rise to symptoms of incarceration and the so-called retroflexion of the gravid uterus is developed.

Retroflexion of the gravid uterus with symptoms of incarceration.— In the majority of cases it is in the way above described that the retroflexed uterus becomes incarcerated. At any rate it is doubtful whether in the normal position of the uterus a retroflexion can be produced by sudden causes at a time when the organ has already so enlarged as to produce immediate symptoms of incarceration.

The first symptoms which a retroflexed uterus produces, as soon as it becomes too large for the pelvic cavity in which it is contained (this usually occurs in the fourth, sometimes in the third, and rarely in the fifth month) are those on the part of the bladder and the rectum. The micturition is always impeded, sometimes so completely that not a drop of urine can be passed. More frequently, however, it occurs that when the bladder is enormously distended, through the incessant desire of micturition, the woman passes a small quantity of urine with very great pain, and in such a state of the bladder most frequently complaint is made, not of retention of urine, but of vesical tenesmus and incontinence. The urine may, as a sequence of this, accumulate in the ureters and the pelves of the kidneys.

The disturbance on the part of the rectum consists either in complete constipation or in great pain at defæcation, especially when the stools are hard. Persistent constipation may then give rise to vomiting and to all the other symptoms of ileus. There are, also, severe pains in the loins and in the abdomen, with an insupportable sensation of a downward pressure. Even now the spontaneous ascent of the enlarged uterus may take place or abortion, and then the tormenting symptoms terminate. In a case reported by Von Haselberg spontaneous abortion occurred though the fœtus was five months old and the uterus was so completely retroverted that its vaginal portion, in the absence of all flexion, looked directly upwards. But in other cases severe inflammation of the incarcerated uterus and its serous covering follows, which increases the swelling of the organ and causes severe pains in the abdomen, meteorismus, and fever. A sero-sanguinolent discharge escapes from the cervix, and by reflex irritation of the abdominal muscles the fundus is more and more forced downwards.

In Halbertsma's case the fundus protruded through the anus, which was opened to the size of a half-crown piece, as a tumour, in which small fœtal parts could be felt. It is very rare to find the posterior wall of the vagina ruptured in this way, and the bare uterus protruding through the rent (Major Dubois).

The retention of urine is followed by cystitis and decomposition of the urine; if it has rather slowly developed, the muscles of the bladder hypertrophy. In consequence of the alkaline urine and the greatly distended bladder, diphtheritic inflammation follows, and, as in the case reported by Schatz, the whole mucous membrane, together with the greater half of the muscular coat, may mortify and be thrown off, so that it forms a second sac, situated within the bladder and almost entirely separated from it. In other cases the bladder ruptures, followed by infiltration of urine and gangrene of the surrounding connective tissue or by an immediately fatal peritonitis. This latter may, however, result from the spreading parametritis, and death follows with uræmic symptoms or from ileus.

In most of the cases the diagnosis is not attended by any difficulty. Simple inspection of the abdomen shows the characteristic shape of the bladder, which may reach even up to the umbilicus. As a rule, fluctuation is very distinct, but sometimes, on account of the great tension, the bladder may give a more hard sensation. The application of a catheter may be difficult, but almost always succeeds if a silver male catheter is used and if special attention is paid to the regular deviation of the bladder to one side. Often an enormous quantity of urine is thus drawn off.

By an internal examination the vagina is found filled by a large soft tumour, which is very sensitive. The vaginal portion of the uterus lies close behind the symphysis, sometimes rather high up, but rarely so high that it cannot be reached. After emptying the bladder, if the abdomen is not too sensitive, it may be found by a combined internal and external examination that a tumour representing the normally situated uterus is absent from the abdomen, and therefore the tumour felt in the posterior vault of the vagina must be the enlarged uterus. If the sensitiveness be too great, a careful examination must be made under chloroform. This condition will be readily distinguished from a hæmatocele or from an incarcerated ovarian or fibroid tumour, since with them the uterus is felt in its usual place.

The treatment consists first of all in emptying the urinary bladder. In somewhat recent cases the reposition of the uterus is easy through the vagina or rectum. If the application of the catheter should not succeed, the reposition may be cautiously tried whilst the bladder is still distended, yet the attempt to introduce the catheter must be continued. If this proves fruitless, puncture of the bladder now becomes necessary. If the reposition of the uterus is otherwise impossible, it may be tried in the lateral or knee-elbow posture. A finger is introduced into the rectum, and the uterus pushed upwards along the sacro-iliac synchondrosis, and this manipulation is a second time repeated if it should not immediately succeed. The reposition is then only successful when the vaginal portion has assumed its normal place in the pelvis. Frequently whilst the fundus falls forwards the vaginal portion is placed behind in the excavation of the sacrum. To obviate a relapse, which is especially liable to occur when the uterus is not yet sufficiently enlarged, and when the reposition has been easy, a common Mayer's caoutchouc ring is placed in the vagina so as to prevent the deviation of the vaginal portion forwards, and by it the dislocation of the uterus.

If the reposition does not succeed in any way the uterus is to be emptied of its contents. As it most frequently occurs that a sound cannot be introduced through the cervix, it will be necessary to puncture the ovum through the uterine wall, either from the vagina or from the rectum. By the escape of the liquor amnii the circumference of the uterus is so considerably diminished that a repeated attempt at reposition may prove successful, or that abortion may be rendered possible.

The other symptoms are to be treated on general principles.

Hunter's opinion that retention of urine is the primary cause of retroflexion of the gravid uterus was accepted until very recently by almost all authors. The following were mentioned as causes predisposing to the development of the dislocation:—an emaciated body, a too wide or a too narrow pelvis, too slight inclination of the pelvis, &c. Lohmeier, whom Benninghausen followed, stated the dislocation to be the primary occurrence, and that the other symptoms depended upon the increased size of the organ. He also very correctly supposed that the retroflexion might have existed previously to the occurrence of pregnancy without the patient being aware of it. Recently Tyler Smith has stated that it is his opinion that these cases are commonly nothing else but pregnancy occurring in a retroflexed uterus.

Slight lateral deviations of the uterus regularly occur in the gravid organ. The development of the uterus itself, and its relation to the rectum and the bladder during intra-uterine life are the determining causes of the way in which it arises. The course of pregnancy is not in the least modified by them.

2. *Prolapsus.* Prolapsus of the gravid uterus has most frequently been observed in cases where conception has taken place during the existence of the displacement. It must be very rare to meet with

prolapsus of the gravid uterus in the first few months produced by acutely acting causes. Under such conditions the disturbance of circulation which is suddenly produced would lead to hæmorrhage into the membranes, and consequent death of the fœtus and abortion.

If conception takes place in a prolapsed uterus the organ will reascend into the false pelvis with the increase of its volume, and in this way the prolapsus disappears so long as gestation lasts. When the prolapsus is so great that a larger portion, or even the entire uterus, is placed outside the vulva, symptoms of incarceration of the pelvic organs may be caused by the growth of the uterus, and abortion may be the consequence. There is no authenticated case on record where pregnancy has gone on to term in a uterus completely prolapsed and situated outside the vulva.

A greatly hypertrophied cervix projecting beyond the vulva may give the appearance of an incomplete prolapsus of the uterus when the organ is far advanced in gestation. The fœtal parts are then in the true pelvis and above the brim. Doubtless most of the cases described as prolapsus of the gravid uterus belong to that category.

As a rule, the uterus descends, in the normal pelvis of primiparæ, in the last month of pregnancy, the head and the surrounding uterine segment entering deeply into the true pelvis, and sometimes even down to the outlet.

In all cases this displacement can easily be made out by a careful examination. In the first month the gravid prolapsed uterus is to be replaced and retained in position by a suitable pessary (the best is Mayer's caoutchouc ring). If the latter is well chosen it does not in the least disturb pregnancy. In replacing the uterus attention must be paid to the direction which the fundus takes. By neglect of that precaution the fundus may be pressed against the promontory, so as to be bent beneath it. A retroflexion of the gravid uterus is then produced when means are employed to retain the organ in that position. But the reposition of the uterus may be actually impossible when the prolapsed organ has already too greatly enlarged. The pregnant woman is then to be watched carefully, and artificial abortion induced as soon as symptoms of incarceration appear.

3. *Herniæ.*—Herniæ of the unimpregnated uterus are met with comparatively rarely, and the cases also are very limited in which such a dislocated organ becomes gravid, or where a gravid uterus is found within a hernial sac.

Umbilical and ventral herniæ are the most frequent of all. Of course an umbilical hernia can only contain the uterus far advanced in pregnancy when the recti muscles are widely separated. Such cases are reported by Murray and Léotaud. In both delivery was spontaneous.

True ventral herniæ of the gravid uterus are also very rare. But there are many cases reported where the gravid uterus was found lying in a hernial sac, which was produced by the dilatation of a large cicatrix of the abdominal walls.

By the mere separation of the recti muscles along the linea alba herniæ frequently arise, if they may be considered as such, and the hernial contents are then covered by the fasciæ also as well as by the skin. A slighter degree, that is, when a part of the uterus only lies in the excavation, is very frequent indeed, since a separation of the recti muscles is always connected with a very considerable pendulous belly. The displacement is easily recognised if in the recumbent posture the woman is made to raise the upper part of her body. The uterus then protrudes as a globular tumour in the linea alba, and on each side the contracted bellies of the recti muscles are distinctly visible. The treatment is the same as that of the pendulous belly.

(C) *Lacerations and injuries.*—Lacerations of the uterus, due to its anomalous development or displacement, have already been mentioned, whilst those arising from interstitial pregnancy and new growths of the uterus will be treated of later. Independent of these conditions the gravid uterus only tears after severe contusions, and this is followed by the escape of the uterine contents into the abdominal cavity, causing fatal peritonitis.

Wounds of the uterus are rare, and, as a rule, death follows on account of the grave complications. Small wounds may cause abortion, and the uterine contractions close the wound; or they may heal, and later on normal parturition occur.

(D) *Inflammation.*—1. *Decidual endometritis.*—When the fecundated ovum enters the uterus it is grafted, so to speak, into the uterine mucous membrane, and on account of the irritation exerted upon it this becomes the seat of a hyperplasia by which the normal decidua is formed. Under certain circumstances, *i.e.* in endometritis prior to conception or, perhaps, through syphilis, this irritative condition of the mucous membrane may exceed the normal limits.

The consequences of the increased formative activity vary according to the character and course of the inflammatory irritation.

The very acute inflammatory irritation causes apoplexy into the membranes and subsequent abortion; or it destroys the ovum, and the latter is converted into the so-called blood- or flesh-mole.

A more chronic inflammation may assume a different course, namely:

(*a*) *Diffuse chronic decidual endometritis.*—The proliferation of the cells of the decidua takes place slowly or is diffused over the whole mucous membrane, in which the reflexa also may participate. In the severe form of the disease, which is distinguished from the rest only by its uniform extension over all the decidua, the fœtus becomes stunted and abortion follows. But the abortion may, doubtless, also follow the inflammation of the mucous membrane by the irritation which it produces on the uterine nerves. The thickened mucous membrane shows numerous large decidua-cells, or the proliferation has, especially within the deeper layers, a more cavernous structure. Cysts also occur in the chronically inflamed decidua.

In other cases the inflammation is still more chronic, the nutrition of the ovum is in no way disturbed, pregnancy continues to term, and the inflammation is recognised only by an examination of the membranes after birth. Madame Kaschewarowa found in the thickened membranes of a mature fœtus, not only proliferating connective tissue and decidua-cells, but also newly formed unstriped muscular fibres. The inflammatory thickening is frequently not uniformly extended over the whole periphery of the ovum, but is limited to single spots.

(*b*) *Tuberous and polypous decidual endometritis.*—The characteristic changes of the decidua belonging to this class have been observed in ova expelled at periods varying from the end of the second to the beginning of the fourth month.

As a rule, only the decidua vera is the seat of the hyperplasia, but in the cases reported by Dohrn and Müller, where the decidua vera was absent, the proliferation took place in the reflexa. Especially at the places corresponding to the anterior and posterior walls of the uterus, the decidua was greatly thickened, as much as 4·7 lines; on the surface looking towards the ovum it was studded with large elevations or more polypous growths, whilst the uterine surface was rough and covered with coagula of blood. The growths have either a broad base or are more pedunculated and polypoid; they reach a length of half a line and almost as much in breadth. On their greatly reddened surface there are no openings of the uterine glands, whilst they are very distinct on the other portions of the mucous membrane. The microscope shows that the proliferation takes place in the interstitial tissue of the mucous membrane.

In a slightly fibroid matrix are seen large, at some places even enormous, cells of the shape of a lentil, which on vertical section appear as thick spindle cells (Virchow). The whole mucous membrane, and especially the papillæ, are very vascular.

Without doubt the changes of the abortive ova just described started primarily from the decidua. The villi of the chorion were normal only in one case; whilst in the ovum described by Virchow they represented thick cylindrical knobs with very long but chiefly fine epithelial processes; in that described by Gusserow they terminated with knobby swellings, and in Dohrn's and Müller's cases they showed the commencement of a multiple myxoma of the villi. The embryo itself was well developed in one case only; in another it was stunted and neither allantois nor umbilical vesicle could be discovered; in a third it was entirely absent, whilst twice only a rudiment of the umbilical cord was present.

The etiology of this hyperplasia of the mucous membrane is not quite known. In Virchow's case there was syphilis, but in the others it was not so. One of the women was chlorotic, another previously suffered from endometritis, and a third had conceived very soon after delivery, and it may, therefore, appear that this process of proliferation is somehow connected with an irritation of the mucous membrane prior to conception. The pathological changes of the chorion and of the fœtus are secondary.

(*c*) *Catarrhal decidual endometritis; hydrorrhœa gravidarum.*—Chronic inflammation of the decidua, like that of the mucous membrane of the non-impregnated uterus, may manifest itself by an abnormal secretion as well as by a proliferation of its cells. This apparently accumulates between the decidua and the chorion (in some cases also between the chorion and the amnion), and escapes from time to time after the rupture of the reflexa. Larger quantities, up to one pound and more, of the yellowish serous fluid, sometimes tinged with blood, can of course only escape in the last months, but a dribbling of single drops may occur in the third month. Most frequently hydrorrhœa is observed in hydræmic women, who have a pronounced tendency to serous exudations, and in some cases it is said to have the catamenial type.

By the transudation of the serum the uterine activity is sometimes roused, and pregnancy interrupted, but commonly it reaches the full term.

Hydrorrhœa could only be mistaken for the premature escape of the liquor amnii, but this is always followed by the commencement of labour.

The treatment must be directed to remove all the influences which are known to cause congestions of the uterus; for the rest it is merely symptomatic.

2. *Metritis.*—Inflammation of the parenchyma of the gravid uterus is very rare. It is most frequently diffused over the whole organ, in consequence of displacements, especially of retroflexion of the gravid uterus, or it is partial, limited to some spots. It may then terminate in suppuration or with the formation of abscesses. It is not improbable, however, that in some cases of spontaneous rupture of the uterus during parturition the solution of continuity is favoured by a tissue rendered friable by a circumscribed inflammatory softening.

3. *Perimetritis.*—The hyperæmic condition which exists in the peritoneum covering the uterus during the whole process of gestation often passes into a partial and slight inflammatory process. This gives rise to a circumscribed sensitiveness of the gravid organ, and sometimes also to slight flocculent depositions. A considerable exudation is very rare, and the transition into a general fatal peritonitis only occurs after severe injuries.

(E) *New Formations.*—1. *Fibroid growths.*—Fibroid tumours have, according to their seat, a very varying influence on pregnancy.

Subperitoneal fibroids impede conception or modify the course of pregnancy only when they are very large. They are easily detected by palpation, but may, under certain circumstances, at times be mistaken for small fœtal parts.

Interstitial fibroids greatly impede conception, especially if they are situated close to the internal os uteri, thus narrowing it. Pregnancy, when it has taken place, may sometimes go on to term without interruption; frequently, however, abortion or premature labour occurs, or there may even be spontaneous rupture of the uterus.

Submucous fibroids, with more or less broad bases, allow conception to take place only in very rare cases. If it occurs gestation commonly terminates prematurely by abortion, and only very seldom reaches the normal end. A pedunculated submucous fibroid, especially if the polypus originates from the cervix, may favour the occurrence of very dangerous hæmorrhages during pregnancy.

Interstitial submucous fibroids are in most cases very difficult to recognise, unless there has been an opportunity of making a gynæcological examination before the commencement of pregnancy. At an early period of gestation there is the danger of diagnosing the fibroids when they have come to a considerable size, and to overlook the pregnancy, and at a later period the fibroids mostly escape detection.

Subperitoneal fibroids preserve their normal structure during gestation. The others become, as a rule, more or less softened, and in extreme cases may be converted into a reddish-brown pultaceous mass.

2. *Carcinoma.*—One of the most fatal complications of pregnancy is carcinoma of the uterus, which, when primary, is almost exclusively seated in the cervix. Only in the first stages of its development is conception possible. But by the increased supply of nutritive material to the uterus, the proliferation as well as the simultaneous disintegration of the newly formed masses, is so greatly augmented that the disease makes rapid advances during pregnancy to its fatal termination. Most commonly abortion sets in, or, since the newly formed masses do not possess the property of extending with the growing uterus, rupture of the neoplasm follows. In rare cases, if the disease has not yet too far advanced, pregnancy goes on to term.

Carcinoma can always be recognised, for the cervix is here in the same state as it is when not complicated by pregnancy. In great proliferation the treatment is only symptomatic, and all attention is to be paid to cleanliness. Induction of artificial abortion is not justified.

2. *Anomalies of the Vagina*

The mucous membrane of the vagina is during gestation swollen and the seat of an abundant secretion. Often this swelling is equally diffused throughout the submucous connective tissue, but frequently, and particularly in primiparæ, an hypertrophy of the papillæ takes place, which gives the mucous membrane of the vagina a peculiar rough and granular aspect. A very intense catarrh may be attended by enormous hypertrophic proliferation, or the tips of the enlarged papillæ may grow together, and thus give rise to the formation of numerous small cysts between them. This at least appears to be the most plausible explanation given by Winkel of the cases of "colpo-hyperplasia cystica" described by him.

The secretion of the vaginal mucous membrane may become so abundant as to produce a very tedious blenorrhœa. Sometimes it is of a virulent character also, and then the secretion easily becomes purulent. The secretion, if not dependent upon gonorrhœal infection, is thicker, white, and creamy, or, when the quantity is very great, purely serous. The treatment cannot be very active, in order not to interrupt pregnancy prematurely. Cleanliness and mild injections will generally answer the purpose.

Sometimes the catarrh of the vagina is complicated by mycosis. At some places, especially at the vaginal outlet, whitish or greyish-yellow patches are seen adhering to a red basis and consisting of vast proliferations of fungi. At times they do not give rise to any symptoms, but, as a rule, they cause a good deal of itching. The infusorium, Trichomonas vaginalis, which is commonly found in the vagina, is of an innocent nature.

Hæmorrhages from the vagina or from the external genitals are very rare during pregnancy. They may proceed from ruptured varicose veins, but more frequently they occur independently of these. If the hæmorrhage is external, it must be treated on the established principles; but if the blood is effused into the tissues (*thrombus vaginæ sive vulvæ*) cold must be applied in order to prevent a further bleeding, and when this has succeeded the tumour is to be opened.

3. *Anomalies of the Breast*

The hyperæmia of the mammæ commences during pregnancy, and the secretion of the gland may be so much increased as to cause a milder form of parenchymatous mastitis. The mamma greatly swells, becomes red, tense, and tender. The commencing inflammation can be easily cut short by pressing the milk from the mammæ and by stimulating the action of the intestines. But even without any treatment it usually passes away without any other consequences, and it rarely terminates in purulent mastitis. Although this seldom occurs in pregnant women, yet when it does it is most frequently caused by injuries.

Literature.—Kussmaul, Von dem Mangel u. s. w. der Gebärmutter. Wurzburg, 1859.—Fürst, M. f. G., B. 30, pp. 97 u. 161.—W. Hunter, Medical Observations and Inquiries, vols. iv and v, 1771 and 1776.—Lohmeier, Theden's Neue Bem. u. Erf., &c., 1795, 3 Th., p. 144.—L. van Praag, Neue Zeitsch. f. Geb., B. 29, p. 219.—Tyler Smith, Transactions of the Obstetrical Society, ii, p. 286.—Martin, Neigungen u. Beug. d. Gebärmutter. Berlin, 1866, p. 185.—Säxinger, Prager Viertelj., 1866, iv, p. 53.—Schroeder, Schw., Geb. u. Wochenbett, p. 36.—May, Ueber die Reclination der schwangeren Gebärmutter, D. i. Giessen, 1869.—Barnes, Obstetrical Operations, 2 ed., 1871, p. 242.—Hüter, M. f. G., B. 16, p. 186.—Gusserow, M. f. G., B. 21, p. 99.—Litten, Ueber den Vorfall der schwangeren Gebärmutter, D. i. Berlin, 1869.—Klob, Path. An. d. weibl. Sexualorgane. Wien, 1864, p. 105.—H. Müller, Bau der Molen, p. 80.—Hegar, M. f. G., B. 21, Suppl., p. 12.—Klebs, M. f. G., B. 27, p. 401.—M. Duncan, Researches in Obstetrics, p. 290.—Frau Kaschewarowa, Virchow's Archiv, 1868, B. 44, p. 103.—H. Müller, Bau der Molen. Würzburg, 1847, p. 33.—R. Virchow, Virchow's Archiv, 1861, B. 21, H. 1, p. 118, and die krankh. Geschwülste, B. II, p. 478.—Strassmann, M. f. G., B. 19, p. 242.—Gusserow, M. f. G., B. 27, p. 321.—Dohrn, M. f. G., B. 31, p. 375.—V. Haselberg, Berl. Beitr. z. Geb. u. Gyn., B. I, p. 34.—Chassinat, Gaz. de Paris, 1858, No. 29, &c. (s. M. f. G., B. 15, p. 465).—C. Braun, Zeitschr. d. Ges. d. Wiener, Aerzte, 1858, No. 17, p. 257.—C. Hennig, Der Katarrh der innern weibl. Geschlechtstheile. Leipzig, 1862, p. 48.—Hegar, M. f. G., B. 22, pp. 299 and 437 (Vergl., B. 25, Suppl., p. 64).

CHAPTER II

DISEASES OF THE OVUM

A.—EXTRA-UTERINE GESTATION

Exceptionally the fecundated ovum does not enter the uterus, but continues to develop in the ovary, in the abdominal cavity, or in the Fallopian tube.

To understand the occurrence of this interesting but dangerous anomaly, it is necessary to recal to mind the normal process of the expulsion of the ovum from the Graafian follicle and its migration into the uterus. Since spermatozoa have been found on the ovaries and within the whole pelvic portion of the abdominal cavity of animals, and since in man also the advance thus far of spermatozoa has been demonstrated by the cases of their external transmigration, the possibility cannot be doubted that an ovum may be fecundated outside the uterus and the Fallopian tubes. If this be admitted, then the occurrence of ovarian or abdominal pregnancy has lost its extraordinary character. For the ovum may, instead of being swept out with the serum after the rupture of the Graafian follicle, as commonly occurs, remain within the cavity of the follicle (and this chiefly occurs when the rent in the follicle is small and the serum slowly drains away) and is there fecundated by the spermatozoa which have advanced so far. Or it may happen that the ovum expelled from the Graafian follicle fails for some reason to meet the fimbriated end of the Fallopian tube and is fecundated in the free abdominal cavity. The impediments to the entrance of the ovum into the Fallopian tube are various. Thus, the Graafian follicle may have ruptured at a spot too far distant from the end of the tube, so that the serous current in the abdominal cavity directed towards the latter cannot sufficiently act on the progress of the ovum; or an abnormal movement of the abdominal contents, especially of the intestines, may have forcibly deviated the ovum from its right course; or the ciliary vibrations in the tube may have ceased with the partial or entire destruction of the ciliated epithelium in consequence of a catarrh of the tube; or the end of the tube of the same side is occluded and the fecundating sperm has passed through the tube of the other side into the pelvic cavity. All those circumstances cannot appear so very strange, and it is somewhat surprising that extra-uterine pregnancies do not occur more often than is actually the case. This comparative rarity is, perhaps, due to the fact that the ova exceptionally only find a spot suitable for their development, so that in most cases they perish and are reabsorbed.

A tubal pregnancy is effected when the ovum has entered the Fallopian tube in the usual way but has been impeded in its further progress to the uterus; it then remains within the tube, and continues to develop there. In such a case the tube may have allowed the spermatozoa to pass, but the ovum itself has been retained, either by a fold in the mucous membrane or because the tube was too narrow for it. In Beck's case a small polypus situated at the uterine end of the tube prevented the entrance of the ovum into the uterine cavity; but in Breslau's case the polypus was not apparently the impediment. Or it may occur thus: the left tube, for instance, is occluded at some spot close to its uterine opening, and the spermatozoa have passed through the right tube into the abdominal cavity and there fecundated an ovum derived from the left ovary. Now, if the fecundated ovum enters the abdominal opening of the left tube it is, of course, stopped at the occluded spot. Perhaps a catarrh has destroyed the ciliated epithelium of the tube, by which the progress of the ovum is either entirely impeded or at least so slackened that at the time when the ovum is about to enter into union with the surrounding tissues it is still contained within the tube. Stenosis or atresia of a tube is most frequently caused by a flexure or constriction of it, the sequence of an old perimetritis; such a partial peritonitis is by no means rare. This often causes sterility, or, at least, greatly impedes conception. Clinical observations appear to confirm what has just been stated. For a great number of extra-uterine pregnancies occurred in primiparæ who had lived for some years in sterile marriage, and also in very many pluriparæ, whose extra-uterine pregnancy had been preceded by a long pause in conception. It is shown also by a case of Hassfürther, although this must be pronounced extremely rare, that tubal pregnancy may be due to an internal transmigration of the ovum, that is, the ovum which has normally entered the uterine cavity passes thence into a Fallopian tube and there develops.

Extra-uterine pregnancy occurs more frequently in pluriparæ; its rarest seat is within the ovary, and the most frequent in the abdominal cavity.

Tubal pregnancy is again distinguished into—(1) interstitial or tubo-uterine, (2) ovario-tubal, and (3) tubal pregnancy proper, according to the spot within the tube on which the ovum developed. We shall first consider the latter, on account of its greater frequency.

The fecundated ovum arrested in the tube continues to develop. The mucous membrane of the tube proliferates in a way similar to that of the uterus in utero-gestation, and consequently a normal decidua serotina, often a very thick decidua vera, and even a decidua reflexa, is formed. The chorion forms, as usual, from the ovum; its villi sink into the mucous membrane and form at the spot where the ovum has imbedded itself the placenta the rest of the periphery atrophying. With the growth of the ovum the corresponding part of the Fallopian tube also is stretched, so that at a later period both openings of the tube cannot, as a rule, be found, or the one at least towards the uterus has disappeared. Whether in this case the tubal pregnancy was caused by the impermeability at that spot, or whether the occlusion is only a secondary change in the vicinity of the growing ovum, cannot definitely be decided from the appearances in the dead body. The ovum forces asunder the muscular fibres of the tube, and the sac containing the fœtus consists only of the peritoneum and the tubal mucous membrane, which appears as a hernia-like protrusion from the tube. In the mean time the uterus itself is pretty constantly the seat of the same changes as in the beginning of normal gestation. The whole organ hypertrophies, the mucous membrane is converted into a true decidua, and in the cervix the well-known glassy mucous plug is formed.

The most frequent termination of tubal pregnancy is in rupture of the sac containing the fœtus. This occurs in most cases within the first two months, frequently in the third, and rarely in the fourth or fifth month.

It is generally at the thinnest part of the free periphery of the sac that the rupture takes place, but sometimes also it is at the placental insertion. Often also the rupture is gradual; this becomes complete after the sac has been partially torn for some time previously.

When rupture has taken place the bleeding into the abdominal cavity causes great and immediate danger. In many cases the effusion of blood is so considerable that death follows from internal hæmorrhage, but even where the hæmorrhage is not so great as at once to cause death the consecutive peritonitis, as a rule, proves fatal.

If the ovum escapes into the abdominal cavity death generally follows from peritonitis. But even under such circumstances the ovum may be encapsuled, and if it is at an early period of development it may be reabsorbed to a great extent. If the embryo is older, it may undergo all those changes which will be mentioned when speaking of abdominal pregnancy.

In rare cases the tube ruptures at a place on its periphery where it is not covered by the peritoneum. The embryo then escapes from the tube between the lamina of the broad ligament of its own side, and thus after rupture of the tube it lies outside the peritoneum. Death also follows in these cases from hæmorrhage, laceration of the broad ligament, and peritonitis.

Without doubt the most favorable termination of tubal pregnancy is when the embryo perishes early, and then recovery takes place without rupture. An embryo dying at a very early period undergoes all the changes which will be described as occurring in abortive ova; the embryo itself is reabsorbed, and the ovum greatly altered by hæmorrhagic effusion. If the more fully developed embryo dies, it may be converted into a lithopædion within the tube without rupture of it.

Ovario-tubal or interstitial pregnancy differs greatly from the tubal pregnancy just described. Here the ovum develops in a portion of the tube which is already continuous with the uterine substance.

In the majority of cases rupture takes place here also in the first three months. After rupture of the tubal mucous membrane the ovum may remain between the separated muscular fibres of the uterus, and gradually stretching the peritoneum as it enlarges may continue to develop up to the normal end of pregnancy. In an isolated case observed by Braxton Hicks the uterine opening was so dilated that the fœtus afterwards entered the uterus. Even then rupture is likely to follow on account of the great thinning of the uterine wall at the original insertion of the ovum. But the child may also be born per vias naturales.

The relation of the round ligament to the sac containing the fœtus easily shows in the dead body the difference between this and simple tubal pregnancy. In interstitial pregnancy the ligament passes external to the sac, and in simple tubal pregnancy between the sac and the uterus. A distinction between this and gestation in a perfect or rudimentary horn of the uterus offers great difficulties, since the relation of the round ligament to the sac of the fœtus is the same in both, and the thickness of the septum between the sac and uterus proper, though generally less in interstitial pregnancy, differs only in degree.

If the ova remain in the abdominal opening of the tube, the sac is formed partly by the Fallopian tube and partly by the ovary or portion of the peritoneum. This constitutes tubo-abdominal pregnancy, which, however, does not practically differ from the abdominal form.

In very rare instances the ovum continues to develop in the ovary itself—ovarian pregnancy. This will be the case when the ovum has remained in the ruptured follicle, and has there been fecundated. The ovum continues to develop, as a rule, until, as in tubal pregnancy, the sac ruptures.

In abdominal pregnancy the fecundated ovum develops in the abdominal cavity. An inflammatory new formation of very vascular connective tissue envelops the ovum in a sac, which, capable of great distension, may equal in size a gravid uterus at term. Very rarely this sac ruptures prematurely, and more frequently than otherwise the fœtus is fully developed. It is not probable that the fœtus should be alive more than ten months in the abdomen of the mother; nor has it been proved by any reliable observation. Abdominal pregnancy terminates variously. If the fœtus is alive up to the normal end of pregnancy, generally a kind of labour-pains sets in, and often a decidua is expelled from the uterus. In the mean time the fœtus dies, and after death undergoes a variety of changes. In other cases the fœtus dies in the last months of gestation, and then also undergoes the same changes.

In the majority of cases the body of the dead fœtus irritates the walls of the sac. They inflame, and suppurate, or putrefy under the influence of the disintegrating fœtus. The mother, as a rule, dies of peritonitis due to an extension of the inflammation from the walls of the sac to the rest of the peritoneum, or she perishes with hectic symptoms, exhausted by the profuse suppuration. If the peritonitis, however, does not become general, and the profuse suppuration be supported, the contents of the sac may perforate a neighbouring hollow organ, or be discharged externally, and their elimination from the organism may be followed by recovery.

The perforation is most frequently into the large intestine, and in the course of some months (or sometimes even after many years) bones and disintegrating soft parts of the fœtus are discharged by the rectum. Often the anterior abdominal wall is perforated. Fistulous openings form, discharging an ichorous pus and small bones, and only close after the whole body of the fœtus has been removed, either spontaneously or artificially. Very rarely the whole, or almost the whole body, of the fœtus is at once expelled through a larger opening in the abdominal walls. The vagina also, and the bladder, and in other cases several hollow organs, may be simultaneously perforated by the putrefying fœtus. The process of perfect elimination is always tedious and exhausting; nevertheless it may terminate in recovery, although in some cases rectal or fœcal fistula remains (in one case even a gastric fistula—Raneyer), or communications between the bladder and the intestines, &c. More frequently death takes place during the suppuration, either from exhaustion or blood poisoning.

In some cases the walls of the sac are less irritated by the dead fœtus. The liquor amnii is gradually absorbed, the sac collapses, and is closely applied to the body of the fœtus, and the fœtus itself is greatly metamorphosed. The soft parts undergo fatty degeneration, and are gradually converted in the course of years into a thick, smeary, pultaceous mass, consisting of fat, lime salts, cholesterin, and diffused and crystalline pigment. The pultaceous mass also is partly reabsorbed, and occasionally the contents of the greatly shrunken sac consists only of the bones of the fœtus and a quantity of calcareous plates. A fœtus changed in the way just described is called a lithopædion. A fœtus in such a state may for many years be carried within the body without harm, but in other cases, though in this state, it causes inflammation, suppuration, and even death.

The name of secondary abdominal pregnancy has been given to those cases where the fœtus originally contained in a Fallopian tube, the ovary, or even the uterus, has escaped after rupture of its sac into the peritoneum and remained there. It is certainly doubtful whether a very young ovum expelled from the tube may be so engrafted into the peritoneum as to enter with it into a real placental connection. At any rate the embryo escaped into the peritoneum may there be preserved in a variety of ways. In some cases it is soon encapsuled by the newly formed connective tissue, and degenerates into the common lithopædion, in others there is quite a peculiar preservation of it. This apparently occurs when, after the rupture of the tube, the fœtus alone has entered the abdominal cavity, whilst the placenta has remained in undisturbed connection with its place of insertion. Whilst inflammatory new formations cover the embryo on all sides, it may doubtless continue to live for a short time; and under these circumstances the very vascular connective tissue closely approaches the skin on all sides, and the soft parts of the body are so marvellously preserved after death that many years after they show perfectly their normal structure.

The symptoms of extra-uterine pregnancy are usually not very characteristic. One most constant symptom is a severe pain in the abdomen, the same as in an acute circumscribed perimetritis, probably due to the tension and twisting of the piece of peritoneum covering the sac of the fœtus. Sometimes also the discomfort of pregnancy is greatly increased, and a bloody mucus is discharged from the uterus, with pains like those of labour. In other cases there are no anomalies whatever in the course of pregnancy; a change only occurs when—as a rule in tubal pregnancy—the sac containing the fœtus ruptures. This is attended by symptoms of an acute internal hæmorrhage, or those of acute peritonitis are more conspicuous, and death follows rapidly. The instances are more rare in which partial successive rents give rise repeatedly to the alarming symptoms of an acute internal hæmorrhage or an acute peritonitis, until finally the patient succumbs to the complete rupture. The course of an abdominal pregnancy is similar in the commencement. There is, however, no rupture, and the fœtus comes to maturity or almost so, and dies after a short labour, in which a decidua is expelled. Even the comparatively favorable termination by the formation of a lithopædion, or by elimination through suppuration, is connected with very grave symptoms and dangers.

Diagnosis.—Women with tubal pregnancy, as a rule, apply for medical assistance only when the rupture has already taken place. An accurate palpation is rarely possible on account of the great sensitiveness of the abdomen. But the whole aspect of the case, together with the history that the woman was supposed to be pregnant in the second, third, or fourth month, suffices to make a diagnosis with great certainty.

If the case comes under treatment before the rupture has occurred, or if there be an abdominal pregnancy, it is generally easy in the first months to make sure of an extra-uterine tumour, whilst the diagnosis of a pregnancy is frequently involved in difficulties. At a later period the reverse is the case—pregnancy can easily be diagnosed, but to show that the fœtus is outside the uterus may be very difficult.

If the accoucheur is summoned to a woman supposed to be in the first months of pregnancy and now suffering from severe pains in the abdomen, everything depends upon an accurate combined internal and external examination. If the uterus be found moderately enlarged, and on one side of the uterus an elastic sensitive tumour, this would point to an extra-uterine pregnancy. The first thing, however, would be to be certain whether pregnancy exist at all. Since the signs which are usually furnished by the shape, size, and consistence of the gravid uterus are, of course, absent in extra-uterine pregnancy, this may be very difficult, yet it can almost always be ascertained. If menstruation had formerly always been regular, and if it had suddenly stopped, whilst there is a presumption in favour of conception—and if there are, moreover, the subjective symptoms of gestation, which are of especially high value in pluriparæ—the existence of pregnancy is highly probable. Besides, if the mammæ are swollen and a somewhat large quantity of pigment deposited in the areola and linea alba, the diagnosis is rendered still more probable. The uterus also having been found absolutely too small for the supposed time of gestation, one sufficiently practised in the use of the sound may now try to make out with the aid of this instrument whether the uterus is empty. The bulb of the sound easily and without force enters the empty uterine cavity if there be no flexure at the internal os. There is a certainty of the existence of an extra-uterine pregnancy if in the extra-uterine elastic tumour hard and movable parts are felt. Should the tumour be in any way accessible it may be punctured by means of an exploring trocar. The fluid will be found to be liquor amnii from the very small quantity of albumen and mucine, and at a later period from the thrown-off pavement epithelium and the lanugo.

With the advance of gestation it becomes more easy to decide whether pregnancy exists. The audible sounds of the fœtal heart, as well as the palpable parts of the fœtus, the movements felt and other changes of gestation (as in the mammæ and linea alba), make the diagnosis of pregnancy absolutely certain. All that is to be done at present is to be sure that the fœtus is not contained in the uterus. In many cases palpation alone will show that the uterus is only slightly enlarged and that it is distinctly separable from the tumour. Some difficulty will be experienced if the extra-uterine tumour is very large and the uterus itself drawn backwards and somewhat upwards. But by the aid of the sound it can be shown whether the uterus is empty or full. We will not omit here to draw attention to the fact that even in utero-gestation the sound may pretty easily pass between the membranes and the uterine wall, and that abortion is by no means the usual consequence of a careful introduction of the sound. Nor does the possibility of introducing a sound to the knob or beyond it speak in favour of an empty uterus, but the ease only with which the sound has passed through the internal os. The difference, therefore, is one of degree, the estimation of which requires practice in the use of the sound.

A differential diagnosis between extra-uterine pregnancy and gestation in a rudimentary horn may be quite impossible. This will be easily understood if it is considered that such states, as Kussmaul has shown, are frequently mistaken even in the dead body. Nothing but the direction of the round ligament can definitively decide the question.

The diagnosis of the kind of pregnancy is also involved in great difficulties. The best indications are furnished by the course of gestation. If rupture occurred in the first four months the ovum was contained in a Fallopian tube, but if the ovum has developed beyond that time there is abdominal pregnancy. Interstitial pregnancy is recognised under exceptionally favorable circumstances, and ovarian pregnancy, perhaps, never as such.

With regard to prognosis, a distinction must be made between the mother and the child. The life of the latter has been rarely only preserved by Cæsarean section, whilst the prognosis is more favorable as regards the mother, varying greatly according to the seat of the ovum. In tubal and ovarian pregnancy death followed, at least in the great majority of cases which had been diagnosed, but in abdominal pregnancy the prognosis is more favorable, according to Hecker, 42 per cent. die.

Treatment.—If in the first months the diagnosis is certain and the sac accessible from the vagina, rectum, or abdominal walls, it would appear advisable to puncture it. This most frequently and most easily succeeds from the vagina. A very fine trocar is pushed into the tumour and the liquor amnii allowed to drain off. The fœtus then dies, and according to the time of its development is either re-absorbed or converted into a lithopædion.

More frequently, however, medical aid is sought when the rupture has already taken place. To ward off now the immediate danger, *i.e.* to stop the hæmorrhage, is not always an easy and successful task. Ice applied to the abdomen, and perhaps also, if possible, compression of the abdominal aorta, are the most efficacious means. The peritonitis which results is to be treated as of traumatic origin, on the usual principles. Perhaps the most successful treatment, if the ovum is of a considerable size and if assistance is immediately at hand after the rupture, would be the Cæsarean section, in order to stop the hæmorrhage, remove the fœtus and, if possible, the whole sac which contains it. Although this treatment has formerly been universally rejected, yet from the recent favorable experiences of the Cæsarean section it requires very careful reconsideration.

If menacing incidents occur in extra-uterine pregnancy which has so far advanced that a living child may be expected, or if the extra-uterine pregnancy has gone on to term and the child is alive, the Cæsarean section is unconditionally to be performed. If during the operation the child is alive, there is every chance of preserving it. To the mother certainly the operation is very dangerous, but scarcely more so than waiting. The principal danger lies in the insertion of the placenta. Since after the detachment of the placenta its place of insertion does not contract as in the uterus, there is usually a rapid and fatal hæmorrhage. Attempts may be made to prevent this by ligature or cauterization, on the principles followed in ovariotomy, with a short pedicle. If the whole sac with the fœtus and the placenta can be removed, this method of operation is decidedly preferable, and is to be performed as in ovariotomy. Should this not succeed, and if the placenta be firmly adherent, it is, perhaps, best to let it remain and allow it to be removed by suppuration. Stutter allowed the placenta to remain on account of its intimate connection with the intestines. By injections into the cavity, for the purpose of cleanliness, the placenta came away on the fifth day and the woman recovered.

If the child has died towards the end of pregnancy the treatment is merely symptomatic. The formation of a lithopædion renders every therapeutic interference superfluous. If the fœtus decomposes and suppuration occurs, the effort of nature to eliminate the foreign body is to be assisted by incisions and the removal of pieces of the fœtus from the abdominal walls, the vagina, or the rectum. When extensive adhesions have already formed with the abdominal walls, incisions can safely be made in order to remove, as soon as possible, the contents of the ovum. In other cases the formation of adhesions may be favoured, as in Simon's method of treating hydatid cyst of the liver, by pushing in two trocars and by dividing the bridge between them, thus causing a larger opening. The strength of the woman must, of course, be supported by every possible means.

B.—FAULTY CONDITIONS OF THE APPENDAGES OF THE OVUM

1. *Anomalies of the Membranes*

(*a*) *Hyperplasia of the chorion; myxoma multiplex; vesical mole; degenerating vesical mole; myxoma diffusum; myxoma fibrosum placentæ.*—The connective-tissue stroma, with which the villi of the chorion are supplied by the allantois—the endochorion—may sometimes be the seat of a peculiar proliferation. The anatomical appearances vary according to the number of the diseased villi and the time which has elapsed since the tumour began to form.

The hyperplasia commences in the epithelium of the villi, forming a great number of outgrowths, and their connective tissue (the immediate prolongation of the gelatinous tissue of the umbilical cord) extends into them and there proliferates excessively. The cells are partly converted into mucus, but the enormous thickening of the villi chiefly depends upon the proliferation of the cells and the great accumulation of the mucous intercellular substance. This is, as a rule, present to so great a degree that the hypertrophic villi have the appearance of cysts with liquid contents. The proliferating tissue, however, does not uniformly distend an entire villus, but accumulates rather in large masses at single spots, the intervening portions of the villus remaining almost normal. In consequence of this disposition the degenerated villus acquires a grape-like appearance, and the round cystoid tumours are attached like racemes to fine pedicles. This new formation closely resembles in its nature the papillomata; there is, however, this difference, that the stroma consists of thin mucous, instead of the common connective tissue, and that the abundantly excreted intercellular substance accumulates in masses at some spots.

If that degeneration occurs during the first month, at a time when the villi of the chorion are uniformly developed around the whole periphery of the ovum, the ovum is transformed into a tumour, the whole circumference of which is studded with round vesicles attached by pedicles.

When the degeneration occurs after the placenta has already begun to develop, the villi of the chorion in the remaining part of the periphery of the ovum have already atrophied. It occurs, however, very exceptionally that a villus outside the placental insertion which has not atrophied becomes the seat of the degeneration. More frequently the new formation is limited to the villi of the placenta or to some of its cotyledons. In such cases the greater part of the periphery of the ovum is normal, and only the portion corresponding to the placenta is partially or entirely transformed into vesicles. The vessels of the diseased villi atrophy, of course, when the embryo dies. But if this continues alive a very dense and fine capillary network is formed within the degenerated villi.

This hyperplasia of the chorion may be either primarily due to a disease of the ovum or to an irritation produced by the decidua or the maternal blood and transmitted to the villi of the chorion. A series of undoubted facts renders both processes possible.

Apparently all these cases, where a vesical mole develops by the side of a normal ovum, would confirm the opinion that the primary process is the disease of the embryo. Rather frequently aborted ova are found showing the first commencement of a myxomatous degeneration, and either containing no embryo or a stunted one. As this degeneration is so slight it cannot be supposed to have caused the death of the embryo, and would thus tend to prove that the death of the embryo can give rise to the hyperplasia of the villi of the chorion. For it is certain that after the death of the embryo the fœtal placenta may be nourished by osmosis and continue to grow.

On the other hand, the frequent occurrence of a diseased decidua, as well as the partial myxomatous degeneration of the placenta of a well-developed fœtus and the repeated occurrence of that degeneration in the same woman, render it extremely probable that the exciting cause of the hyperplasia of the chorion may be produced on the part of the mother.

But although it must be admitted that the proliferation sometimes follows the death of the embryo, yet it is not to be doubted that the degeneration of the chorion may be the starting-point of the disease of the whole ovum.

In such cases the fœtus will be variously affected, according to the extent of the degeneration of the chorion. If the whole circumference of the ovum, or at least the place of the placental insertion, degenerates, the nutritive material intended for the fœtus will be used up by the villi of the chorion. The fœtus dies, and it is found within the ovum either stunted or shrunken, or if at an early period, entirely reabsorbed, after having been macerated in the liquor amnii. Sometimes a rudiment of the umbilical cord is preserved. If the degeneration is limited to a small part of the placenta, the fœtus may continue to develop normally, and may be born alive at the normal termination of pregnancy.

The vesical mole is a soft flocculent mass, which is formed of a number of racemes or bead-like coherent vesicles of varying size. If a vesicle is punctured a tenacious fluid with the reaction of mucine escapes.

The fully developed condition of a vesical mole is rare; but the slighter degrees of that formation, from an inconsiderable swelling of the villi to single vesicles microscopically visible, are frequently found in aborted ova.

The symptoms produced by a vesical mole are not quite constant, and do not permit a positive interpretation; a diagnosis, therefore, cannot be made with absolute certainty before the expulsion, or at least before the degenerated ovum is accessible to the touch. It may, however, be supposed when the size of the uterus by no means corresponds to the time of pregnancy, when the uterus is either too large or too small, and when it does not enlarge after a prolonged period of observation, and when at the proper time and under suitable conditions the fœtal parts cannot be felt or the heart's sounds heard, which ought to be according to the duration of the pregnancy, and when repeatedly or almost continually blood escapes from the uterus.

If it is certain that the uterus contains an ovum (which may be made out by the method of exclusion, and the history of the case), the diagnosis between a vesical mole and a putrefying embryo, even in the sixth or seventh month of pregnancy, and after repeated careful examination, may yet be doubtful. It may sometimes be very difficult to distinguish between these two conditions, but the consistency of the uterus containing a mole is softer, and at times almost fluctuating.

The prognosis depends on the amount of the hæmorrhage which occurs during pregnancy or the expulsion; this may be very profuse. The result of recent observation has rendered the prognosis still more unfavorable, since it has shown that moles often assume a decidedly destructive character. In the case described by Volkmann the substance proper of the degenerated villi had grown into the interstices of the hypertrophied wall of the uterus. The villi had continued to proliferate into the maternal blood-sinuses, and by pressure caused atrophy of the tissue of the uterus. The villi penetrated into the fundus of the uterus, reaching the peritoneum, and on section the cavity of the uterus appeared to be divided into two parts by a diaphragm. Only the part situated below was the proper uterine cavity, whilst the larger upper portion was occupied by the proliferating tumour, which had distended the parenchyma of the uterus into a large cavity. Through an opening in this diaphragm of the size of a shilling the tumour descended into the cavity of the uterus. A very similar case is reported by Jarotzky and Waldeyer, only the degeneration occurred before the formation of the placenta, and the whole circumference of the ovum was covered by villi. The extension of the villi into the veins contained within the parenchyma of the uterus, and followed by atrophy, would naturally greatly impede the expulsion of the mole, and render it quite impossible in more advanced cases. Krieger reports a case where the "usur" of the uterus was followed by a fatal peritonitis.

Treatment with regard to the prevention of this degeneration is powerless. All that can be done consists in means to stop the bleeding and to hasten the expulsion of the ovum. It is hardly possible to diagnose with absolute certainty a vesical mole, but in the second half of gestation, at least, the dead product of conception can be recognised, and in such cases, especially if there are urgent symptoms, the artificial interruption of pregnancy is indicated. If the expulsion of the mole has already begun, and if there is no great amount of hæmorrhage, the best treatment would be to place a tampon in the vagina and to give ergot internally, for all depends upon the complete expulsion of the new formation and the good contraction of the uterus after the expulsion.

Since the embryonic connective tissue of the allantois surrounds the whole circumference of the ovum between its serous sheath (the exochorion) and the amnion, this layer also may become the seat of a *myxomatous hyperplasia.* Until now only one such case, that reported by Breslau and Eberth, is known and described under the name of *diffuse myxoma.* Between the exochorion and the amnion, around the whole circumference of the ovum, a layer was found from 4 to 5 mm. thick, like the gelatine of the umbilical cord, the placenta being normal. Here, therefore, the degeneration occupied the connective tissue belonging to the chorion, which had proliferated around the circumference of the ovum, whilst the villi of the chorion had atrophied in the usual way. It appears from the case of Spaeth and Wedl that the rest of the allantois situated beneath the placental part of the amnion may also be the seat of a hyperplasia without the myxomatous degeneration of the villi of the chorion, which form the placenta. For close to the insertion of the thick umbilical cord, immediately beneath the membranes, a stratum of young gelatinous connective tissue 1 mm. thick was situated and extended towards the edge of the placenta. Rokitansky also mentions amongst the new formations of the placenta an accumulation of gelatinous connective tissue on its concave surface.

The myxoma presents a very different appearance to the vesical mole, where the homogeneous, thinly mucous, intercellular substance becomes richer in fibrous constituents, so that the tissue becomes firmer and more like connective tissue, as in the peripheral layers of the umbilical cord. Virchow first described this as *myxoma fibrosum placentæ*. Between the normal cotyledons a degenerated one was found, the child being healthy. Upon the thick and firm knobs, which were the size of a pigeon's egg (the stems of the villi), secondary and tertiary prolongations were seated in the shape of smaller knobs as large as a hazel-nut or a hemp-seed. Hildebrandt, who has observed a very similar case, attributes that degeneration to the congestion of the efferent veins of the diseased cotyledon.

(*b*) *Anomalies of the placenta.*—Whilst speaking of the diseases of the chorion we have already mentioned the hyperplasia which is met with at the insertion of the placenta under the forms of myxoma multiplex and myxoma fibrosum. The other anomalies of the placenta are—

1. *Anomalies of formation.*—The size of the placenta is subject to considerable variation. The older writers speak of an entire absence of the placenta, which, however, is nothing else than the arrested development of the villi of the chorion at the situation of the decidua serotina. In vary rare instances the villi of the chorion retain their blood-vessels round the whole circumference of the ovum, or at least the greater part of it, and consequently the thick placenta proper does not form. The ovum is in communication with the decidua in the whole periphery (a condition similar to that found in the pachydermata). Stein, senior, called this formation placenta membranacea.

More frequently the placenta is abnormally large. This occurs in cases of uncommonly strong children, but chiefly in cases of atrophied fœtus together with hydramnios. In these cases the hyperplasia is due to an inflammatory condition of the uterine mucous membrane, and the great proliferation of the decidua serotina and of the villi of the chorion, which cause the profuse exudation of the liquor amnii, by which the fœtus is deprived of a part of its nutriment.

Sometimes the placenta is divided even in simple pregnancy. Two or more, even seven placentæ, have been observed, and at the side of a larger several smaller placentæ succenturiatæ occur. These formations can easily be explained from the development of the membranes. Some of the villi of the chorion not inserted at the place of the decidua serotina retain their vessels and enter into vascular communication with the decidua vera. If this does not take place the enlarged villi form the so-called placenta spuria.

The shape of the placenta is more or less round or oval, very rarely that of a horseshoe, in which case the placenta is inserted close to the internal os uteri, which it surrounds with the two crura.

An anomalous seat of the placenta, of very great practical importance, is the placenta prævia. It will be spoken of under "hæmorrhage during parturition."

2. *New formations of the placenta.*—The myxomata have already been mentioned, and other tumours scarcely ever occur. Clarke observed a case where a knot within the placenta weighed about fourteen ounces, and looked internally like firm muscle, and Löbl described one the size of a child's head as a fibroid. Both were probably nothing else but myxoma fibrosum. Hyrtl described two tumours imbedded in connective-tissue capsules, which latter consisted of newly formed connective tissue in all stages of development, and called them sarcomata.

In not very rare instances cysts of various sizes form on the concave side of the placenta; the connective-tissue layers of the chorion and amnion are raised like cysts, and are lined by a flat epithelium, whilst the portion upon the placenta has a shaggy appearance, and is covered by fibrinous deposits. Those cysts are thin transparent vesicles, which contain a yellow or reddish opaque and thin fluid. They are probably always formed from apoplectic centres.

Calcareous deposits of a moderate size are not rare in the placenta. They are found on the tips of the villi (also in the maternal tissue), and when the deposition is more abundant also on the pedicles. The infiltration is either more diffuse and starts from the epithelium of the villi, or, what is more frequently the case, it commences in the walls of the capillaries and in the finer vessels of the villi.

3. *Inflammation of the placenta—placentitis.*—Inflammation of the placenta proceeds from the cells of the decidua serotina, also according to Maier, in some cases from the adventitia of the fœtal arteries. Young connective tissue is formed, which extends between the cotyledons of the placenta and afterwards begins to shrink. By the proliferation of the connective tissue the interjacent maternal blood-vessels are compressed, the villi of the chorion which dip into them become atrophied and undergo fatty degeneration. The collateral hyperæmia may also lead to extravasations of blood on the fœtal surface of the placenta.

The disease is chiefly caused by metritis or endometritis, which has previously existed. If there are sufficient villi to carry on their necessary functional activity the fœtus may fully develop. The hyperæmia, however, which is always connected with that inflammation, very easily leads to considerable hæmorrhage from the maternal vessels, and consequently to abortion. The inflammation of the placenta has also some influence upon the delivery of the afterbirth. For on account of the greater brittleness of the recently proliferated tissue, pieces of the placenta may easily tear off and remain attached to the uterine wall, especially when the cicatricial contractile tissue has already entered between the villi.

A similar wedge-shaped proliferation which passed between the cotyledons is described by Virchow and Stavansky as being of syphilitic origin. It consisted of firm, large-celled connective tissue, in which here and there younger cells were abundantly accumulated. The cortical layer of the nodes was whiter and fibrous, and the internal substance was yellowish and soft.

Inflammation terminating in suppuration is certainly extremely rare. Bouchut found once two abscesses of the size of a pigeon's egg beneath the fœtal surface, and in another case several small ones surrounded by the hepatized tissue. Jacquemier also mentions purulent centres, and Cruveilhier speaks of a form of purulent infiltration of the placenta.

(*c*) *Anomalies of the amnion and of the liquor amnii.*—The liquor amnii may, from various causes, accumulate in the amniotic cavity in such excessive quantity as to constitute hydramnios. This is seen in disturbances of the circulation of the maternal organism, and is followed by œdema and dropsy in other organs, but it exists also in cases where the latter is perfectly absent or is apparently a (secondary) consequence of the enlarged uterus. In such cases the placenta is commonly hypertrophied, the decidua in a state of great proliferation, and the villi of the chorion thickened and enlarged into knobs. The fœtus is either dead or much atrophied, and sometimes it does not weigh more than the placenta. The inflammatory proliferation of the membranes is in such cases apparently the primary process; it is followed by an enormous secretion of the liquor amnii, and secondarily by atrophy of the fœtus, so that the membranes consume the greater part of the nutritive material which is intended to nourish the fœtus. In other cases the fœtus is sometimes found of normal conditions.

McClintock has found that hydramnios is far more frequent in pluriparæ than in primiparæ (twenty-eight to five), and that amongst his thirty-three cases there were twenty-five female and eight male children born.

By the great distension of the ovum the hydramnios gives rise to the same disturbances which occur when the volume of the uterus is enlarged from other causes. There is discomfort or pain in the abdomen, and, as symptoms of pressure, œdema and neuralgia of the lower extremities. The diaphragm pushed upwards, especially in primiparæ, gives rise to disturbances of respiration, and sometimes even to attacks of impending asphyxia. Often the great distension of the uterus causes a premature interruption to pregnancy.

The higher degrees of hydramnios at the end of pregnancy can almost always be easily recognised. The abdomen is enormously distended, the uterus very tense, elastic, and rarely showing distinct fluctuation; sometimes when the tension is very great the uterus feels as hard as wood. Usually the fœtus cannot be distinctly felt, at least the smaller parts cannot be made out by palpation, whilst ballottement shows one or two large parts remarkably well. The latter and the sounds of the heart very easily change position, sometimes almost continually. By internal examination the lower uterine segment is felt elastic and greatly distended, but no fœtal part is distinctly perceptible behind it. But even when the child is dead, and neither the sounds of the heart can be heard nor the fœtal parts felt, the tumour containing the fluid must always be recognised as the enlarged uterus; and such an accumulation of fluid in the uterus does not occur in any other condition than in hydramnios. In the lesser degrees of hydramnios, and at an earlier period of gestation, the diagnosis of pregnancy is not beset with so great difficulties as that of hydramnios.

The prognosis as regards the mother chiefly depends upon the cause of the hydramnios and the amount of the fluid. As regards the child the prognosis is doubtful, according to McClintock, of thirty-three children nine were stillborn and ten died within the first few hours of extra-uterine life.

There is no remedy known to limit the accumulation of fluid. If the discomfort is very great premature labour may be induced, but only when the mother is in imminent danger, before the thirty-second week; for children born before the thirty-sixth week are always very weak.

Too small a quantity of liquor amnii is only dangerous at the commencement of the formation of the fœtus. If the amnion in the course of its development is not separated from the body of the fœtus by the necessary quantity of serum, this anomaly may be followed by folds in the amnion, by abnormal connection between it and fœtal parts, by the union of single cutaneous spots (Simonart's bands), by the formation of fissures, and the deficiency of whole organs (self-amputation).

In the latter period of pregnancy the amnion may rupture, whilst the chorion preserves the integrity of the ovum. The active movements of the fœtus may cause the amnion to be rolled up in the ovum, and form cords which compress the umbilical cord and lead to the death of the fœtus.

2. *Anomalies of the Umbilical Cord*

The length of the umbilical cord varies very considerably. It may be so short as to be immediately attached to the placenta, or have a length of seventy inches. The quantity also of the Whartonian jelly varies greatly. It sometimes accumulates at those places where the vessels form a kind of loop in thick lumps, and constitutes the so-called "false knots."

True knots arise when the fœtus in its active movements passes through a loop of the umbilical cord, which is afterwards drawn together. It is, however, extremely rare that these knots form an impediment to the circulation. Hecker found them once in 266 cases, and Elsäser once in 202 cases.

Very frequently the umbilical cord is twisted round the trunk, the neck, or the extremities of the fœtus. But it is certainly a very exceptional occurrence that this causes the death of the fœtus during pregnancy.

More frequently torsion of the umbilical cord leads to the death of the fœtus during gestation, since the umbilical vessels are occluded when the cord is twisted at a place where the jelly is deficient. The cord, however, may be twisted in a great many places without there being any great impediment to the circulation.

It is only the absence of the Whartonian jelly which renders the twistings of the cord so sharp that the vessels are compressed.

In the majority of instances the umbilical cord is inserted into the middle of the placenta, though not quite centrally. But frequently also its insertion is marginal, and in some cases the cord goes to that part of the chorion which is free from villi, so that the vessels pass through the membranes to the placenta. In such cases both umbilical arteries usually blend into one longer or shorter common trunk.

Schultze explains this anomaly, the velamentous insertion, in the following way:—In every ovum the fœtal vessels are carried by the allantois to a point in the circumference of the ovum, and it is relatively rare for them to meet exactly the future place of the placental insertion. At first the vessels enter into all the villi of the chorion, and those which do not belong to the placental portion of the circumference of the ovum are obliterated; a vascular communication remains only at one spot. During the progress of the growth of the ovum the fœtus is normally so turned that the vessels have a straight course towards the place of the future placenta. This twisting of the fœtus may be prevented by abnormal adhesions, which one of the constituents of the future umbilical cord forms with that place on the periphery of the ovum which first of all the vessels had reached. Then the umbilical cord remains inserted at one spot in the membranes outside the placenta.

There are numerous anomalies in the arrangement of the vessels of the umbilical cord, but they are without practical importance.

C.—DEVIATIONS IN THE DURATION OF PREGNANCY

1. *Premature Interruption of Pregnancy (Hæmorrhages during Pregnancy); Abortion; Blood-Mole; Fleshy Mole; Premature Labour*

The normal course of pregnancy may be suddenly interrupted by some acutely acting cause. In most cases, however, material alterations of the ovum or pathological conditions of the mother give rise to a peculiar predisposition to abortion. Under such circumstances an external impetus which would otherwise be quite ineffectual will suffice to cause premature uterine contractions and lead to the expulsion of the ovum. A distinction has therefore to be made between the conditions which constitute the peculiar and remote cause, the predisposition to abortion, and those which determine the occurrence of abortion.

The tendency or predisposition to abortion may proceed either from the ovum or from the mother. On the part of the ovum every disease of the chorion predisposes to abortion, but by far the most frequent cause is the death of the fœtus, the result of a variety of conditions. They are congenital malformations, internal diseases, and exceptionally injuries; then all conditions which diminish or entirely cut off the supply of nutritive material. Chronic or acute anæmia of the mother may prove fatal in this way by causing asphyxia of the embryo, as we had the opportunity of observing in a case of profuse haemorrhage in consequence of rupture of a varicose vein in a case of twin pregnancy. In febrile diseases of the mother the fœtus may succumb to the abnormal accumulation of heat or to the influence of a constitutional disease as syphilis.

We have already said that the fœtus often dies, and thereby not only becomes the most frequent predisposing cause of abortion, but also determines with great certainty the premature expulsion of the ovum. In the absence of other changes the expulsion never occurs immediately upon the cessation of life, but a shorter or longer period afterwards in consequence of secondary changes in the ovum. These may be various, but essentially they act by loosening the connection between the ovum and uterus, and thus cause uterine contractions. When the fœtal circulation ceases the villi of the chorion are obliterated and undergo fatty degeneration, so that the decidua also is involved in the same change. But almost always the cavity of the ovum collapses after the death of the fœtus, and this diminution of the volume of the ovum only fails to occur when dropsy or tumours of the ovum (hydramnios and myxoma chorii) are the causes of the death of the fœtus. In the other cases the diminution of the volume of the ovum is followed by a change in the connection between the circumference of the ovum and the wall of the uterus, so that the villi of the chorion are drawn out of the maternal blood-vessels. A solution of continuity takes place, and gives rise to hæmorrhage from the over-filled maternal vessels into the cavity of the uterus, which now is under an altered pressure. The hæmorrhage into and between the membranes is still more considerable when, as often occurs in the first months of gestation, the membranes rupture, and the liquor amnii and fœtus are expelled, whilst the empty ovum remains a still longer time within the uterus. These hæmorrhages are one of the most frequent sequences of the death of the fœtus. We shall have to speak afterwards of the anatomical changes caused by the varying seat of the hæmorrhage and by the metamorphosis of the effused blood.

On the part of the mother the abnormal conditions which predispose to abortion are anomalies of the decidua; a too scanty development or atrophy of the mucous membrane has only a pernicious influence on the growth of the ovum when the process takes place at its periphery, that is, if the decidua serotina or reflexa is implicated.

When the decidua serotina is not sufficiently developed an abnormally small placenta is formed. This in itself does not exert a noxious influence on the growth of the ovum, but it becomes of importance when the decidua reflexa also happens to be less developed than normal. If the latter membrane is either too soft and thin, or if its proliferations have not covered the ovum at all, or only imperfectly, the ovum may, partly by its own gravity, partly by premature uterine contractions, be forced towards the internal os, and occasionally also into the cervix. The ovum, then covered only by the chorion, hangs by a long drawn-out pedicle, formed by the uterine glands of the decidua serotina, or it is entirely separated from its place of insertion, and hangs by a pedicle formed by the decidua reflexa.

The changes in the decidua vera cause either the death of

the fœtus or give rise to primary apoplexies between the membranes, and are in this way chiefly concerned in the induction of abortion. Acute endometritis is especially apt to give rise to apoplexy.

Besides the changes in the mucous membrane there are other pathological conditions of the uterus which easily induce abortion, and which have been mentioned previously.

Of the changes in the position of the uterus the chief are retroflexion and prolapsus. Anteflexion alone is not a direct cause of abortion, but if it occurs it is usually due to the chronic metritis with which the anteflexion is complicated. Metritis itself is decidedly a frequent cause of abortion, as are also new formations of the uterus (fibroid and carcinoma) and other conditions, by which the enlargement of the uterus is impeded, as large tumours of the abdomen, exudations enveloping the uterus, &c.

There can be no doubt that, independently of these demonstrable pathological conditions of the maternal organism, some individuals are of such an irritable constitution that they have a peculiar predisposition to abortion. In virtue of which, even when the ovum and the maternal organs of generation are healthy, incidents, such as slight bodily or mental excitement, which are borne without any bad consequences by the great majority of women, cause a premature interruption of pregnancy. In those individuals the motor nerves of the uterus react on stimuli which are commonly quite powerless.

Though pathological conditions of the ovum frequently predispose to abortion, it rarely happens that they determine its immediate occurrence. On the side of the ovum rupture of the membranes and evacuation of the liquor amnii constitute the sole cause of an immediate interruption to gestation; here the effect is very precise, and the more so the farther pregnancy has advanced.

Far more frequently the direct causes of abortion proceed from the maternal organism, and the chief agent is the anomaly in the quantity of blood in the gravid organ.

It is doubtful whether hyperæmia of the gravid uterus without rupture of the vessels can cause uterine contractions by irritating the nerve-fibres. So much is certain, that hyperæmia greatly favours hæmorrhage, and that this, if at all considerable, leads to uterine contractions by solving the connection between the ovum and the uterus. Every condition, therefore, which leads to a hyperæmia of the uterus may be the starting-point of an abortion.

Thus, general plethora alone suffices to produce hæmorrhage from the greatly proliferating mucous membrane, and acute diseases of the mother attended by a more considerable febrile disturbance cause lacerations of the decidual vessels. All those conditions which cause active congestion of the genital organs act in the same way, such as inflammation of these organs, too frequent and too excited coitus, hot footbaths, greatly exciting pleasures or reading, and spirituous drinks. Then, again, those conditions which cause passive congestions, viz. change in the position of the uterus, diseases of the heart, lungs, and liver, &c. The cases in which vessels are directly torn by injuries are very rare; it more frequently happens that concussions of the body from a fall or a jump, even from violent coughing, sneezing, or vomiting, cause lacerations of the vessels of the decidua.

It can also be experimentally shown that contractions are produced by intense anæmia of the uterus. And although chlorosis and anæmia, resulting from various kinds of cachexiæ, rarely have that direct influence, yet pregnancy is easily interrupted by a general and sudden loss of blood. It is in this way, probably, that the action of ergot of rye can be explained by causing contractions of the arterial walls, which are followed by an acute anæmia.

Contractions of the unstriped muscular fibres of the uterus can also be caused by irritations acting directly on the nerves. Thus, the uterine nerves may be irritated as by friction of the organ through the abdominal walls. Premature delivery in consequence of too great distension of the uterus, especially in twins and hydramnios, is, perhaps, due to this cause; or sympathetic fibres of other nerve-centres may be excited by reflex action, and thus the activity of the uterine nerves be aroused. Thus, irritation of the vagina, vulva, and mammæ very frequently causes contractions of the uterus. It is as yet unknown in what way mental emotions, especially fright, cause uterine contractions (paralysis of the spinal inhibitory nerves?).

The causes of abortion are to be distinguished into—(1) the predisposing, or those which prepare the ground, so to speak, and (2) the determining causes, those which induce abortion. Although this distinction is of the greatest importance in practice, yet these two causes cannot be recognised and distinguished in every case.

Where a predisposition exists, *e.g.* the presence of a dead fœtus, and degeneration of the decidua has already somewhat advanced, it may gradually go on to the production of uterine contractions without an acute cause being demonstrable, and, on the other hand, some determining cause, as particularly hæmorrhage into the membranes or direct irritation of sympathetic nerve-fibres, may induce abortion even under quite normal conditions.

Amongst the causes of abortion hæmorrhage has been said to be the most prevalent, and in cases where bleeding has not previously taken place it always attends the act of abortion. Of course, in a given case it may be very difficult to trace with accuracy the connection between the death of the fœtus and apoplectic effusions into the membranes. The death of the fœtus may give rise to hæmorrhage, which either immediately induces abortion or which first produces changes in the ovum, which in their turn lead to abortion, so that it is either caused or attended by hæmorrhage. But a primary bleeding into the membranes may so much interfere with the nutrition of the fœtus that death follows. Death of the fœtus, then, is the consequence of the hæmorrhage, and the dead fœtus may again give rise to hæmorrhage. It may therefore be extremely difficult or impossible to determine the succession and causation of the changes observed in an ovum. What renders this still more difficult is that nothing definite is known as to the time which is required for the various changes usually undergone by the fœtus. This much is certain, that a varying length of time is required for the various changes in fœtuses of different periods of gestation. It thus becomes extremely difficult to say from the changes how long the fœtus has been dead.

Similar difficulties are met with in determining the age of hæmorrhagic effusions. Quite recent effusions can be easily distinguished from older ones, and these again from very old ones, but it is impossible to determine, from the metamorphosis of the effusions themselves, the time when an apparently older hæmorrhage has taken place. Quite recent effusions are fluid, an older one is coagulated, almost black, and firmly adherent between and to the membranes. After a still longer period the coagula become discoloured, and assume a light reddish coriaceous appearance.

The seat of the hæmorrhage may also vary greatly. The blood (with the very rare exceptions already mentioned) is always derived from the maternal vessels.

If the entire decidua vera be expelled its uterine surface is always covered with fresh blood, and frequently old, black, or even discoloured coagula are found. Within the tissue of the decidua vera itself small apoplectic centres are sometimes seen, and this is especially the case when the decidua is more than usually developed.

When the extravasations of blood into the decidua vera are more copious they completely destroy it, and are effused between the vera and reflexa. Before the decidua serotina can become the seat of the hæmorrhage, the fold which forms the vera and reflexa has to be broken through. Blood may also be effused between the reflexa and the chorion. The effusion has then either to rupture the reflexa, or, when derived from the placental insertion, to extend thence along the chorion, separating it from the decidua reflexa.

These extravasations of blood occur very frequently and unless the effusion has been very slight, give the ovum a characteristic appearance. The chorion and amnion are pushed inwards, forming thick projections (Velpeau's *bosselures*). On section they are seen to be formed by coagula of blood, which are either recent and black or older and discoloured. As long as the ovum remains healthy the pressure under which the extravasation of blood takes place hardly suffices to narrow the amniotic cavity. Probably these projections form after the rupture of the membranes, or at least when, after the death of the fœtus, the ovum has collapsed. In such ova (which were formerly called blood-moles when the extravasation of blood was still recent, and flesh-moles when older and of a colour like liver) the fœtus is usually found atrophied, or has entirely disappeared, either by maceration and reabsorption or in consequence of premature expulsion.

It is exceptional for the chorion to be ruptured by the extravasation of blood, so that it is separated from the amnion for any distance. Even the amnion may be ruptured, and hæmorrhagic effusions of various kinds are found within the cavity of the ovum.

The period of gestation is also of importance with regard to the place where the blood is extravasated. After the formation of the placenta this is by preference the seat of the hæmorrhage, whilst at the commencement of pregnancy it takes place with equal frequency in the whole circumference of the ovum.

The quantity of blood effused necessary to produce immediately contractions of the uterus varies greatly. Here also the time of pregnancy has a great influence, for at an earlier period a slight extravasation of blood may be followed by expulsion of the ovum, whilst at a later period and under favorable conditions a comparatively profuse loss of blood may be well borne. The consequence of the hæmorrhage materially depends upon the state of the ovum, whether this is healthy or not. The ovum having collapsed after the death of the fœtus, very considerable loss of blood may well be borne for some time, but all large hæmorrhages, whilst the ovum is healthy, as a rule cause an immediate interruption to pregnancy. Hæmorrhage which is not too profuse may remain without endangering the life of the fœtus or interfering with the course of pregnancy. Especially when the source of the hæmorrhage is close to the internal os uteri so that the blood can flow out of it, and one flooding during gestation may pass by without any noxious consequence. When the seat of the bleeding is farther from the os the effused blood undergoes the usual changes which can be recognised in the mature ovum.

An important question to decide is, at what time, after the proper cause of abortion has begun to act, the actual expulsion of the ovum takes place? This time varies greatly. It has already been said that some causes are followed by immediate expulsion; after others days, weeks, and even months may elapse. Often a long time elapses when the death of the fœtus has been the cause of the interruption of pregnancy. As a rule the fœtus is then found in a peculiar condition of maceration, which will be afterwards described; it is only in rare cases that the embryo which died during gestation retains a fresh appearance. Should the death of the embryo occur at a very early period of gestation the ovum may continue to be nourished. Such nutrition of the fœtal placenta also is effected through the decidua serotina by osmosis. The placenta is then found to be more developed than corresponds to the fœtus, which may be still in the ovum or have been expelled a long time previously. The placenta is by no means putrid; it looks quite fresh, and differs from a healthy placenta only by the strikingly pale colour of the villi and by its somewhat firmer consistence. If the placenta grows in the way just described, the ovum may for months be carried in the uterus without a trace of a fetid discharge.

Abortion must be frequent, although for obvious reasons no accurate statistics on the subject can be given. Hegar, who has extensively studied the pathology of the ovum, is of opinion that to eight or ten births at term there is one abortion in the first months of pregnancy, and this frequency is hardly over-estimated. Experience also shows that, without demonstrable general causes, abortions occur at times in large numbers. Pluriparæ abort far more frequently than primiparæ (twenty-three pluriparæ to three primiparæ); this is probably dependent on the far more frequent occurrence of endometritis, chronic metritis, and change of position of the uterus in pluriparæ.

The commencement of abortion is, as a rule, indicated by a loss of blood. This usually commences, without preceding symptoms, with a few drops of blood, or at once with a larger quantity. In the first months of pregnancy there are usually no distinct labour-pains, but only a dull pain in the hypogastrium. Gradually the cervix dilates, and as soon as it allows a finger to pass the tip of the ovum is already felt within it. The internal os is the first to open, the ovum is pushed into it, and thus the cervix is more and more dilated. Bleeding may hereby become very profuse. Gradually the ovum is pushed through the external os; it then falls into the vagina, and is finally expelled by the latter. The decidua comes away with the ovum either as a whole or in shreds, or it is at first retained within the uterus and expelled some days afterwards. If only a portion of the ovum is expelled, which frequently occurs in the third or fourth month, the remainder contained within the uterus may give rise to obstinate hæmorrhage. The hæmorrhage is the more profuse the larger the portion of the ovum remaining in the uterus, but its complete expulsion is more easily effected.

The interruption of pregnancy by the expulsion of the ovum is called an abortion when it occurs at a time when the fœtus is as yet unable to live outside the uterus, but a premature birth when the fœtus has acquired such a degree of development that it is able to live independently. There is, however, no accurately defined limit between the two. Usually the twenty-eighth week is considered that limit, and this is correct so far, as it is certainly extremly rare that fœtuses born at an earlier period are kept alive, whilst the majority of those born immediately after the twenty-eighth week also die, and are kept alive only under especially favorable circumstances.

Within the first three months the ovum is usually expelled intact, without laceration of the amnion. Sometimes, however, it bursts at the commencement of the abortion, the as yet small embryo escapes unnoticed with a small quantity of liquor amnii, and later the empty ovum is expelled separately. From the third month the placenta developes distinctly, and about the fourth month the usual process is that the membranes present and rupture, then the embryo is born, and that afterwards the placenta is expelled with the membranes. The farther pregnancy has advanced the more the expulsion of the ovum resembles a normal birth. It is only requisite to mention that, although in premature births head presentations are still the most prevalent, yet the frequency of pelvic and transverse presentations considerably increases, and the more so the earlier gestation is interrupted.

The diagnosis of abortion is not difficult if the process can be fully observed. But when in the commencement all other symptoms except hæmorrhage are absent, the diagnosis may be very difficult.

It may often be difficult to attribute hæmorrhage from the uterus in the earliest period of gestation to pregnancy; and in the first month, especially when the menses have not yet ceased, this may be quite impossible. However, abortion at that time is very rarely observed, because the very young ovum passes away unnoticed with the slight hæmorrhage, and the bleeding itself is considered the returning catamenia. Hæmorrhage after the cessation of the menses must always be thought of as being due to abortion, and, though pregnancy cannot be distinctly diagnosed, treatment must be instituted as if it existed.

If, however, pregnancy is recognised it will certainly be in the interest of the patient if every hæmorrhage is considered pathological, inducing abortion, and treated as such, for we have already mentioned above that the return of the catamenia during pregnancy is decidedly an extremely rare occurrence.

If the abortion has already progressed so far that the cervix is dilated the diagnosis is easy, since the finger can, as a rule, be introduced, and the tip of the ovum felt. Sometimes, however, it may be difficult to distinguish it from a polypus, and it may even be impossible until after the expulsion of the tumour. This is especially the case when the ovum has previously ruptured, and when the thickened membranes enveloped in coagula of blood are felt in the cervix (there being no fluctuation in the cavity of the ovum), and when on account of previous hæmorrhages the history of the case is unreliable.

From an examination only once made it may also be difficult to decide whether the ovum is still within the uterus or whether it has been entirely or partially expelled. If the cervix is pervious the finger can easily be introduced into the cavity of the uterus, and by external pressure with the other hand the uterus can be brought against the examining finger, and thus it can be ascertained if parts of the ovum are still contained in the uterus. But in impervious cervix an accurate diagnosis must depend upon a repeated examination. The history only shows that clots have passed away, but whether they were blood or constituents of the ovum remains doubtful; and the uterus also, even after the expulsion of the ovum, remains enlarged. The uterine sound, cautiously used, can only show that there is not an intact ovum within the uterus, but it is useless for the distinction between coagula of blood and remains of the ovum. In such cases, when after repeated hæmorrhage attended by severe lumbar pains the examination is repeated, the cervix will often be found pervious. Or the hæmorrhage renders the tampon necessary, in consequence of which the cervix is dilated and the remains of the ovum expelled.

It is important in regard to treatment to know whether the fœtus is alive or not. In the first months of gestation this is, as a rule, impossible. Death of the fœtus may be supposed in all cases where for a long time sero-sanguineous discharges have been met with, and where the constitutional changes previously mentioned exist. But not even this is quite reliable, and it is safest if even in such cases abortion is treated as if the fœtus were alive.

The prognosis is naturally distinguished into that for the fœtus and for the mother.

With regard to the former, it has already been said that in many instances abortion is caused by the death of the fœtus. In all those cases there is, of course, no question of a prognosis. If the ovum is healthy, and if it is expelled in the first seven months of gestation, the fœtus is always lost, even if born alive. What has to be considered is whether the expulsion can be prevented. It is an important point in the treatment never to give up the idea of preventing abortion as long as a part of the ovum has not already passed out of the external os. It happens in some cases that the ovum has already partly descended into the cervix, but that it is again retracted into the uterus, and that at the normal end of gestation a healthy child is born. In the last three months of pregnancy the prognosis is the more favorable to the child the more the premature birth approaches the normal end of gestation.

The mother is never quite free from danger. The hæmorrhage may sometimes even in the beginning of the abortion, more frequently after the expulsion of the ovum, be so considerable that her life is endangered. But experience teaches that the hæmorrhage of abortion may go on to the highest degrees of anæmia, syncope, and pulselessness, but is very seldom followed by death. As a rule the bleeding is stopped by the syncope. The quantity of blood also which a woman can lose without danger to life varies greatly in different individuals. In some cases very little, at others enormous quantities of blood may be lost. If the expulsion of the ovum has been imperfect the remains of the ovum may for a long time cause obstinate hæmorrhage, followed by lasting illness. Ichorous and putrid infection may also be the consequence of retained portions of the ovum. Unless the necessary care is taken after abortion it is easily followed, as in a normal birth, by inflammation and change in the position of the uterus.

The treatment of abortion ought first of all to be prophylactic. All causes must be removed which primarily lead to premature expulsion of the ovum or to the death of the fœtus. Under normal conditions, that is, in healthy primiparæ or women who have not previously aborted, general dietetic rules are sufficient. But the case is different with the so-called habitual abortion, that is, where regularly at a fixed period of gestation abortion takes place. No general rules can be given to remove this disposition to abortion. The actual causes must be found out. Most frequently it lies in local diseases of the uterus, and consequently the treatment must be directed to them. We must not fail to remember that syphilis, though latent, in one of the parents, is a very frequent cause of the death of the fœtus, and this is especially the case in the last months of gestation. This point must be inquired into, and if it is so the parents should be put on a course of mercurial treatment; healthy children will then be born.

Under such circumstances it may be necessary to keep the woman in the recumbent posture for weeks. In the discussion on a case of the kind related by Cuthbert before the Edinburgh Obstetric Society the use of chlorate of potash was advocated. Simpson suggested that it should be administered in large doses, about fifteen grains three times a day.

The first thing to be done when called to a pregnant woman suffering from hæmorrhage is to endeavour to prevent the threatening abortion. An exception can only be made in those cases, very rare, at least in the first months of pregnancy, where the death of the fœtus can be recognised with certainty. Otherwise, unless the bleeding becomes too profuse, no active means need be employed. The woman ought to remain constantly in the dorsal posture, and a few full doses of tincture of opium should be given by mouth or rectum. The recumbent posture should be maintained for some days after the hæmorrhage has ceased. Even if the cervix is already dilated, and the ovum felt penetrating into it, the hope of stopping the abortion must not be given up. Here, also, if the hæmorrhage is not too profuse, the treatment just mentioned may be employed. It undoubtedly occurs that the cervix closes again, and that the embryo is carried to the normal end of gestation.

Other treatment is necessary in those cases where profuse hæmorrhage threatens the life of the mother. Then no time must be lost in employing uncertain and mostly ineffectual remedies, such as mineral acids and ergot internally, and vinegar and cold water applications to the abdomen, but immediately the tampon should be applied to the vagina. (Injections also of cold water against the cervix are not very certain.) Amongst the various kinds of tampon that is the best which is sure to stop the hæmorrhage without exciting too much uterine action. The caoutchouc tampon can therefore not be recommended, because it does not stop the hæmorrhage with certainty if only moderately filled, whilst by great distension it causes severe pains and uterine contractions. The hæmorrhage is most certainly stopped by pieces of lint which are pressed against the bleeding spot. They firmly adhere to it, and thus prevent any further bleeding. It is therefore unnecessary to fill up the whole vagina, and often a small tampon suffices to stop the most abundant hæmorrhage. We have always had the best results from the following very easy way of applying the tampon:—A speculum as wide as possible is introduced into the vagina, so as to get a full view of the bleeding cervix. Then a large piece of lint is evenly placed over the external opening of the speculum, and upon it other and smaller pieces of lint are placed so as to fill up the speculum to the bottom. Whilst the tampon is pressed against the cervix with a long rod the speculum is withdrawn over it. There remains then within the vagina a closely packed tampon, contained in a sac of lint of about the thickness of the speculum. If a speculum is not at hand single pieces of lint must be pressed against the bleeding cervix. A tampon also so constructed renders good services; it is, however, more inconvenient to remove, since each piece has to be withdrawn singly from the vagina. After the removal of the tampon it can easily be seen that it very effectually stops the hæmorrhage, for only at the spot which was pressed against the cervix is a coagulum of blood found, whilst the rest of the tampon is moistened by serum only. On the removal of the tampon after six to twenty-four hours, according to the strength of the contractions of the uterus (it cannot remain too long in the vagina without producing a disagreeable odour), the ovum is frequently found expelled from the uterus and situated behind the tampon. If the ovum have not advanced, and the bleeding persist, the tampon must be renewed. Even after the application of the tampon all hope of stopping abortion need not be given up. The method of using the tampon just described by no means always increases uterine action; it occasionally happens that after its removal the hæmorrhage ceases, the os uteri is again somewhat contracted, or at least not more dilated, and pregnancy takes its normal course without further disturbance.

When the greater part of the ovum is found to be in the cervix it may be removed provided that this can easily be done without lacerating it. To effect this the uterus is fixed from outside; one or two fingers are introduced into the cervix at the side of the ovum, and this is then withdrawn. When the ovum is too high up, so as to render the removal difficult, the attempt must be given up, for every care must be taken that portions of the ovum be not left behind in the uterus. Instead of making further attempts at extraction it is preferable to employ the tampon immediately. A few hours afterwards the ovum will be found expelled as a whole, and situated behind the tampon. Höning recommends for the removal of the ovum the compression of the uterus by combined manipulations. When, as is usually the case, the uterus lies somewhat anteflexed, two fingers of the one hand are brought into the anterior vault of the vagina, and there placed against the body of the uterus, whilst the other hand presses from outside upon the posterior wall of the uterus. Or the contents of the uterus may be pressed out, when the organ is pressed from outside against the posterior wall of the symphysis. When the uterus is retroflexed the fingers introduced into the vagina are brought into the posterior vault. This method of expulsion succeeds easily and perfectly.

Portions of the ovum retained within the uterus can be removed only with difficulty, especially when a long time has elapsed since the abortion. If the cervix is not yet pervious for one finger, it is to be dilated by a sponge or laminaria tent. If the cervix is pervious the uterus is fixed from the outside and the index and the first two fingers introduced into the vagina (the introduction of the whole hand is unnecessary when the uterus is well fixed from the outside). If the portions are loosely adherent in the cavity of the uterus they are in this way very easily removed by introducing two fingers. If this does not succeed they must be extracted by means of long forceps. The forceps are introduced along the fingers, the uterus being fixed by an assistant.

After the complete removal of the ovum the hæmorrhage usually stops spontaneously or by friction of the fundus or injections of cold water. In other cases, when hæmorrhage continues, if the abortion has occurred in the early months, a loss of blood equal in volume to the expelled ovum can be borne without danger by the woman. We may plug the vagina, and by it the necessary contraction of the uterus will also be obtained. When the uterus has been more distended, so that it can contain a greater quantity of blood, the use of the tampon is inadmissible, since by it the hæmorrhage would only be converted into an internal one. The treatment is here the same as in the period of the afterbirth in normal parturition.

When the abortion and the attending hæmorrhage are over, the patient, even if she feels perfectly well, is to be treated as a parturient woman. For some time she must lie still, else inflammation and change in the position of the uterus will be the consequence.

If the ovum was putrefied or decomposing shreds of the decidua are found, injections of tepid water or infusion of chamomile should be used two or three times daily.

It is well that the medical attendant himself should attend to this, for nurses often only inject into the vagina.

2. *Abnormally Long Duration of Pregnancy (Partus Serotinus).*

Those cases in which a fœtus situated outside the uterus or within an atrophied cornu, and being encapsuled, is converted into a lithopædion, or gradually putrefies, and is thus retained within the abdomen an abnormally long time, cannot be reckoned amongst those of an abnormal duration of pregnancy; for such a fœtus is considered a foreign body enclosed within the cavity of the abdomen. Nor do those cases properly belong to this category where a degenerated ovum continues to develop within the uterus beyond the duration of pregnancy, or where a fœtus contained within the uterus has not been expelled, but undergoes a similar change to that in a fœtus situated outside the uterus and retained in the abdomen.

We shall only speak of those cases where an intra-uterine fœtus is born after an abnormally long time, either alive or only recently dead.

The cases of this kind which in the course of time have been made known are comparatively valueless, because the normal duration of gestation in the human female is undoubtedly subject to great variations, and the calculation is in almost all cases of pregnancy not very accurate.

For from the calculation alone, and not the appearance of the child, can a late birth be decided. It is known from experience that the size, weight, and the whole development greatly varies in children at term, and, on the other hand, some children who from calculation have expressly been stated to have been born very late are said to have been small, whilst others, on the contrary, have been, if accurately weighed, fourteen pounds in weight. The closure of the sutures, the small fontanelles, the strong voice, and the long hairs, are of little importance for the present, for sometimes small children, even those not quite at term, have all these signs, and, on the contrary, the strongest children may have remarkably wide sutures and large fontanelles.

A late birth can only be decided from a calculation of the term of gestation. It has already been stated that the birth takes place, on an average, 278 days after the last menstruation. Whilst gestation prolonged for some days is quite a common occurrence, cases where birth occurred 300 days or later after the last catamenia are, at any rate, very rare; they only are correctly called "late births."

From analogies in the animal kingdom we may conclude that such cases occur (the duration of gestation in the cow is, on an average, 282 days, and may exceptionally be prolonged to 321 days); and this is confirmed by a number of accurate observations; but their frequency must not be overrated, and only after mature consideration of all the conditions respecting it can such a case be pronounced a late birth.

Literature.—Heim, Horn's Archiv, N. F., 1812, I. 1 (s. Wittlinger's Analecten, I. 2, p. 321).—Meissner, Frauenzimmerkrankheiten, III, 1 and 2 Abth. Leipzig, 1847.—Kiwisch, Klinische Vorträge, &c., 2 Aufl., Prag., 1852, II, p. 233. —Hecker, M. f. G., B. 13, p. 81.—Czihak, Scanzoni's Beiträge, B. IV, p. 72.—Klob, Path. Anat. d. weibl. Sexualorg., p. 159.—Mme. Boivin, Nouv. rech. sur la môle vesical. Paris, 1827; Deutsch, Weimar, 1828.—Mikschik, Zeitschr. d. Ges. d. Wiener Aerzte, July to Sept., 1845.—H. Müller, Abh. über den Bau der Molen. Würzburg, 1847.—Gierse, Verh. d. Ges. f. Geburtsh. in Berlin, 1847, H. 2, p. 126. —Mettenheimer, Müller's Archiv, 1850, p. 417, T. IX u. X.—G. Braun, Wiener Medicinalhalle, III, Jahrg. 1 u. 3.—Graily Hewitt, Transactions of the Obstetrical Society, I and II.—Hecker, Klinik d. Geburtsk., B. 2, p. 20.—Virchow, Die krankhaften Geschwülste, I, p. 405.—Bloch, Die Blasenmole. Freiburg, 1869.—Ercolani, Mem. delle malattie della placenta. Bologna, 1871. (S. das Referat von Hennig, Arch. f. Gyn., B. II, p. 154.)—Simpson, Monthly Journal of Medical

Science, February, 1845, p. 119; and Select Obstetrical Works, 1871, vol. i, p. 134. —Spaeth and Wedl, Zeitschr. d. Ges. d. Wiener Aerzte, 1851, 2, p. 806.—Klob, Pathol. Anat. der weibl. Sexualorgane, p. 542.—Hyrtl, Die Blutgefäss d. menschl. Nachgeburt. Wien, 1870.—Whittaker, American Journal of Obstetrics, vol. iii, p. 193.—Ercolani, Mem. delle malattie della placenta. Bologna, 1871. (S. das Referat von Hennig, Arch f. Gyn., B. II, p. 454.)—Simpson, Edinburgh Medical Journal, April, 1836, p. 274.—Scanzoni, Prager Vierteljahrsschr., 1849, 1.—Hegar, Die Path. u. Ther. der Placentarretention. Berlin, 1862.—Matthei, Gaz. des hôp., 1864, No. 98.—Hegar and Maier, Virchow's Archiv, 1867, März, p. 387. Maier, M. f. G., B. 32, p. 412, and Virchow's Archiv, 1869, B. 45, p. 305.—Busch and Moser, Handbuch der Geburtskunde. Berlin, 1840. Artikel Abortus.—Whitehead, Causes and Treatment of Abortion, &c. London, 1847.—Dohrn, M. f. G., B. 21, p. 30.—Hegar, M. f. G., B. 21, Suppl., p. 1.—Verdier, Apoplexie plac. et les hematomes du placenta. Paris, 1868.—Hoening, Scanzoni's Beiträge, B. VII, p. 213.

CHAPTER III.

APPENDIX.

A.—THE DEATH OF THE FŒTUS DURING PREGNANCY.

THE causes of the death of the fœtus in the uterus are very various, and have already been mentioned in the chapter on abortion. They are constitutional diseases of the mother, diseases of the uterus, of the membranes, and lastly of the fœtus itself. If the fœtus dies very early it is soaked in the liquor amnii, softened, and may be entirely reabsorbed. The cavity of the amnion then contains the cloudy, turbid serum, and sometimes the remains of the umbilical cord. If the fœtus (at the time of death) was somewhat older, it is found to be soft and pulpy. At a still later period it undergoes a peculiar transformation. The fœtus is then commonly said to be putrid (*todtfaul*), but this condition is best considered a kind of maceration. The whole body is softened; if placed for some time upon an even surface it is found quite flattened at those parts of the body on which it rested. The flaccid abdomen falls to one side. The fœtus has not properly a putrid smell, but is of a peculiar, stale, sweetish, disagreeable odour. The epidermis is raised in large patches, especially on the abdomen and the face, and the reddish-brown corium is denuded at those places. The umbilical cord is withered, discoloured, and reddish-brown from the diffusion of the blood. The cranial bones are loose, the skin of the head withered, flaccid, and too large for it. The sutures between the cranial bones are loosened, easily movable, or the bones are entirely separated, and contained in the integuments as within a sac. The internal organs are variously altered. The brain is mostly changed. It is converted into a reddish-brown pulp, in which formed material can no longer be recognised.

The muscles and connective tissue preserve their external shape; the transverse striæ of the muscles are frequently distinctly recognisable, although the primitive fibrils are filled with a finely granular fat. Of the organs contained in the thoracic and abdominal cavities, the latter of which contains, as a rule, bloody serous transudations, the liver is most of all altered. The cells are disintegrated, and their membrane contains only finely granular fatty detritus and pigment. The uterus, and lastly the lungs, are most perfectly preserved, and the latter can still be distended. In all the organs the blood has disappeared from the vessels and become suffused into the surrounding tissues. All the organs show a finely granular opacity of their parenchyma, and very commonly also crystalline fat and pigment. Sometimes crystals of margarin and cholesterin accumulate in such large masses that some partly well-preserved organs are covered by a whitish grey pulp, and this condition is called "lipoid metamorphosis" (Buhl).

If one of twins dies at an early period it may be so much flattened by the pressure of the other growing ovum that it presents a layer as thin as a sheet of paper (fœtus papyraceus).

Experience does not as yet enable us to determine with any certainty from the changes of the fœtus the time which has elapsed since its death. From causes with which we are unacquainted the changes appear to progress with varying degrees of rapidity. Sometimes a fœtus, which, on the examination of the woman a short time before, showed undoubted signs of life is expelled in a high degree of maceration; at other times one which has been dead for weeks is expelled comparatively little altered.

B.—THE DEATH OF THE MOTHER DURING PREGNANCY.

The fœtus continues to live for a very short time after the death of the mother. Every legislation, therefore, commands that on the death of the mother, if the presumption is in favour of the fœtus being alive, means are to be employed to save it by the Cæsarean section, that is, by opening the abdominal cavity and the uterus, and by extracting the fœtus through that opening. But the greatest obstacle lies in the uncertainty of the actual death of the mother. For the surest sign of death is putrefaction, and if this is waited for the fœtus is in the mean time certainly dead; but if the death of the mother is not certain there is great risk in performing the very dangerous operation of Cæsarean section on a woman who is only apparently dead. Children extracted ten minutes after the death of the mother are only very exceptionally kept alive. After the actual death of the mother the fœtus soon becomes asphyxiated, and this takes place the more rapidly the more mature the fœtus is. The fœtal heart may continue to act for some time, even after asphyxia has set in, and as long as the action continues there is still hope of keeping the extracted child alive by means of suitable treatment. It is therefore not impossible to save a child a quarter of an hour or even somewhat longer after the death of the mother.

Reported cases where a living child had been extracted hours after the death of the mother are, if worthy of credit at all, those of apparent death. After accidents which are undoubtedly followed by the death of the mother it is easiest to save the child by the Cæsarean section rapidly performed. Even if the death of the mother is beyond doubt, the operation and the succeeding dressings are to be made *lege artis*.

Literature.—Reinhardt, Der Kaiserschnitt an Todten, D. i. Tübingen, 1829.—Heymann, Die Entbindung lebloser Schwangerer, &c. Coblenz, 1832.—Lange, Casper's Woch., 1817, No. 23—26.—Schwarz, M. f. G., B. 18, Suppl., p. 121.—E. A. Meissner, M. f. G., B. 20, p. 40.—Ferber, Schmidt's Jahrb., 1863, B. 117 p. 179. (Referat über die Verh. d. Pariser Academie.)

PART VI

GENERAL PATHOLOGY AND THERAPEUTICS OF PARTURITION

It has already been mentioned in the physiology of parturition that the favorable mechanism of labour essentially depends upon two factors. They are (1) the due activity of the expelling forces; and (2) that the obstacles opposed to the forces are normal.

These two factors may overstep the normal limits in two directions. The expelling forces may be too strong or too feeble, and the obstacle too great or too small. Under proper supervision of labour even very strong expelling forces have no other effect than to terminate rapidly the act of parturition, whilst a very slight obstacle under the same conditions facilitates it without exerting any harmful influence.

The pathology of parturition, therefore, is formed (1) by disturbances produced by too feeble pains; and (2) by those produced by a too great obstacle. To these may be added a third pathological factor, which includes those accidents which do not interfere with the mechanism of labour itself, yet exert an unfavorable influence upon the mother or the child or upon both.

The resistance opposed to the expelling forces is determined by the relation of the object to be expelled—that is, the presenting part of the fœtus—to the parturient passages. Consequently a too great resistance may be due either to abnormal conditions of the parturient passages or of the child.

We have, then, the following scheme for the pathology of parturition:

I. Too feeble action of the expelling forces.

II. Too great obstacles. These may be caused either by—

1. The too great resistance of the parturient passages, either of—

   a. The soft parts; or,

   b. The hard parts.

2. Anomalies of the ovum, which, as a rule, depend on—

   a. The anomalous condition of the fœtus. These are—

      α. Too great a development;

      β. Abnormal position and attitude;

   b. abnormal condition of the fœtal appendages.

III. Dangerous accidents, which do not interfere with the mechanism of parturition. They are chiefly—
1. Pressure upon the umbilical cord when prolapsed.
2. Laceration of the parturient passages.
3. Hæmorrhage.
4. Inversion of the uterus (in the afterbirth period).
5. Eclampsia.

The treatment is in general on the same principles as in other branches of medicine. Careful attention is to be paid to prophylactic measures in order to avoid dangerous accidents. Anomalies of the expelling forces are chiefly prevented by a proper dietetic supervision during pregnancy, occasionally assisted by suitable therapeutic measures, for primary weakness of pains rarely occurs in women who at term are in perfect health. Too great resistance on the part of the soft parturient passages can during pregnancy be prevented by emollient warm injections. More successful still are prophylactic measures in contraction of the hard parts of the maternal passages. For although the origin of pelvic distortion be rarely and only in a slight degree under control, there is a valuable remedy in the induction of premature labour at a time when the disproportion between the head and the pelvis does not exist at all or only in a slight degree. Prophylactic measures in regard to some of the disturbances of parturition will be mentioned in the special pathology.

Of the means at our disposal to remove existing disturbances and limit at least their unfavorable influences, the mildest, if of the same efficacy, ought to be preferred. Thus, no powerful or dangerous measure should be taken if there is reason to expect the same result from a milder one. If, *e. g.*, the head has deviated from the pelvic inlet it should not be replaced by introducing the hand into the parturient canal if the altered position of the woman or external pressure will do the same. But, on the other hand, no valuable time is to be lost by employing mild or inefficient measures when more powerful ones, if immediately used, would have the desired effect, and which used after the mild ones have been tried come too late.

The obstetrician has also the whole list of remedies used in internal medicine at his disposal. Exceptionally only very important results are obtained by internal remedies; occasionally, however, they may prove very useful. Thus, in eclampsia parturientium measures purely obstetrical have only a noxious influence upon the disease, whilst complete narcotism cures it with absolute certainty.

The measures most frequently applied and most efficacious are operations more or less peculiar to midwifery. They are required by the peculiar conditions met with during parturition, which conditions differ widely from any other known state, and are seldom employed except in the treatment of diseases of women. Thus, the forceps may be very usefully employed in the treatment of large polypi, and the artificial dilatation of the os uteri is as often practised in diseases of women as in midwifery.

Obstetric operations are performed partly with the unarmed hand and partly by means of surgical or peculiar obstetrical instruments. It has here to be remarked that the hand is the most perfect instrument, and that in all cases in which the unarmed hand suffices instruments are to be banished.

Some operations, however, can only be performed by means of instruments, and it is necessary that the obstetrician be provided with them and carry them with him when called to a case. It is best to carry them in a leather bag, which is capable of containing also a few necessary drugs.

Besides a stethoscope, a male silver catheter (for a female catheter is insufficient in difficult cases), a pointed and a probe-pointed bistoury, a needle-holder, needles and sutures, every obstetric bag should contain the following instruments:

A pair of forceps of medium size.

A few elastic catheters of not more than 3·5 mm. diameter (for the treatment of asphyxia neonatorum), a few colpeurynters of caoutchouc to be used as tampons, with a suitable syringe, which can also be used as a uterine syringe.

Two slings for turning.

A scissor-shaped perforator.

A slightly bent and strong pair of Siebold's scissors.

Other instruments will usefully find a place in the obstetric bag, particularly when the case is somewhat distant from the residence of the accoucheur. They are—

A cephalotribe.

A blunt and half-blunt crotchet.

A pair of bone forceps.

A Braun's hook.

An instrument for replacing the umbilical cord.

Also the following medicines:

Chloroform, tincture of opium, morphia made up in doses, and fresh ergot of rye.

Before we consider the special pathology of parturition we think it necessary to say a few words on the various methods of obstetrical treatment, especially the operations. They may be arranged, like the pathology of parturition, under three heads, so that there are measures to be employed—

1. When the expelling forces are too weak.
2. When the obstacles are too great.
3. Against dangerous accidents which do not, however, interfere with the mechanism of parturition.

The treatment of the third section may be twofold. It may be directed against the dangerous accidents directly, or it may be employed to terminate delivery as rapidly as possible, in order not to allow time for the dangerous occurrences to exert their injurious influence.

The measures of the first class belong to the special therapeutics of the corresponding complications; those of the second are identical with those employed in weakness of the expelling forces, for, in point of fact, the expelling forces are too feeble to terminate delivery in the short time required for the safety of the mother or the child. The difference does not consist in an absolute anomaly of the expelling forces, but only in their insufficiency in the special case to terminate delivery in so short a time as to escape the dangerous influence of the complication. It is therefore a relative insufficiency of the expelling forces, and if their power cannot in a sufficiently short time be increased to the necessary degree the forces of nature must be replaced by others.

We have therefore only to consider the first two sections in the general obstetrical therapeutics.

When the expelling forces are too feeble, the most perfect treatment is to increase their activity, for which purpose there are external and internal remedies. If these do not succeed in strengthening the pains the necessary force is to be obtained from one or more other sources.

The effect of the pains consists in (1) dilating the soft parturient passages, (2) expelling the fœtus, and (3) expelling the afterbirth. There is only one way of replacing the triple effect of the pains, namely, pressure applied to the fundus uteri, the so-called "method of pressure." (By means of the Cæsarean section, by which a new parturient canal is opened, the fœtus and afterbirth are artificially extracted, and thus the expelling forces replaced in their triple effect). Pressure acts very uncertainly and slowly on the dilatation of the soft parts, somewhat better on the progress of the child, but with great certainty on the expulsion of the afterbirth. If this method is inapplicable, or what is more frequently the case insufficient, each individual effect of the expelling forces must be replaced by a definite operation with a view to that end. If the pains are unable to dilate the soft parts, artificial, manual, or instrumental dilatation of the os uteri is to be practised. If the pains are insufficient to expel the fœtus, the wanting or insufficient *vis à tergo* is to be replaced by traction applied to the presenting part of the fœtus—when the head presents traction is, as a rule, made by means of the forceps (exceptionally also when the pains are weak, after the head has been perforated for other reasons, by means of the cephalotribe)—when the breech presents traction is made by means of the bent finger or the blunt hook—when one or both feet present traction is made on them. In some cases, especially when the head is high up and movable, the forceps cannot be used. Since no suitable traction can be applied to the head in any other way, another operation, turning by the feet, is to be performed. This is therefore done in order to extract by the feet. The third effect of the expelling forces, the expulsion of the afterbirth, can be replaced by artificial removal with the hand.

All these measures have already been said to be applicable, not only when the pains are absolutely feeble, and consequently the delivery too much delayed, but also when the pains are normal and accidents occur which, threatening the life of the mother or that of the child, urgently demand the rapid termination of labour. Treatment is then instituted, not because the pains are abnormal, but because they are insufficient to terminate labour with the rapidity desirable in the special case.

We shall now consider the measures to be employed when the obstacles are too great, and shall begin with those caused by the maternal passages.

If the obstacle is caused by the soft parts we shall have to employ the same means as when the pains are not strong enough to dilate the normally constituted soft parts. In this case the soft parts only are altered, manual dilatation is as a rule useless, and it must be performed by means of instruments.

If the dystocia is due to too great narrowness of the hard passages their dilatation would be the most plausible treatment. This, of course, cannot be done without an operation, or only in very far advanced stages of osteomalacic softening of the pelvis, at the stage of the disease when the bony parts have just ceased to be hard. The contracted pelvic outlet may also be dilated in consequence of too great mobility of the pelvic joints; this has been observed in the case of a kyphotic, funnel-shaped, contracted pelvis. Independently of these extremely rare cases, attempts have been made to dilate the pelvis by dividing the joint of the symphysis pubis. This operation, however, though based on perfectly correct principles, has been entirely abandoned, because the dilatation of the pelvis by symphysiotomy is only very inconsiderable, and attended by great dangers and serious consequences to the mother.

Thus, the bony pelvis cannot be sufficiently dilated without doing serious harm. It is therefore necessary to lessen the disproportion between the pelvis and the head by diminishing the size of the latter. The methods by which we endeavour to attain this object are principally determined by this fact—whether the life of the child can possibly be saved or whether it must be sacrificed. It is evident that the first method, if applicable, is to be preferred.

The diminution of the size of the child's head is usually accomplished by the strong pains gradually forcing the head through the narrow parts of the pelvis. We may try to produce this effect in cases of feeble pains by an artificial increase of the uterine action; this, however, rarely succeeds to a sufficient degree. Unfortunately, the natural process can only be imitated in extremely rare cases. The forceps certainly compresses the head of the child in that diameter in which it has been applied, and the compression is to a certain degree without harm to the child. One circumstance, however, is adverse to the general application of the forceps for this purpose. In the immense majority of contracted pelves it is the conjugate diameter of the inlet which is shortened, and the forceps can only be applied at the inlet in the transverse diameter. It is therefore inapplicable for the compression of the head contained in the conjugate diameter. In very rare instances, where, with sufficient length of the conjugate, the transverse diameters of the pelvis are too small, the forceps may occasionally be applied successfully, and without risk, to compress the head in that direction. Those cases are, however, extremely rare, and there are usually no means of compressing the head in its diameter corresponding to the conjugate.

Another mode of treatment has been proposed to avoid the disproportion. It is to prevent the head of the child acquiring its usual size by putting the mother on a debilitating diet. This plan is now abandoned, since we are fully persuaded that the desired object is not attained by it, and that the mother is debilitated in the highest degree at the time of her delivery.

The interruption of pregnancy at a time when the head has not yet attained its usual size and the child is viable, the induction of premature labour, is based upon sound principles, and is therefore often and usefully applied in contractions of the parturient passages.

This last-mentioned measure is only prophylactic, and is no longer applicable when the physician is called to a parturient woman. Here all hope of saving the child has to be given up, and perforation is the only means by which the head of the child can be diminished. Ordinarily this operation will be resorted to when the child is already dead. In some cases cephalotripsy may be practised for that purpose even in a pelvis with a contracted conjugate diameter. For when this instrument has been applied in the transverse diameter and screwed up, it may be, owing to its small pelvic curve and its smaller size than the forceps, so turned that it, together with the compressed head, comes to lie within the contracted conjugate diameter. More frequently cephalotripsy is practised when, after previous perforation, the pains are insufficient to expel the head.

The obstacles opposed to the passage of the child may be so great that delivery per vias naturales is quite or at best almost impossible. A new way has to be opened up for the delivery of the child by dividing the abdominal walls and the uterus—the Cæsarean section.

The soft parts, too, when the pains are very slight, may also greatly impede labour. If by this the life of the child is brought into imminent danger, as long as the mother is alive the Cæsarean section will not be practised, because the operation is very dangerous to the mother and can rarely be performed in so short a time as to save the life of the child, which is already threatened. Matters are, however, quite different when, after the death of the woman, no special regard need be paid her, and even then it is an exception that the life of the child is saved. We have already spoken above of performing the Cæsarean section on the dead.

Far more frequently the Cæsarean section is practised in cases of greatly contracted pelvis. It must be practised when the child can neither alive nor in pieces be delivered per vias naturales. If the accoucheur is fully convinced that the still living child cannot pass through the pelvis without fatal injuries, the Cæsarean section may be performed to save mother and child.

Anomalies of the ovum, and especially of the fœtus, whilst the maternal passages themselves are perfectly normal, may also be the causes of considerable obstacle to delivery.

Those special manipulations required in certain anomalies of the fœtus or its appendages, and the treatment in diseases, malformations, faulty positions as prolapse of extremities of the fœtus, and the artificial rupture of the membranes, will be considered in the special pathology of parturition.

We shall here only mention the way in which we may help in faulty position of the fœtus. Since, as a rule, the fœtus can only pass through the pelvis when the cephalic or pelvic extremity presents, all deviations of those parts from the inlet, as well as complete transverse positions, require our aid unless nature herself rectifies the position.

It is evident that as long as it is possible to correct a position we must endeavour to do so. The position is rectified by turning the child so as to make head, breech, or one or both feet present.

If the correction of the position is impossible, the child in a transverse position cannot, as a rule, be extracted intact. The delivery of the mother per vias naturales is then only possible by cutting up the child (embryotomy).

The following will be a classification of the general treatment of parturition:

I. Treatment of (absolute or relative) insufficiency of the expelling forces.
  1. By the increase of too feeble pains—
    a, by internal remedies;
    b, by external manipulation.
  2. By the replacement of too feeble pains—
    a, by the method of pressure;
    b, by imitating some of the effects of the pains;
      α, artificial dilatation of the soft parts;
      β, extraction of the child by traction on the presenting part—
        1. In head presentation by means of the forceps;
        2. In breech presentation—
          a, by one or both feet;
          b, by the breech;
      γ, artificial removal of the afterbirth.

II. Treatment of too great obstacles.
  1. Anomalies of the maternal passages—
    a, of the soft, by incisions;
    b, of the hard;
      α, with the preservation of the fœtal life by induction of premature labour;
      β, without preserving the fœtal life—
        1, by artificial abortion;
        2, by craniotomy.

Appendix. Opening of a new passage by the Cæsarean section.
  2. Anomalies of the fœtus in its faulty position.
    a, correction of the position—
      α, by turning by the head;
      β, „ „ the breech;
      γ. „ „ the feet;
    b, extraction of the fœtus in faulty position rendered possible by embryotomy.

CHAPTER I

TREATMENT OF THE INSUFFICIENCY OF THE EXPELLING FORCE

I.—Increase of Too Feeble Pains

A.—*By Internal Remedies*

Ergot of rye undoubtedly occupies the foremost place amongst the medicines to which an influence on uterine action is attributed.

Its action, however, upon the expulsion of the uterine contents is not the same as that of a normal pain. It rather causes a uniform persistent and rigid contraction of the uterus, uninterrupted by any relaxation.

Schatz has shown by means of the tokodynamometer that after the use of ergot the internal uterine pressure is continuously and greatly increased during the intervals, and that the pains become more frequent but less efficient, until at last they entirely cease.

It is known that it is just this persistent contraction which brings the child into danger, for during each contraction of the uterus, as during a normal pain, the diffusion of gases between the maternal and the fœtal blood is not entirely stopped, but greatly limited, so that during a normal pain the sounds of the fœtal heart are less frequent; in very powerful and rapidly succeeding pains the child cannot recover in the short relaxation, but becomes asphyxiated and dies.

This process is very similar to what takes place in the energetic action of ergot. By the uniform contraction of the uterus, interrupted by no pause, the diffusion of gases at the placental insertion is impeded, and thus asphyxia is produced.

Besides, the uniform tension of the uterus does not materially favour either the dilatation of the maternal soft parts or the progress of the ovum, and therefore ergot is useless for the expulsion of the child and may be in some cases even dangerous to it.

But the case is quite different in the afterbirth period. Here all depends upon the placenta being separated from the uterus, and that, immediately after the separation, the uterus should remain so contracted that not much blood is lost from the open vessels; the expulsion of the separated placenta is, as a rule, effected without difficulty.

We have learnt in the "Physiology of Parturition" that the placenta is separated from the inner surface of the uterus when the place of its insertion is greatly diminished by the contraction of the uterus.

This effect, and still more the occlusion of the open vessels, is most certainly obtained by a persistent uniform contraction of that organ, and this action completely corresponds to that produced by ergot. It is consequently indicated when there is reason to suppose that the uterus, after the expulsion of the child, will badly contract. It may be given in all those cases where the expulsion of the child is almost completed, but it must never be given at a time when labour cannot be terminated at any given moment.

It has already been stated that ergot is hardly able to contribute to the expulsion of the child, and, besides, towards the end of labour we have far more efficient means of acceleration. Nevertheless, when used in the afterbirth period it is a valuable remedy, because its action is rather rapid, generally causing contractions after ten minutes.

The use of ergot after expulsion of the placenta, in order to stop or to prevent bleeding, belongs to the pathology of the lying-in state. It is usually given in powder in doses of from sixteen grains to half a drachm (one to two grammes), or it may be given as an infusion or decoction of one drachm to two ounces (four to sixty grammes), in doses of a table-spoonful. Kept for a long time it loses its efficacy.

Borax and tincture of cinnamon have also been recommended for their influence on the uterine action, but their efficacy appears to be doubtful.

Opium deserves to be far more recommended, especially when feeble pains are due to great nervous excitement. Experience is decidedly in favour of it. After one or two doses of one grain of opium the effect is often extremely evident and precise, and strong and uniform pains follow. If the spinal sacral nerves are considered the inhibitory nerves of the uterus, the action of opium may be explained by their paralysis. This explanation can be only hypothetical, but this much is certain, that in many cases uniform and marked pains are produced by no remedy better than by opium.

Chloral also appears to act in a very similar way. Gerson da Cunha has given it in tardy and exhausting labours. After a few hours' sleep, on awakening the labour was very rapidly terminated by powerful pains. We have also observed that by the use of chloral in cases where the uterine action was very painful without being efficacious, the labour assumed an instantaneously rapid course, although the intervals between the pains had considerably increased in duration.

B.—*By External Manipulation*

Almost all the means, by which during pregnancy uterine action can prematurely be produced, may also be employed during labour for the same purpose. Foremost stands the introduction of an elastic catheter between the uterus and ovum. The catheter thus introduced produces during labour uterine contractions with the same certainty as it does when employed to induce abortion. The effect is especially evident as long as the membranes are intact, but we must agree with Scanzoni that this may also be the case when the liquor amnii has partially drained off. At any rate, as long as the membranes are still intact the catheter deserves diligent application, because its action is certain and not injurious.

The warm douche applied to the lower uterine segment is also a very valuable remedy. It is especially efficacious at the commencement of the period of dilatation, and at the same time it relaxes the soft parturient passages. It will be especially useful in those cases where, to the despair of the relatives, the first stage of the period of dilatation is protracted, and thus, through the premature rupture of the membranes, the obliteration of the cervix and the dilatation of the os have to be waited for a very long time.

The artificial rupture of the membranes is in many respects an excellent proceeding to hasten the progress of labour, for the contractility of the greatly stretched uterine muscles may have been impaired by the membranes being overfilled with liquor amnii, so that efficient pains set in only after a part at least of the liquor amnii has drained away.

Still more important, however, is the rupture of the membranes by the mechanical influence of the expelling forces upon the presenting fœtal part. We have endeavoured to show that the internal uterine pressure before the rupture of the membranes only acts on the expulsion of the entire ovum, and not on the progress of the presenting part, and that the efficacy of the form-restitution power is impeded by the intact bag of waters. But if after rupture of the membranes the presenting head itself forms the lowest part of the ovum, it is, as such, moved forwards by the internal uterine pressure, whilst the form-restitution power expels it from the ovum. These two forces, acting on the expulsion of the fœtus, can develop their full power only after rupture of the membranes.

From this consideration the precise effect of the rupture of the membranes will be understood in those cases also where the ovum is by no means overfilled, and the artificial rupture of the membranes therefore deserves to be considered as a principal agent in accelerating tardy labour. This operation may always be performed, though the os be only slightly dilated, when the lower uterine segment is firmly applied to the head. If this is not the case, on account of the possibility of a prolapse of the funis, the membranes can then be only ruptured when, in case of need, labour can be terminated artificially. Sometimes too firm adhesions between the membranes and the lower uterine segment may impede the retraction of the latter over the ovum, and only when the bag has ruptured can the os together with the membranes recede over the presenting part. In such cases the external membranes, decidua and chorion, which are firmly adhering to the lower uterine segment, rupture within the internal os; so that these two, together with the lower uterine segment, recede over the ovum, the continuity of which has been preserved by the amnion alone, the latter only forming the projecting bag of waters. In the expelled ovum it is often found that the amniotic sac is separated to a very great extent from the inner surface of the chorion.

Plugging of the vagina by means of the colpeurynter is very useful when there is reason to fear the premature rupture of the membranes. Plugging retards rupture by counter-pressure against the bag of waters, and at the same time it excites uterine action.

Friction of the fundus uteri may also be employed to augment the activity of the uterus, and it appeares especially efficacious either when the head of the child appears in the vulva, or in the afterbirth period. But since it is usefully combined with pressure, we shall not speak more of it here.

Other means, such as electricity, friction, and dilatation of the os by means of the fingers, are far more uncertain; moreover, they are painful, and even not without danger, so that their use is best entirely omitted.

II.—The Replacement of Too Feeble Pains

A.—*By Imitating the Effect of the Pains by the Method of Pressure*

The method of delivering the child by pressure consists of the following manipulation.

The woman lying on her back, and the operator sitting by her side, the uterus is brought as near as possible to the anterior abdominal walls, and any intestinal coils which are between are gently pushed away. Then the fundus uteri is grasped with both hands, so that the thumbs are placed in front and the palms as deeply as possible on the posterior surface, with their ulnar sides directed towards the pelvis. Gentle friction is gradually continued into strong downward pressure, which is to last from five to eight seconds, and then gradually decrease, as in a natural pain.

After a pause of a half to three minutes, according to the necessity of the case, pressure is again exerted in the same way at a different place, and this compression repeated ten, twenty, or forty times if required.

By external pressure made in this way strong expulsive pains are as easily obtained as by mere friction of the fundus, and, on the other hand, the pressure acts in a very similar way to the contractions of the abdominal muscles. This method, therefore, is to be considered as a replacement or increase respectively of all the forces acting on the expulsion of the child. Generally this method is to be applied when the expelling forces are abnormally weak, when the uterine action has not sufficient effect, or when with rather feeble pains the activity of the abdominal walls is for some reason entirely wanting or insufficient. By it it is intended gradually to terminate labour in just the same way as nature does. Of the effect of the pressure one may very easily be convinced when the os is dilated and the head is within the pelvis, if pressure is made by one hand, and the finger of the other is placed on the head. The latter is distinctly felt to move downwards with each compression, and to press on the pelvic floor. Of course, it recedes again when the pressure is discontinued.

The effect, therefore, is quite similar to that produced by the abdominal pressure. This procedure has not only the advantage of being easily applied and perfectly safe, but also that it imitates as closely as possible the process of nature. On the other hand it acts slowly, often causes pain, and only succeeds in the milder cases. Most probably the pain complained of is due to actual labour-pains, which are excited by the procedure. The women attribute it to the manipulation, and object to it for the same reason as they unwillingly suffer the finger to remain in the vagina during a pain. The procedure is, of course, more easily carrried out in women with thin and flabby abdominal walls, than in those in whom the walls are fat or tense. The method is seldom employed for the dilatation of the os, for it acts too slowly, and, if continued too long a time, greatly fatigues the operator. It will, therefore, as a rule, only be practised when there is a prospect of a speedy termination of labour.

In cases in which, on account of danger to the mother or child, artificial delivery must be effected as rapidly as possible, it is equally contra-indicated, because the various methods of extraction, without doubt, give the same results in much less time. Although Kristeller, in a case of breech presentation, in which the child, already born up to the shoulders, was in danger of being asphyxiated, completely delivered it in two minutes by means of five artificial compressions, yet it must not be forgotten that the same result could have been obtained by manual extraction in less than half a minute.

However, in a case of head position in which instantaneous delivery is indicated and there are no forceps at hand, the energetic application of the method of pressure is then to be recommended.

It will prove highly useful to combine pressure with extraction in true breech presentations, for in at all difficult cases the effect of the traction is slow and uncertain, and by the application of the *vis à tergo* the arms remain applied to the chest, which by mere traction are easily turned upwards.

Pressure upon the fundus will often, doubtless, be usefully combined in difficult extraction by means of the forceps, and this combination is especially recommended in the extraction of the aftercoming head, where it renders excellent service just in the most difficult cases.

In disproportions of the pelvic cavity the procedure is not strong enough. It deserves, however, a far more extended application, and the so-called Credé's method is especially useful for the squeezing out of the afterbirth.

B.—*By Imitating the Individual Natural Effects of the Pains*

(*a*.) *Artificial dilatation of the soft passages* (*accouchement forcé*).—We shall speak here only, as is shown already by our classification, of the artificial dilatation of the normally constituted os uteri.

The normal os pretty rapidly dilates if the pains are at all strong, and never impedes the further progress of labour. In cases, however, where a rapid delivery is indicated, and where the pains, though strong, are not sufficient to end labour in the time necessary for the preservation of the life of the mother and the child, we may be forced to have recourse to artificial dilatation of the os as the first step in the performance of delivery (*accouchement forcé*).

Artificial dilatation of the os, unless we rashly tear or cut the soft parts, is impossible before the setting in of labour, at which time the os is only slightly dilated, if at all, and the cervix not yet obliterated. There is, moreover, no indication which would require the performance of such an operation upon a pregnant woman. Eclamptic convulsions, which disappear with certainty under chloroform, are violently increased by operations on the os, and in a pregnant woman who has just died the Cæsarean section is the only way to save the child, since for the most part the child dies a long time before it is possible to extract it by the natural and yet undilated passages

If profuse hæmorrhage threatens the life of the mother it is absurd to attempt the *accouchement forcé*, for the bleeding still continues during the time of the operation. The os then can only be dilated in a short time, when nature has already taken the initiatory steps.

In considering the applicability of the operation attention must first of all be directed to whether the woman is a primipara or a pluripara. In primiparæ it is an absolute necessity that the cervix be completely obliterated, that the internal os be entirely dilated, and that the edge of the external os be sharp. If, as is frequently the case, the lower uterine segment is already greatly thinned, artificial dilatation of the still impervious os may be performed in case of need. Of course the operation is the easier the more the os is dilated and the thinner the lower uterine segment.

In pluriparæ also the cervix must be obliterated, at least to a great extent, but here a still very puffy, tumid os sometimes dilates comparatively easily. Since the operation is only performed to allow the extraction of the child, and since the advancing fœtal parts themselves continue the farther dilatation, the operation is finished as soon as the presenting part can be drawn through. In breech presentations the dilatation is therefore sometimes already sufficient when one foot can be drawn through the os ; in head position, when the blades of the forceps can be passed through it. As soon as extraction is commenced the os is dilated by the part passing through it.

The operation is performed either by the hand or by means of instruments. In pluriparæ, with the os still puffy but extensible, the first method is the best ; in primiparæ, where the edges of the os are very sharp, the latter. Sometimes both methods are usefully combined. For manual dilatation as many fingers as possible are introduced into the os, and in this way it is dilated. For instrumental dilatation a blunt-pointed bistoury is necessary, or, still better, a pair of scissors curved on the side are introduced with the blunt blade between the ovum and the os. The latter is incised in two, three, or more places. The incisions need not be too deep, but should be numerous and in different places.

It may be necessary to perform the operation in the interest of the mother or the child. The first indication is almost exclusively given by hæmorrhage, and we shall speak about it in the special pathology of parturition. In the interest of the child it becomes necessary when in a still insufficiently dilated os any circumstances bring the life of the child into danger. This indication must be very carefully considered, whether the *accouchement forcé* can be performed with the necessary rapidity without risk to the life of the mother. In the case of the sudden death of a parturient woman the *accouchement forcé* is indicated when there is reasonable hope that by it the extraction of the child can be more rapidly effected than by means of the Cæsarean section. Otherwise, since through the death of the mother the life of the child is solely in question, the Cæsarean section is preferable.

The prognosis in cases of hæmorrhage is favorable for the mother, for it is then only practised when it can be successfully performed. It is known from experience that the incisions do not tear more, nor are they attended by any great loss of blood.

Manual dilatation is, under the circumstances above mentioned, almost always easy and does no harm. The operation itself does not at all affect the child.

(β) *Extraction of the child by drawing on the presenting part.*—1. *In head presentations by means of the forceps.*—The forceps consists of two blades or arms which cross one another, the upper half of which, when applied and locked, is intended to grasp the head on two sides like two thin iron hands. The lock of the blades is so arranged that they may easily be separated and again brought together. Each blade has an upper portion, the "bow" (applied to the head of the child), and a lower portion, the handle. The blade which is introduced with the left hand, and comes to lie on the left side of the mother, is called the left blade, and the other the right blade. In German forceps the lock is on the left blade. A good forceps must have the following qualities:—The blades must be neither too long nor too short. If too short the lock comes to be within the genital fissure, and when the head is not very low down this renders the locking of the instrument difficult. Traction also with a very short handle is inconvenient and not practicable without great force. The too long forceps is heavy and unwieldy, and its great length is unnecessary, since the forceps is advantageously applied only when the head stands in the small pelvis. The blades must be made of good steel, and are usefully provided with fenestræ, the sides of which must be free from sharp edges and slightly convex inwardly; the fenestræ make the instrument lighter.

A great deal depends upon suitable head and pelvic curves. As regards the head curve of both blades, the tips of the bow must not meet when the handles of the forceps are brought together, but should be distant more than 1 centimètre or $\frac{1}{3}$ of an inch, and the greatest distance between the bows must be about 7 centimètres, or 2·7 inches; greater or less compression of the handles allows a firmer or looser grasp of the head contained within the forceps. But the handles should by no means be too firmly compressed, and it is always to be borne in mind that by a close approximation of the handles a medium-sized head is exposed to a compression which only very rarely will not be injurious. The pelvic curve need not be very great, and is gradually to increase from the lock upwards to the tip of the bow. The kind of lock also is of very great importance. In English forceps the union is effected by a projecting bar, and they are distinguished by the facility with which the blades can be locked. At the same time, however, they can easily slide along when locked from above downwards. In French forceps the locking is difficult, because the hole in the one blade has to be brought into accurate opposition with the pivot of the other. Once locked, however, the blades are firmly connected.

In German forceps, as designed by Brunninghausen, the advantages of both locks are united. It is very much like the English lock, only on the left blade there is a pivot surmounted by a flat knob which fits into a notch in the other arm. By that contrivance the forceps can quite easily be locked, and the locking is secured by the projecting knob and by the pivot fitting into the notch.

The handles are best made of metal covered with a thick layer of wood. A groove at the lower end and two hook-shaped projections near the lock, as shown in the Nægele forceps, greatly facilitate its use. The weight of the forceps is not to be more than a pound and a half.

The forceps, by traction applied to the head, is intended to replace the expelling force which in normal labour pushes the head forwards. For that purpose the head contained within the forceps must preserve as much as possible the faculty of accommodating its shape to the pelvic space, and the forceps must compress the head as little as possible. A firm grasp of the head is, of course, not obtainable without a certain amount of pressure; but, after all, compression is only a secondary effect, and the ideal forceps would be such as allowed a sufficiently firm hold without exerting the least pressure upon the head.

Where the progress of the head by means of the forceps is not opposed by too great obstacles, traction alone in the direction of the pelvic axis perfectly suffices to deliver the head. The forceps itself does not act as a lever, but simply represents an artificial prolongation of the head on account of the handles by means of which the head is extracted. In difficult cases the efficacy of traction can be considerably increased when lateral oscillations are made with the forceps; then the forceps ceases to be a simple handle, and becomes a lever. For if we suppose the forceps to be immovably united to the head standing in the pelvis, the left blade, for instance, being fixed against the left wall of the vagina, and by simultaneous traction of the handles the forceps are drawn over to the left side of the woman, the right blade must leave that part of the vagina where it before lay and descend somewhat lower down. The forceps, then, represents a one-armed angular lever, one part of which is formed by the greatest diameter of the forceps, and the other is about in the direction of the right blade. The fulcrum is situated at the spot where the left blade is applied to the vagina, and the fulcrum is fixed by simultaneous traction on the handles. The weight acts at the place where the right blade lies against the vagina, and the force is applied to the handles of the forceps. If the fulcrum is steadied by traction the other blade descends by the movement of oscillation towards the side of the fulcrum. In the corresponding movement to the opposite side the point where the weight before acted now becomes by traction the fulcrum and *vice versâ*, so that now the forceps, and with it the head, descends along the other side of the vagina. In this way the forceps, and with it the head, descends somewhat lower down by continuing the oscillations from the one to the other side of the vagina.

Rotatory movements have a very similar effect, and they are made by describing a circle with the handles of the forceps. These leverage movements, however efficacious, are always to be executed with the greatest care, for the maternal soft parts at the side to which the handles are pressed are exposed by them to great pressure.

It is therefore well to disregard, in a case in which moderate traction suffices, the leverage power of the forceps.

Another action of the forceps is the dynamic, arousing uterine action. Although it cannot be denied that in some cases, immediately after the introduction of one blade, the pains until then feeble, become more efficacious, yet in the majority of cases that effect is wanting, or is at least so slight as not to be of any use.

To make a proper use of the forceps, that is, to effect the extraction of a healthy child without danger to the mother and with the greatest possible security, it is necessary to attend to the following conditions:

1st. The maternal soft parts must be sufficiently prepared, *i. e.* the cervix uteri must be obliterated, and the os so far dilated that the head can easily pass through it.

2nd. The membranes must have already receded beyond the head, and the liquor amnii in front of the head must therefore have escaped. For if the forceps be applied to a head still within the membranes, the insertion of the ovum would be dragged upon, and thus premature separation of the placenta would be caused.

3rd. The head must have approximately the normal size and firmness, since the curvature of the forceps is constructed for that. The instrument cannot be applied to a hydrocephalic head, nor to the head of an immature or of a softened or putrid child. The head also artificially lessened by perforation cannot well be grasped by the forceps.

4th. The head must have already entered the true pelvis, and stand, so to say, ready for the application of the forceps.

5th. The small fontanelle must be directed forwards.

If all these preliminary conditions are fulfilled the application of the forceps, and the extraction by means of it, in the great majority of instances, will not present any difficulty, and the result will be equally favorable to the mother and the child. Cases, however, may occur where all of these five conditions cannot be complied with, and where extraction by means of the forceps appears to be extremely desirable. The use of the forceps is then not contra-indicated in all cases, but what is always to be kept in mind is, that extraction is made with some difficulty under the exceptional circumstances which increase the responsibility of the operator.

The second condition, that the membranes must have previously been ruptured, is easily complied with when the waters have not already spontaneously escaped. As regards the first condition the forceps may be applied in an urgent case as soon as the blades can be passed through the os. For its advantages are extensive; the os will be dilated by the traction itself. If this is not the case, or if the opening of the os is still smaller, and the application of the forceps urgently indicated, the soft parts must previously be artificially dilated.

The third condition must be strictly adhered to, for the forceps very easily slips over abnormally shaped heads, and the young practitioner does well when he strictly adheres also to the fourth condition. It requires a great deal of obstetric experience to estimate with any certainty the difficulties which are opposed to extraction when the head is still high up.

In this respect it is of great importance to know the relation of the head to the pelvis. If the head is not too large, and the pelvis not contracted, sometimes the forceps may be advantageously applied to the head, which still remains movable above the inlet; it is well understood that the application can only be made with any safety when the head is fixed, but the still movable head may easily be fixed externally by the hands of an assistant, so that in this respect the application of the instrument does not offer any great difficulty. But even under such conditions violent traction must be avoided, for the forceps can only lie transversely in the inlet, and the head, if possible, always so turns within the forceps that the blades lie on each side. The still movable head then so turns that the sagittal suture almost runs in the conjugate diameter, and one frontal bone comes to lie close to the promontory.

Unless the dimensions of the pelvic cavity are uncommonly great the frontal bone of the child may, if powerful traction be made, be broken against the promontory of even a normal pelvis. Whilst the head is still immovable above the brim version is easy, and this should always decidedly be preferred.

The fifth condition is very desirable, but not absolutely necessary. Though the small fontanelle may be directed to one side, or even more backwards, extraction by means of the forceps is still possible without much difficulty.

Whether the above conditions are strictly adhered to or not entirely depends upon the reasons on account of which the operation is undertaken.

The indications for the use of the forceps are—

1st. When, with not abnormal obstacles, the pains are so feeble that they cannot effect the expulsion of the child at all, or at least not in so short a time as is desirable for the safety of the mother or the child.

2nd. In cases where dangerous accidents occur which require immediate delivery in the interest of the mother or the child, and where the application of the forceps is easy.

It can easily be seen that in the second case also the uterine action is unable to terminate labour in the short time desired, and the indications for the application of the forceps may shortly be considered as absolute or relative feebleness of the pains, *i.e.* the pains are either too feeble in themselves or for the special condition of the case to terminate labour in the short time desired. Abnormal obstacles rarely indicate the use of the forceps. Those caused by the os or by a too narrow genital fissure are best removed by an incision. The forceps may occasionally be used with success in general narrowness of the vagina, as is often met with in primiparæ, for by the traction of the forceps the dilatable vagina is distended. Still, the application of the forceps in a too narrow vagina requires caution, for lacerations of the vaginal membrane easily occur.

In too great obstacles of the hard parturient passages, since they cannot be dilated by application of the forceps, this instrument can only have a favorable influence when it can be so applied that the diameter of the head situated in the contracted space of the pelvis can be diminished. In the great majority of cases of contracted pelvis (the flattened) this is never possible. In the extremely rare cases of transversely contracted pelvis the forceps may, perhaps, be occasionally applied with advantage. However, it would of course require very great caution, because the iron blades of the forceps would then lie at the narrow part, just between the pelvis and the head, and would exert a very unfavorable influence upon the maternal soft parts, which have already been greatly compressed.

The first of the above indications may be variously met with in practice. However, it is too much limited by those who would only allow the application of the forceps when there is evidently danger to the life or health of the mother or child. Where the expulsive stage is too much protracted, the loss of strength and the excitement of the woman are greatly increased, moreover—since the chances of the child are, *cæteris paribus*, greatly lessened by the protracted labour—we deem it justifiable not to postpone the application of the forceps until the life of the mother is seriously endangered, or until the frequency of the fœtal heart sounds has permanently diminished, but at once to fulfil the happiest duty of the physician by abbreviating the enervating labour and the pain of the mother by employing the forceps, provided always that the abbreviation itself can be made without danger to the mother or the child.

The young practitioner may have difficulty in accurately estimating the circumstances, while the more experienced operator grasps them with greater ease. At any rate, under such conditions, the least practised can only apply the forceps when the head is very low and all the other preliminary conditions are complied with.

As regards the operation itself, we shall only give here a general outline of the method of application. The more intimate acquaintance with the individual manipulations is obtained only on the dummy, and sufficient dexterity and practical experience by repeated operations on the living subject.

When the bed on which the woman lies is accessible on both sides, in most cases where the head is low down the operation may be performed in the dorsal posture, the nates being somewhat raised by a pillow. If after the application of the forceps extraction becomes very difficult, the woman may slide down somewhat towards the foot-end of the bed, and extraction may very easily be made by the operator standing at the foot.

If one side of the bed stands against the wall, and cannot be well removed from it, extraction may also be made from above the foot-end of the bed, where the woman is somewhat obliquely placed, so that the foot lying over the free edge of the bed is placed upon a chair, or the forceps may be applied with the woman in the lateral posture. Under such circumstances the dorsal posture is unsuitable for the application of the forceps, and when the head is very low down only possible with enormous difficulties. A cross bed is only necessary exceptionally, when great difficulties are expected in the application of the forceps, and afterwards in extraction. The blades of the forceps are to be slightly warmed and well oiled.

The operator prepares some clean towels, and again accurately examines the position of the head. This is a golden rule, and ought never to be neglected, for in the mean time the head may have altered its position, and occasionally also previous errors in the diagnosis may be discovered.

In the dorsal posture the operation can be performed without any assistance. It is, however, more convenient to have an assistant at hand. If the operation is performed on the cross bed two assistants at least are required, the one to hold the legs and the blade already introduced, the other to attend to the woman.

When the head is low down, and the small fontanelle directed forwards and to one side, the forceps is applied in the following way.

The operator stands at the right side of the bed, and takes the left blade in his left hand like a pen. In the interval of a pain he introduces two fingers of the right hand into the vagina, and passes them up on the left side of the pelvis along the head. Guided by the fingers of the right hand the blade is then introduced by the left from above the symphysis, and passed along the head towards the sacro-iliac articulation (if the edge of the os can still be felt, of course, between the head and the os). Under no condition whatever is the blade to be pushed violently upwards. If it is gently pushed forwards, and the hand is not too soon lowered, the blade usually passes without difficulty along the head so that the latter comes to lie in the concavity of the blade. Now the operator comes to the other side of the bed. He takes the right blade in the right hand, introduces the left hand into the right side of the pelvis, and proceeds in the same way as just described. Both blades are now situated somewhat behind, and in order to lock the forceps either both or at least one of the blades must be brought forwards; in the first head position the right-hand blade. This succeeds easily when the handles are depressed towards the perinæum. Usually the locking of the forceps then becomes easy. Otherwise each handle is grasped and the blades accommodated to each other; by slight traction it may be ascertained whether the forceps is well applied, and now the first part of the operation, viz. the application of the forceps, is terminated.

The head is extracted by the operator making steady traction upon the forceps, and by means of it upon the head. Since the course of the genital canal from above downwards is curved forwards, traction, if it is to be efficacious, must be made according to the position of the head and the various apertures of the pelvis in different directions. If the head is still high up the direction of the traction must be accordingly downwards, and the more the head approaches the genital fissure the more must the handles of the forceps be raised corresponding with the pelvic axis, so that the handles will be above the symphysis when the head appears in the vulva.

Traction must never be sudden, but gradually increasing, persisting for a time in uniform intensity, and then again gradually decreasing. The greater number of lacerations in obstetric operations are due to jerky traction, the degree of which is not able to be controlled. A steady and continued traction in the same line is the most effectual.

In difficult cases gentle oscillations are permissible, only it must be remarked that they as well as the rotations are not without influence upon the maternal soft parts.

The rapidity with which extraction is performed essentially depends upon the cause which called for the use of the forceps. If the life of the mother is in danger, or, as is more frequently the case, that of the child, extraction must be performed as rapidly as possible, and, therefore, continued traction must be made upon the forceps.

If there is no immediate danger, or if the forceps has only been applied on account of deficient expulsive force, it is best to imitate the natural course of labour as much as possible.

Traction is to be begun at the commencement of a pain and gradually increased; as the pain passes away the traction is diminished. The hold on the forceps is to be maintained in the intervals of the pains, especially when the passage of the head is arrested by a narrow vagina and the elastic pelvic floor, for otherwise the head recedes during the pause. After some time, and if possible with a fresh pain, traction is again made and repeated at intervals until the head appears in the vulva.

It will be shown below that the pressure of the forceps upon the child's head produces respiratory movements, and may thus cause asphyxia; and it is therefore necessary, when extraction is slowly performed, to examine from time to time the fœtal heart sounds.

We must hasten the extraction, not only when the frequency of the heart sounds is considerably decreased, but also when it is increased up to 160, 180, or more.

The pressure upon the handles in traction need only be very slight. The stronger the traction the greater also must be the compression of the handles, in order to prevent the sliding of the forceps; but here, again, it is to be remembered that the hand which compresses the handles at the same time compresses the child's head. When the head appears in the genital fissure, lacerations of the perinæum can be more easily avoided than in natural labour, because the head grasped by the forceps can be completely controlled. If the operator stands on the right side of the mother he holds the instrument with the left hand, and if a strong pain is about to bring the head through the ostium vaginæ, he presses it back into the vagina. When the pain has passed away the edge of the vulva is pushed backwards over the head with the right hand, so that the head now appears in the vulva during an interval of the pains. If the edge of the vulva is very much stretched it is best to make one or two incisions in it.

After the birth of the head the blades are removed by drawing the handle of each blade somewhat downwards; the rest of the labour is conducted in the ordinary way.

Neither the application of the forceps nor the extraction materially differs when the head is high up, provided the pelvis be normal. Since the forceps can then only be applied in the transverse diameter, it may therefore happen that the head is grasped by the forehead and occiput. In the contracted pelvis the use of the forceps is contra-indicated until the head has passed through the contracted brim.

Procedure when the large fontanelle is directed forwards:—As long as the large fontanelle is directed forwards the forceps are not to be applied, for, by their application in the opposite oblique diameter, the rotation of the head would be prevented. We must, therefore, wait until the small fontanelle has rotated forwards, or, at least, until the sagittal suture runs in the transverse diameter. An exception to this must only be made when the life of the mother or the child is in danger. The forceps are then applied in the oblique diameter of the pelvis, in which the sagittal suture does not run, that is, when the small fontanelle is directed backwards and to the left, in the right oblique diameter. If traction is now made the forceps lies more transversely, and the small fontanelle rotates more backwards. By this the rotation of the small fontanelle forwards is no longer possible. The head is then extracted with the large fontanelle directed forwards in the same way in which it is exceptionally expelled by the forces of nature. Extraction is always here more difficult than when the occiput is directed forwards. It is, however, facilitated by traction being made downwards until the forehead appears beneath the pubic arch, and then only raising the forceps in order to make the vertex and occiput sweep over the perinæum.

In face presentations the forceps ought never to be applied when the face is so high up that it is still possible that turning can be performed. Even in head positions, in cases in which both operations are practicable, turning is always to be preferred; this is still more so the case with face presentations.

When the face has entered the pelvis the chin turns, as a rule, forwards. The application of the forceps to both sides of the face in an oblique or the transverse diameter of the pelvis does not under such circumstances meet with any difficulties. In extracting the handles of the forceps must not be raised too early. When the chin appears beneath the pubic arch, the head is then raised over the perinæum.

As long as the longitudinal axis of the face is in the transverse diameter the application of the forceps is to be avoided, if at all possible. If applied in urgent cases the concavity of its pelvic curve must be directed to that side of the pelvis to which the chin looks, that is, when the chin is directed to the right in the right oblique diameter. When the face has descended deeply into the pelvis it will easily turn within the forceps.

It is rare for the chin to remain directed backwards after the head has completely entered the pelvis. Extraction in that position is much less possible than its expulsion by natural efforts. If delivery is absolutely necessary at that time and the child is alive, we may try to turn the face by means of the forceps in the way recommended by Lange or Scanzoni. The manipulations are, *mutatis mutandis*, just the same as in head positions with the occiput backwards.

As, in general, the greatest caution is required in the use of the forceps in face presentation, so it is especially in the cases just mentioned that forcible attempts must be avoided. If delivery is demanded for the sake of the child, it perishes during the operation unless the rotation is easily and rapidly accomplished. Then the natural rotation may be waited for, or, if necessary, the dead child may be perforated. But if the mother is in danger, and delivery must be immediately accomplished, certainly long-continued attempts at rotation will greatly increase the danger. The sole chance in such cases of saving the mother consists in perforating the living child unless the rotation of the face succeeds relatively easily and at once.

The prognosis is decidedly favorable for the mother and the child when the forceps is skilfully applied and extraction cautiously made in those cases, which are the ones most commonly met with, where the head is low down and the os dilated.

Occasionally an accurate examination may show, even after a cautious use of the forceps, that the mucous membrane of the vagina has been slightly torn. These rents, however, are due to the great distension of the vagina by the head; they occur also in natural deliveries, and heal without producing ill effects. Lacerations of the vulva may by skilful manipulation of the forceps be more easily avoided than in the usual management of spontaneous labour.

Often after very slight operations and the cautious use of the forceps the head of the child has distinct impressions of the blades, and even slight abrasions of the skin are very frequent. Yet both injuries are not of the slightest consequence to the child. Even in paralysis of the facial nerve, which is occasionally produced by the pressure of one blade, the prognosis is decidedly favorable.

The results of a forceps operation are bad, even in a normal pelvis, when great force is required for extraction, and when the blades have to be applied somewhat in the oblique diameter. The soft parts of the mother are thereby subjected to very severe contusions, and the effect is the more injurious since in such cases usually the soft parts have been exposed to pressure for a long time previously. The child too is in danger as soon as considerable force is employed in extraction. Increased traction necessitates a firmer grasp and greater compression of the handles lest the forceps slips off the head. Consequently the compression to which the head of the child is exposed is greater. The injuries which the head in such cases suffers externally (superficial dermatitis, sugillations, and even circumscribed gangrene) are known to greatly endanger the child's life. But the changes within the head after too great compression are of a far more unfavorable prognostic significance. Lacerations of the venous sinuses and of other blood-vessels of the cranial contents may result, and doubtless also the brain may be exposed to such a pressure by the blades of the forceps as to produce a considerable retardation of the pulse by irritation of the vagus. The effect of this upon the placental circulation may be so injurious that the child becomes profoundly asphyxiated and dies.

According to Hecker children may rapidly die from the pressure of the tip of the blade upon the umbilical cord twisted round the neck, or upon the large cervical vessels. Besides, Pappel's statistics show that the use of the forceps renders the prognosis unfavorable. Of 102 children on whom the forceps were applied without any other complication 61 were born perfectly healthy, 36 asphyxiated (of whom 6 died), and 5 were still. This shows a death-rate of 10·8 per cent. of the children where, without any complication, labour had been terminated by means of the forceps.

The application of the forceps in a contracted pelvis may be followed by the saddest consequences to the mother and the child. Under such circumstances Hugenberger says, 70 per cent. of the mothers became ill and 30 per cent. died. Children whose heads are exposed to a double compression, laterally by the forceps and from before backwards by the contracted pelvis, also die very readily. We have already shortly mentioned the reasons why we do not recommend the use of the forceps when the conjugate is narrowed. Nor can the forceps be applied to a head which has entered a pelvis whose size is generally diminished without the worst results, especially if forcible traction is required The pressure may cause extensive gangrene of the maternal soft parts. Caries of the pelvic bones, obliteration of the vagina with consecutive hæmatometra and vesico-vaginal fistulæ, may result. Nay, if very great force is employed the joints of the pelvis may be torn asunder and the bones broken.

Another danger, the slipping of the forceps, can only occur from the head being too high up if the forceps are at all carefully handled; in such cases we must distinctly dissuade from the use of that instrument. The forceps may slip in a vertical and in a horizontal direction; the latter is likely to occur when the head is movable above the brim, and the former when the head is low down and the handles are not sufficiently compressed. If the instrument suddenly slips and is forced out of the vagina it may inflict very severe injuries upon the mother.

Such lesions as piercing the vaginal walls by the introduction of the forceps cannot occur at the present time, when obstetricians receive a scientific education.

2. *In breech presentations.—a. Extraction by one foot or both.*—The conditions which give both mother and child the surest prospect of a favorable issue when extraction is made by the feet are—(1) that the soft parturient passages be sufficiently dilated; and (2) that there be no disproportion in size between the child and the pelvis.

Under those conditions manual extraction succeeds in a very short time, and the child is born without having prematurely made respiratory movements (of course, provided always that at the commencement of the operation the child has been perfectly healthy). The result of the operation is still more rapid when traction on the presenting part is assisted by strong pains.

However desirable the above conditions may be they are not absolutely necessary for the success of the operation. Of the soft parturient passages the insufficiently dilated os alone gives rise to any difficulty in extraction. But, if necessary, extraction can be made as soon as one foot can be passed through the os, especially when its edges are thin and extensible. The body then forms a wedge, increasing in thickness from below upwards, which when drawn upon dilates the os. Of course the child cannot always be extracted in the short time desired.

It is also highly desirable that the bony pelvis should not oppose the passage of the after-coming head. But the head may at times be drawn, through even a very narrow conjugate without harm to the child. The disproportion between the head and the pelvis has here the same inconvenience as in labours where the head presents.

As regards the indications for extraction by the feet it has to be mentioned that in breech presentations the life of the child is, without the assistance of art, much more endangered than in head presentations.

In head presentations the most voluminous and the least compressible portion of the child is born first, so that the passage of the trunk and extremities is easy, and the elasticity and contractibility of the vagina suffice to terminate labour. But the reverse is the case with footling presentations. Feet and breech pass without difficulty through the parturient passages, whilst shoulders and head require strong contractions of the uterus. But when the uterus is emptied to a great extent, the pains commonly recur only after long pauses. Accordingly, the expulsion of the child after the birth of the presenting part takes a much longer time in footling than in head presentations. As a rule, though only a small portion of the child be contained in the uterus, and though it firmly contracts in order to expel this, the child feels the want of respiration; for the placenta is partially or completely detached by the considerable diminution of its place of insertion. In head presentations the child can easily satisfy the want of respiration as soon as the head is born, and with the first breath it begins its extra-uterine life, though part of its body is still within the maternal genitals. When the breech is born, and the placental circulation disturbed, the mouth not being accessible to the air, the first breath is followed by an inspiration of foreign bodies, and consequently by asphyxia. Most frequently asphyxia is produced in the way just described, but it may also be caused by pressure upon the funis.

It is thus evident that in breech positions the life of the child is in danger when after the birth of the breech the rest of the body does not immediately follow.

At that time, that is, when the powers of nature would have expelled the rest of the body in not too long a time, manual extraction of the child does not usually offer the slightest difficulties, and is completely free from danger to both mother and child. We, therefore, do not wait till asphyxia has actually begun, but always extract when, after the birth of the lower half of the body, the upper half does not immediately follow spontaneously.

The reason for not waiting until asphyxia has begun is not the fear of not being able to resuscitate the child, but because by the first inspiration foreign bodies enter the air-passages, which may there produce a fatal inflammation.

Independently of the indication just mentioned, manual extraction, as long as the feet are still within the maternal passages, is also necessary in those cases where, as in head presentations, an instantaneous delivery is indicated, that is, when the safety of the mother or of the child is threatened. Since in footling presentations the danger increases with the descent of the feet, the sounds of the fœtal heart must all the time be carefully watched by auscultation.

It has already been said that cases occur where the child cannot be extracted by the presenting head, when a speedy delivery is indicated. A footling presentation must then be artificially produced in order to allow extraction. Turning by the feet then becomes a preparatory operation to manual extraction.

The operation itself is performed in the following way:—

The woman is to lie close to the edge of the bed, either upon a cross-bed or at least in an oblique position, one leg outside the bed upon a chair, the other within the bed, for circumstances may require the trunk of the fœtal body to be greatly bent backwards.

This position can only be given up when extraction is not attended by any difficulties, as in the case of a second twin. The lateral position, which is decidedly advantageous for turning, is inconvenient for extraction, and cannot, therefore, be recommended in difficult cases. A few warmed towels must be at hand and an elastic catheter, to be applied, if need be, to the air-passages of the child. It is also desirable to have a sling ready.

If the whole hand is to be introduced into the vagina, chloroform should be given in all cases where there is sufficient time to do so. It is unnecessary, however, when the feet are accessible.

If the footling position is complete, the hand, with the back well oiled, is introduced into the vagina. Then both feet are grasped, so that the middle finger comes to lie above the ankles, and the other fingers at the sides of the feet. One foot thus lies between the middle and the index finger, and the other between the middle and the ring finger. They are now drawn upon until they can be seen outside the genitals and in order to have a more secure hold of them they are enveloped in a towel. The operator now takes a foot, the toes of which are looking backwards, in each hand, so that the thumb lies upon the calf, and the other fingers upon the dorsum of the foot and the anterior surface of the leg. Firm traction, especially in a direction backwards, soon brings the thighs and breech into the genital fissure. The more the child becomes visible the higher up it is grasped, so that for the extraction of the breech both thighs are taken into the whole hand; if only one foot presents this is drawn upon with both hands as soon as the space permits it, and when the breech has appeared in the vulva, the index finger is put into the hip nearest the coccyx of the mother in order to draw upon it. If the funis passes between the thighs (" the child rides upon the funis"), it must as early as possible be pushed over the nates corresponding to the foot which is still within the genitals. After the birth of the breech the second foot comes easily out of the vagina, and extraction is then continued as in complete footling position. Now, the breech is so taken hold of that both thumbs lie upon the nates, whilst the index fingers are supported upon the crest of the ilium. Strong traction downwards is made until the thorax appears outside. The umbilical cord is somewhat loosened in order to avoid pulling upon it. If it is greatly stretched and cannot be loosened, it may be rapidly divided, and the fœtal end of it is compressed by an assistant. (The same is done when in complete footling position the child rides upon the funis, and the latter cannot be brought back over one foot.)

If the child is not yet asphyxiated, and has been slowly extracted up to the breech, now the greatest speed is necessary. For, although the cord be not at all compressed, the contents of the uterus have so greatly diminished that, as a rule, the placenta is at least partially detached. The following steps are therefore to be performed with the greatest possible speed consistent with the safety of the mother and the child. If now one arm appears by the side of the trunk, it is taken hold of and drawn upon until the corresponding shoulder gradually descends; otherwise the arms must be liberated artificially. The arm situated more backwards is first brought down, because the necessary manipulations can be more easily performed in the sacral excavation than in the space between the thorax and the symphysis. For that purpose, when the back of the child is directed forwards and somewhat to the left, the feet are raised, and the trunk carried towards the right side of the mother. This causes the right shoulder to descend, and facilitates the liberation of the arm. The shoulders may be also brought dowh by pressure upon them in the way described by Rosshirt, and similarly by Baudelocque. They recommend the liberation of the arm by introducing two fingers into the bend of the elbow, and by pressure upon it the whole arm is brought down past the face. The liberated arm is now drawn upon in order to make the shoulder descend, and to obtain more space for the release of the other arm, the trunk of the child is by a gentle rotation on its longitudinal axis brought to the opposite side. Thus the other shoulder is brought downwards and somewhat backwards, and in this way the arm of that side is released.

Immediately after the liberation of the arms the head is extracted by means of Smellie-Veit's manipulation. For that purpose the hand corresponding to the abdominal surface of the child is introduced into the vagina, and having reached the lower jaw, two fingers are put into the mouth of the child so as not to press upon the floor of the mouth, but upon the alveolar process of the jaw. The trunk of the child is supported by that arm, as if the child were riding on it, whilst two fingers of the other hand are hooked over the nape of the child. If now traction is made with both hands upon the nape and lower jaw at the same time (though of course more upon the former than upon the latter), the head descends, and is born by raising the trunk over the perinæum.

In this way the child is usually extracted without difficulty. There are, however, many deviations from it.

When the back of the child looks directly forwards, it may be doubtful which arm is the posterior. In such cases it is well to try to liberate both arms, and to take that first which is freed the easier.

Considerable difficulties may be met with when one arm lies behind the nape. We may then try to rotate the trunk on its longitudinal axis, at the same time slightly raising it so as to bring the shoulder which corresponds to the arm in the unfavorable position more to one side; or, if this fails, we may try to liberate the impacted arm by strong pressure. Cases, however, occur, and especially easily, in which it is quite impossible to release the arm impacted between the head and the symphysis without inflicting some injury. This is most often met with when the attempt to free the arm has been made after the trunk has been delivered by great traction, and the head by this has been firmly drawn down into the brim; then an attempt must be made to draw the head through the pelvis with the arm in its unfavorable position. If this does not succeed, and it is not possible to liberate the arm without some injury, or to deliver the child with the arm in this faulty position, it then becomes the duty of the physician to save the life of the child at all costs, and to release the impacted arm were he even forced to fracture the humerus.

Although commonly the feet appear in the vagina with the toes directed backwards, yet it is not so very rare to find the toes directed forwards. In the further progress of a labour with such a position, the back almost always turns at least forwards and to one side. This rotation is favoured by drawing more upon the foot which is about to come forwards. In extraction by one foot this always turns forwards, because the resistance is here the least. All attempts at forced artificial rotation must be avoided, since often quite unexpected results follow, only it is proper to assist the rotation which has naturally commenced.

If the rotation of the back forwards has not taken place, the liberation of the arms may be very difficult. We may then try to push one shoulder backwards in order to obtain in this way room at the side of the pelvis to allow the arm to pass. If this also fails, the elbow-joint is pressed backwards, and the arm drawn down by the forearm.

The extraction of the after-coming head is often a task of great difficulty when the chin is directed more or less forwards. Frequently also in these cases one or two fingers can be put into the mouth of the child, and thus the face can be easily drawn into the side of the pelvis, or by pressure externally upon the head this may be made to turn so that it is placed transversely. If this also has been tried in vain, the position of the chin must be accurately made out, and by two fingers hooked over the nape (the so-called Prague manipulation) traction is made on the child, but in such a manner that the head and trunk retain their natural position. This manœuvre succeeds in bringing the head to one side, and the mouth becomes accessible.

Where the extraction of the after-coming head offers special difficulties, the traction may be assisted by pressure upon the head externally. The pressure may be very great without doing harm.

The prognosis of extraction by the feet is decidedly favorable to the mother. All the various steps of the operation are devoid of danger to her; only when there are disproportions in size between the head and the pelvis are the maternal soft parts compressed by the head passing through them. This pressure, acting once and only for a very short time, is known not to have any bad effect, and as a whole extraction is at times more favorable to the mother than delivery with the head presenting. It must be mentioned also that when the after-coming head is impacted, occluding the genital canal, a flabby uterus may be the seat of a profuse internal hæmorrhage by the detachment of the placenta. In view of this possibility it is advisable to exert pressure externally upon the after-coming head.

As regards the child, the prognosis is not quite so favorable, but more so than when a footling position is left to nature exclusively. The later the extraction is performed the better are the prospects of the child. Yet even when we have to deliver in those cases in which the feet are still within the uterus, and where there is no disproportion between the head and the pelvis, the operation is always so rapidly performed that the child cannot make any, or at least very few, premature respiratory movements. But since the foreign bodies drawn into the air-passages may, within the first few days after birth, produce lobular pneumonia, the prognosis of the extraction depends to some extent upon the success of the treatment of premature respiration. The existence of a disproportion between the head and the pelvis, whereby the extraction of the head is obstructed, may be of serious consequences to the child. Yet even in a greatly contracted pelvis where, if the head presented, a live birth could hardly be expected, we may succeed in extracting a living child. On the whole, therefore, extraction performed by a practised operator gives by no means unfavorable results.

(*b.*) *Extraction by the breech.*—When the breech presents, extraction is only indicated when there is danger to the mother or to the child. However, the fact is to be remembered that no conclusion can be drawn as to commencing asphyxia of the child from the passage of the meconium, which, as a rule, is due to the compression of the abdomen.

A breech presentation is, if possible, to be converted into a footling presentation, because extraction by the feet is generally easy, and that by the breech very difficult. As long as the breech is movable above the brim that can easily be done; but we must not despair of being able to bring down a foot even when the breech lies firmly in the inlet, or has partially descended into the pelvis.

To accomplish this the woman is to lie on that side to which the feet of the child are directed, that is, in the first breech position, on the right side. By this the fundus uteri with the head falls over to the right side, whilst the breech is inclined to deviate to the left. Now the left hand is introduced, the breech pushed upwards, and somewhat to the left, and if this succeeds that foot is grasped which lies foremost. The use of chloroform greatly facilitates in difficult cases the bringing down the foot. There is not, however, in all cases the necessary time to use it. Extraction is then continued in the way previously described.

If the foot cannot be brought down, or if the breech has already so far entered the pelvis that it cannot be pushed up without great force, it will be necessary to extract by the breech. When the breech is low down in the pelvis, even though the hand can be passed by the breech, no attempt must be made to bring a foot down; for the thigh-bone is too long to allow of its being drawn through the pelvis by the side of the breech without being fractured.

The index finger of one hand is now brought into the hip situated in front, and hooked over it; the wrist of that hand is grasped by the other hand, and with both firm traction is made downwards. In easy cases the breech is in this way brought down rapidly. If the breech is so low down that an index finger can be placed in each hip, the extraction does not offer any great difficulties. In other cases the impaction of the breech resists all efforts; during a pain it may be brought a little lower, but in the interval it stands as firm as a rock.

We may then try to replace the index finger by a blunt hook. This, of course, gives the operator much greater power, but its application is not quite free from danger. When the extraction is urgently indicated, the child must undergo that danger. Guided by the fingers the blunt hook is introduced between the anterior wall of the pelvis and the breech. Its free end is to be directed towards the knee, and to be so much raised that it can be slipped over the thigh between the two legs. It is then pressed into the hip, and after its position has been ascertained extraction may be begun.

The use of the blunt hook is certainly not free from danger to the child, and Hecker's recommendation to extract by means of a sling deserves consideration. He states that the sling can be applied without much difficulty, and that the traction is effectual. It is at any rate less dangerous than the blunt hook. Pappel recommends the application of the sling by means of an instrument constructed like Bellocq's sound.

The artificial removal of the afterbirth will be described in the special "Pathology of Parturition," under the chapter on Hæmorrhage.

Literature.—Feist, M. f. G., B. 3, p. 241.—West, Transactions of the Obstetrical Society, vol. iii, p. 222.—Mayerhofer, Wiener med. Presse, 1868, No. 1, 3, 5.—Kristeller, Berl. Kl. W., 1867, No. 6 and M. f. G., B. 29, p. 337.—Ploss, Zeitschr. f. M., Ch. u. Geb., 1867, p. 156.—Abegg, Zur Geb. u. Gyn. Berlin, 1868, p. 32.—Playfair, Lancet, 1870, vol. ii, p. 465.—Levret, Observ., &c., p. 82 *sequ*, and Suite des observ., &c., p. 154 *sequ.* Smellie, A Treatise on the Theory and Practice of Midwifery, third edition. London, 1756, vol. i, p. 248.—Baudelocque, L'art des acc., 8 ed. Paris, 1844, t. ii, p. 133 *sequ.*—Wigand, Beiträge zur Geburtshülfe, H. 2. Hamburg, 1800, p. 27.—Boër, Natürliche Geburtshülfe, 3 Band. Wien, 1817, p. 75.—F. B. Osiander, Handbuch der Entbindungskunst. Tübingen, 1830, p. 245.—Mme. Lachapelle, Prat. des acc., t. i. Paris, 1821, p. 60.—G. W. Steind, j., Lehre der Geb., th. ii, 1827, § 606, u. s. w., Siebold's Journal f. Geb., &c., B. vi, p. 481. Gemeins. deutsche, Z. f. G., 1829, 4 Bd., p. 374 and an vielen anderen Stellen.—Kristeller, M. f. G., B. 13, p. 396.—Spöndli, Die unschädliche Kopfzange, &c. Zurich, 1862.—Dieterich, M. f. G., B. 31, p. 262.—Mauriceau, Traité des mal. des femmes grosses, 6 éd. Paris, 1721, chap. xiii, p. 280.—Portal, La pratique des acc., &c. Paris, 1685.—De la Motte, Traité compl. des acc., &c. Paris, 1722.—J. von Hoorn, Die zwo u. s. w., Weh-Mütter Siphra and Pua. Stockholm and Leipzig, 1726.—Puzos, Traité des acc. Paris, 1759, p. 184 *sequ.*—Levret, L'art des acc., ii ed. Paris, 1761, p. 122, *sequ.*—Baudelocque, L'art des acc., 8 éd. Paris, 1844, p. 513 *sequ.* Deleurye, Traité des acc. Paris, 1770, übers. von Flemming, Breslau, 1778, p. 186 *sequ.*

CHAPTER II

TREATMENT OF TOO GREAT RESISTANCE

I.—IN ANOMALIES OF THE MATERNAL PASSAGES

(a) *Of the Soft Parturient Canal*

The treatment of pathological narrowness of the soft parturient passages will be given in the special therapeutics of parturition.

(b) *Of the Hard Parturient Canal*

a. Preservation of the life of the child; induction of premature labour.—The induction of premature labour consists in the artificial interruption of gestation at a time when the fœtus is able to live outside the uterus. It is hereby intended to avoid the dangers which, under certain circumstances, threaten the mother or the child, or both, where pregnancy is allowed to proceed, or delivery to take place at term.

In the definition just given the time is approximately stated when premature labour may be induced. Since children born before the twenty-ninth week invariably perish, the operation cannot be performed prior to that period. Nor is the fact to be lost sight of that the great majority of children born between the twenty-ninth and thirtieth week die within a few days after birth.

The induction of premature labour is indicated in the following cases—

I. Where the pelvis is so much contracted that a fully developed child of medium size can only pass with the greatest difficulty and danger, whilst one not fully developed, yet viable, will do so without endangering itself or the mother.

Now, however clear this indication may appear in theory, in practice it may be difficult to decide whether the operation is applicable to a considerable degree of pelvic contraction, and still more so to determine the time at which the operation should be performed.

First of all, the preliminary consideration, how far the woman is advanced in pregnancy, is full of uncertainties. Under normal conditions it is possible to determine pretty accurately by an examination the period to which pregnancy has advanced. But under abnormal conditions, and especially where the contracted pelvis, by impeding the descent of the head, obscures, so to speak, the signs indicative of the time of pregnancy, it may be extremely difficult to do so; and these are just the cases in which, as regards the induction of premature labour, that question has to be accurately decided. Therefore the calculations of the woman herself are of great importance, and are not to be disregarded unless they are evidently incorrect. Of course the highest degree of certainty will be obtained if the calculations of the woman agree with the result of the examination.

When the period of pregnancy has been determined as accurately as this is possible, there is still another question to be decided, namely, in which week premature labour should be induced. The earlier this is done the less the risks of injuries to the mother, and the greater the chances of the child being born alive, but not of living; and *vice versâ*, the later this is done, the greater the dangers to the mother and the child, but the greater the prospects of the child if born alive. Then the time has to be decided upon when the child can pass through the pelvis without danger to the mother or itself. To be able to fix the time for the operation with approximate accuracy a just estimate of the size of the pelvis and that of the head is required. As in the great majority of instances it is the flat pelvis, that is, one in which the conjugate diameter is contracted either alone, or, at least, more so than the other diameters, which demands the induction of premature labour, it can easily be seen that very much depends upon the most accurate possible measurement of that diameter. A pelvic examination is therefore to be made, strictly adhering to the rules which will afterwards be given. In this variety of contracted pelvis the proportion between the pelvis and the head is the easiest and simplest to estimate, because one diameter only is narrowed. But in the generally contracted, or in the irregularly contracted, pelvis, and especially the osteomalacic, this becomes a matter of much greater difficulty. Here the calculation of one or more diameters does not suffice, but several fingers or, if possible, the whole hand is to be introduced in order to obtain some idea of the pelvic space.

With regard to the size of the child's head there are still greater uncertainties. It is not accessible to direct mensuration, and its size can only be approximately determined by palpation. All that can be done is to have recourse to the known average measurements of the fœtal head at the different periods of gestation.

It is evident that an accurate calculation can never be made of the proportion between the pelvis and the head. This especially refers to primiparæ, in whom there is no indication of the most suitable period for the performance of the operation. In pluriparæ the course and results of previous labours deserve the most careful attention, and it is especially valuable to know the measurements of the heads of the children previously born. If the course of a previous labour was difficult and unfavorable, it is well to induce premature labour though the pelvis be only slightly contracted. It can also be only approximately stated to what degree of pelvic contraction premature labour is applicable. Since we are mostly concerned with the flat pelvis, a conjugate diameter of 7 cm., or 2·75 in., or at least 6¾ cm., or 2·64 in., would be the extreme limit, for a head with a transverse diameter of 8 cm., or 3·2 inches, could not pass through such a pelvis without considerable compression. In the irregular contracted pelvis no single measurement can be given as the extreme limit.

II. The induction of premature labour may ccasionally be attended with success in women whose children have been observed always to die at a certain period of pregnancy, provided that period be not too far distant from the normal term of gestation, and that all other remedies to avert the intra-uterine death of the children have proved of no avail in the previous pregnancies.

III. Where morbid conditions which endanger the life of the mother prove refractory to all other treatment, and if there is reason to expect that those conditions will either disappear entirely after delivery, or at least be then attended by less suffering and danger.

Of course this indication has a very wide range, and its applicability greatly depends upon the period of gestation. Whilst between the twenty-eighth and thirtieth week of gestation a child has very little chance indeed of living outside the uterus, there need hardly be the same degree of anxiety about the life of a child born after the thirty-sixth week, for experience has shown that children born at this period thrive just as well as those born at term. Now, since the operation itself is entirely free from danger to mother and child, it is indicated at that time when actual dangers can be averted from the mother, or when she can be spared intense sufferings for weeks to come.

To avoid post-mortem Cæsarean sections Stehberger induced premature labour in two cases of serious illness of pregnant women, where death was with certainty expected before term. Both times delivery was quick and easy, and the children at least were born alive. This indication may very rarely present itself, yet it requires to be fully considered.

As a rule premature labour will only be induced when the child is known to be alive. But this is no absolute condition of the third indication, as the operation is required for the sake of the mother, and the state of the fœtus is to be entirely disregarded. It must as yet remain undecided whether it would not be preferable to interrupt pregnancy even when there is a certainty of the death of the fœtus. Although the presence of a dead fœtus within the uterus has not the same injurious influence which older authors feared, it will nevertheless be admitted that its earliest possible removal is desirable when it can be done in a harmless way.

Nor is it at all necessary that the child should present by the head; for, even when the child presents with the breech, cephalic version can be made in most cases where labour is prematurely induced; and, on the other hand, with suitable assistance breech positions have scarcely a more unfavorable prognosis than head positions; nay, in the contracted pelvis a labour with the breech presenting is decidedly more favorable for the mother, and at times also for the child. Transverse positions by no means contraindicate the operation. Head or breech can almost always be made to present, and if by the use of a plug the membranes are prevented from rupturing too early, version can be performed after the os is dilated.

Prognosis.—The operation performed in suitable cases, and after approved methods, is favorable for the mother; and in contraction of the pelvis it gives still better results the earlier labour is induced. The same cannot be said for the child; its chances are worse the earlier the operation is performed. If only suitable cases are selected for operation, and all attending circumstances correctly estimated, the great majority of the children ought to be born alive. Under such circumstances, however, not only a slight increase of the size of the head may prove fatal to the child, but even those children if born alive are by no means kept alive. The more immature the child the less chances it has of living. The external circumstances of the mother have the greatest influence upon this. Almost all prematurely born children of unmarried women, who go as wet nurses, die. But when the loving and tender kindness of the mother watches over the child, and when her station in life enables her to devote all her attention to her offspring, to procure what is required to invigorate it, the chances of life are greatly increased.

It appears that the preservation of the child, the sole object for which the operation is frequently recommended, is often enough not obtained. But we lay special stress upon the fact that in the contracted pelvis the operation is to be performed not so much for the sake of the child as for the sake of the mother, whom it is intended to spare the dangers of labour at term.

With regard to the methods of operation a great number of them have been given. We shall shortly describe and examine them.

1. The introduction of an elastic catheter or bougie (Krause's method) deserves, in our opinion, the preference in almost all cases.

An elastic catheter without a stilet is guided along the fingers into the cervix, and is gently pushed upwards for some inches between the uterine wall and the membranes. It is allowed to remain there, and its slipping out may be prevented by a napkin put before the vulva, or by tying it. In introducing it care must be taken not to rupture the membranes. It not infrequently happens that after the introduction of the catheter drops of liquor amnii drain off, whilst the bag of waters appears to be intact. The membranes have then been ruptured, not near the os, but somewhat higher up. This is only to be considered an advantage, for labour is more speedily provoked by it, and the gradual escape of the liquor amnii by drops has no injurious consequence whatever. In pluriparæ pains often set in immediately; in primiparæ within a few hours at least.

This method satisfies all that can be reasonably expected from it. It is very simple, requires no complicated apparatus, is easily performed, is certain, and comparatively rapid in its effect, and has no inconveniences or dangers. If uterine action is exceptionally delayed the following method can easily be employed in addition:

The introduction of a bougie of laminaria digitata instead of a catheter on account of its swelling up is unnecessary, and is much less safe. If it has to remain there for some hours it becomes fetid, and causes decomposition of the secretions.

2. Injections between the uterus and the ovum (Cohen's method). An instrument constructed for the purpose, cone-shaped, in order to dilate the os, or a common elastic catheter, is introduced between the uterine wall and the membranes. Tepid water is then injected through it until the woman feels an increased tension. By the injected fluid the membranes are detached to a great extent from the uterus, and the uterine action is quickly roused.

This procedure is sure and rapid in its effect, but it is more inconvenient than the previous, and far more dangerous. A good many deaths have resulted partly from it and partly from simple vaginal injections, probably due to the entrance of air into the uterine veins.

3. Tarnier's method by means of the "Dilatateur intra-uterin." Tarnier recommends a new instrument of the shape of a bladder, which is to remain above the internal os. It consists of a caoutchouc tube, very much like an elastic catheter, the upper end of which has very thin walls in one place. The tube is introduced on a stilet into the uterus. If water is now injected into it under a great pressure, the thin walls situated above the os internum are distended into a bladder, which remains there when the stilet is withdrawn.

Spiegelberg has tried this method in seven cases, and strongly recommends it. The results he obtained with regard to the mother and the child are not very encouraging, but it must be admitted that they were not the fault of the method. We have only applied it in one case, and are therefore not in a position to give a definite opinion. Only this much is clear, that it never will supersede the simple catheter in practice. Though uterine action may quickly set in, nevertheless it often ceases after the expulsion of the caoutchouc ball, as Tarnier himself states, and delivery is scarcely terminated earlier than by the use of the elastic catheter. In our case the pains completely ceased after the expulsion of the ball, and we had to introduce the catheter, which quickly produced efficient expelling force. If the caoutchouc tube is only moderately injected (to the size of a walnut) the ball is very early expelled. If injected to the size of a hen's egg it very easily bursts, either from the mere pressure of the pains, or from coughing, vomiting, &c. The stopcock also gets easily unscrewed, and then the bladder collapses. It may therefore be well to follow Tarnier's advice to take the stopcock away altogether, and to ligature the end of the tube. On the whole the procedure is inconvenient in every respect, and requires a special apparatus, so that it will certainly not be applicable for general practice. When a very speedy termination of labour is urgently indicated Tarnier's instrument may be used, and after its expulsion Barnes' dilator is to be applied.

4. Mechanical dilatation of the cervix, as recommended by Brünninghausen, Kluge, and Barnes.

When the cervix is still completely closed, and before using the uterine douche or plugging the vagina, Brünninghausen and Kluge recommend the introduction of a sponge tent into the cervix up to the internal os, retaining it there by means of a plug of cotton-wool. The sponge-tent, however, soon becomes fetid. It should, therefore, be removed within twelve hours at the latest, and if uterine action has not yet set in it is to be replaced by a new one. Instead of a sponge-tent, which greatly irritates the cervix and becomes very fetid after a short time, a laminaria cone is suitably employed, but this also if retained a long time favours the decomposition of the secretions. Not only are the tents sometimes difficult to introduce, but the method itself is, in regard to the rapidity and persistence of its effect, far inferior to those above named, and, therefore, does not deserve to be recommended.

Formerly special instruments were employed for the dilatation of the cervix uteri. Such dilators have been constructed by Osiander, Busch, Mende, and Krause. It need hardly be said that the dilatation of the cervix by means of iron instruments is a rough treatment and not to be advised. Another far milder instrument has been designed by Schnakenburg, and called by him "Sphenosiphon." It is the bladder of an animal tied to a syringe, which when injected with water mechanically distends the cervix.

R. Barnes' procedure is based upon the same idea. He recommends the dilatation of the cervix by fiddle-shaped india-rubber bags, which are of three different sizes. They are introduced by means of a sound, and when injected remain within the cervix, because their middle portion is narrower than the two ends. As the dilators are rather thick they can only be used when the cervix is permeable to two fingers, that is, when labour provoked in another way has already begun. Elliot recommends them very strongly.

5. Plugging the vagina recommended by Schöller, Hüter, and Braun.

The colpeurynter—a caoutchouc ball with a brass stopcock—is put into the vagina, and injected with water. The effect of it, especially after a moderate injection, is slow and uncertain, but when fully distended it is very inconvenient, and even causes great pain. In particular cases this method may be useful, especially when there is hæmorrhage, or when after the provocation of labour the pains cease, and an increase of the uterine action as well as a counterpressure against the membranes about to burst is wished for. The rupture of the membranes is prevented with great certainty by the colpeurynter.

6. The ascending uterine douche (Kiwisch's method).

A jet of water of a temperature of 30—35° R. is directed against the lower uterine segment for 10—15 minutes, and if a somewhat precise effect is expected, this has to be repeated every two or three hours, until sufficient uterine action has set in. This method is not very reliable, and not quite free from danger, but at times it may be used to favour the dilatation of an entirely closed cervix, and to prepare it for the application of another method.

7. The puncture of the membranes, Scheel's method, or that of Hopkins and Meissner.

According to Scheel's method the membranes are ruptured either by means of the uterine sound, or by one of the numerous pointed instruments designed for that purpose. The effect follows with great certainty, though sometimes rather slowly. On account of the disadvantages, which may accrue from the premature rupture of the membranes, this procedure had well nigh been abandoned until lately. Rokitansky, jun., has again strongly recommended it after numerous observations in Carl Braun's clinic in Vienna. The puncture is made by a pointed quill, which is introduced on the uterine sound. According to Rokitansky the liquor amnii drains off slowly, and gives better results to mother and child than any other method.

Meissner has constructed a long curved trocar to puncture the membranes higher up. After half an ounce of liquor amnii has been evacuated the instrument is withdrawn. Gradually the liquor amnii drains off by drops, and this is followed by uterine contractions. This procedure, also, is almost entirely abandoned, because the operation is often very difficult to perform, and requires a special instrument.

Other methods which provoke labour slowly and with great uncertainty, and which besides are not quite free from danger, have only a historical value. They are—the method of Hamilton in which the membranes are detached from the uterine wall by the fingers passed round the internal os; friction of the fundus uteri after d'Outrepont; friction of the os after Ritgen; ergot of rye recommended by Ramsbotham; irritation of the mammæ by suction, Scanzoni's plan, and also his douche of carbonic acid by which there has been one fatal case; galvanism (recommended by Schreiber); the interrupted current (recommended by Hennig); hot baths, irritating clysters, and many more.

(β) *Without preserving the life of the child.—a. Artificial abortion.*—Artificial abortion, that is, the induction of labour at a time when the fœtus is not yet able to live outside the uterus, is an operation which, without doubt, is to be practised when the sole hope of saving the life of the mother depends upon it. It is evident that this operation is justifiable. In such cases the physician has only the alternative of saving the mother by artificial abortion or of allowing her to die. The death of the mother is necessarily attended by that of the fœtus, so that the death of the fœtus is equally certain in both cases. The alternative, therefore, is either to save the mother or to let her die and with her the fœtus. In such cases it is not only permitted the physician, but it becomes his duty to save the life of the mother by inducing abortion.

The indication is doubtless correct in principle; the great difficulty in practice is to select the suitable cases. Evident danger to the life of the mother may possibly and probably be averted by artificial abortion.

1. When the gravid, retroflected, or prolapsed uterus is strangulated. (Scanzoni induced abortion in inflammation of a gravid uterus contained in a femoral hernia.) If all attempts at reposition have failed the sole hope of saving the mother depends upon emptying the uterine cavity. Here even the escape of the liquor amnii produces a considerable diminution of the uterus and the immediate recovery of the woman.

2. In all other diseases of pregnant women, when life is in immediate danger, which all other remedies have failed to avert. Obstinate vomiting during pregnancy may be mentioned as a disease that most frequently demands the emptying of the uterine cavity, whilst this is extremely seldom the case in diseases of the heart, lungs, or kidneys with acute and threatening symptoms.

It is more doubtful whether artificial abortion is admissible when the danger to the woman is not instantaneous, or need not necessarily be averted, but the circumstances are such that her life would be endangered during parturition. This may especially be the case when the pelvis is by some peculiar abnormality or by large tumours so much obstructed that the child cannot pass through it in any way. Here the doubt is chiefly due to the fact that there are measures which, taken at the normal termination of pregnancy, may possibly save both mother and child. If the Cæsarean section be performed, the life of the child is with great probability saved, whilst the chances of the mother are certainly very bad, though not absolutely fatal. But it must rest with the mother to decide at the commencement of pregnancy as well as at term whether for the sake of her offspring she will place her own life in danger; and if she declares that she will not permit the Cæsarean section to be performed, we are justly entitled to induce abortion, the more so as frequently the bodily and spiritual welfare of a whole family of children depends upon the preservation of the life of the mother.

In such cases, where the indication is somewhat doubtful, the practitioner will do well in the interest of the patient as well as his own to consult with a colleague before taking any decisive step.

Many more indications have been given which, in the opinion of some, would justify artificial abortion. But close examination will show that the reasons are insufficient.

In hæmorrhages there is no necessity to induce abortion. For if the hæmorrhage cannot be completely stopped it can with certainty be limited sufficiently to avert the danger, spontaneous abortion always following in the course of persistent hæmorrhage. Artificial abortion is never admissible in the following cases:—1. In Eclampsia (this, however, is extremely seldom met with at the time in question), because it completely disappears under chloroform narcosis. 2. In carcinoma of the uterus. Here the birth of a mature and live child is possible, and it is unwarrantable to sacrifice the life of the child when that of the mother, absolutely lost, is prolonged only for a few weeks. 3. In contractions of the vagina, since they never form an absolute impediment to delivery.

An operation equivalent in principle to abortion is the puncture of an extra-uterine growing ovum. In such cases the fœtus is always as good as lost, and the mother herself being in serious danger the destruction of such an ovum is decidedly beneficial.

The operation for the induction of premature labour is usually very simple. A uterine sound is introduced up to the internal os so as to pierce the membranes and allow the escape of the liquor amnii. In retroflexion of the gravid uterus this operation may be very difficult and even impracticable, so that it may be necessary to puncture the ovum through the posterior uterine wall, and thus induce abortion.

b. By craniotomy.—Under craniotomy are comprised all those operations which serve to reduce the bulk of the child's head, and to extract it after that reduction.

The bulk of the head is reduced by perforation, which consists in opening the cranial vault by means of properly constructed instruments, and allowing the cranial contents—the brain—to escape.

All the diameters of the head are very considerably diminished by the operation, which proves so far effectual as to lessen the disproportion between the head and the pelvis. Of course the operation implies the sacrifice of even a living child.

The indications for perforation depend to a very great extent upon whether the child is already dead or is still alive.

The dead child is always to be perforated when a disproportion greatly obstructs the passage of the head through the pelvis. An exception is only made in hydrocephalus, where the cranial contents consist of serum which can easily be evacuated by a simple puncture. The reason for doing so is evident. When the child is dead the safety of the mother is then the sole object of attention. To her, delivery under an existing disproportion is, beyond doubt, more injurious than after its removal. But since the disproportion can be removed in an innocuous way by perforation, it is the duty of the obstetrician to procure the mother that relief. There can be no question of sparing the dead child when, from its mutilation, the mother can profit. The same right which sanctions the opening, in post-mortem examination, of the skull of a dead body, also allows the same to be done on the fœtal body contained within the uterus. That all unnecessary dismemberment is to be avoided is self-evident. It is necessary also that the perforated child should be cleaned after the extraction, and that the injuries inflicted should be hid from the eyes of the mother, for the same reasons as after a post-mortem examination the body is sewn up.

If the child is alive perforation is then to be practised, when delivery must be terminated in the interest of the mother; this, however, being impracticable in a way harmless to the child (as by forceps or version), and the mother declaring against the Cæsarean section. In all such cases the preservation of the living child is, if not always absolutely impossible, yet in the highest degree improbable, whilst procrastination places the mother in imminent danger of her life. The practitioner who acts upon that principle will be able to save the lives of many mothers, whilst the one who rejects perforation, or limits it to the extremest cases, will sacrifice one maternal life after another, and scarcely preserve in return the life of one child.

Perforation of the after-coming head is always indicated when the head cannot be extracted by manipulation. Nor can this operation be replaced by the forceps; for the same power can be exerted by manual extraction as through the forceps. Moreover, the head accommodates itself far more easily to the brim by it than when compressed on both sides by the blades of the forceps. Cephalotripsy, which could replace perforation, is more dangerous to the mother and unnecessary, the extraction of a perforated head not being subject to any difficulty. A live child in head-last labours ought never to be perforated, for the firm impaction of the head rapidly causes death.

The time when the operation is to be performed naturally depends upon the indications. It can safely be asserted that in general it is almost always performed too late. When the child is dead, and perforation necessary, it need not be delayed for a moment. If the child be alive perforation will certainly not be decided upon without the presence of grave symptoms, only care must be taken not to wait too long, or the mother will not derive benefit from it.

To perform the operation the head must be immovable above or within the brim. Perforation may also succeed when the head is movable, but then it has to be fixed by the forceps or by an assistant externally. In all those cases preference is to be given to turning by the feet, and extraction after perforation of the after-coming head if necessary.

In the course of time a large number of instruments for the performance of the operation have been devised. At present two kinds are in use, viz. the scissor-shaped, and the trepan-shaped perforator.

Of the former, Naegele's is the most useful. It is locked when introduced, and a clasp prevents its premature opening; it cuts outwards, and is brought into action by compressing the handles. Levret's perforator is also useful. It also cuts outwards, but its blades are crossed and the handles must be opened, and this makes it more inconvenient and more difficult to use.

The woman is to lie across the bed on her left side, with the nates a little beyond the edge of the bed. The scissor-shaped perforator can also easily be used in the dorsal posture; but the escaping blood and brain cannot well be prevented soiling the bed, whilst in the position above indicated they can easily flow into a vessel standing at the bedside. The left hand, and with it the closed perforator, is introduced into the vagina. Guided by the fingers the perforator, with the handles greatly depressed, is applied to the head (if possible, but not necessarily, to a suture); it is then pushed in up to the shoulders, the clasp is removed, and the instrument opened. Having thus made a longitudinal wound, the perforator is again closed and turned half round, again opened, and afterwards withdrawn. When the head has been firmly compressed in the brim, and its contents, therefore, exposed to great pressure, cerebral substance will immediately escape from the opening. To facilitate this, any instrument (such as the uterine sound or a silver catheter) is introduced into the opening, and the brain broken up. The injection of warm water into the cranial cavity will also facilitate the escape of its contents.

Now the operation is finished; its object, to reduce the bulk of the child's head, has been attained. The expulsion of the reduced head may generally be left to nature—almost always so if the operation has been performed early.

Instead of the scissor-shaped perforators, trephine-shaped have recently been employed. The most useful of them are Leissnig-Kiwisch's instrument, which has a movable pyramid, and large, well-catching teeth; and Carl Braun's, which is provided with a pelvic curve. The trephine is applied to the hairy scalp, the pyramid well fixed, and by turning the handle the corresponding portion of the head is trephined.

The object of the operation, to make a hole into the skull large enough for the escape of the brain, is easily and safely obtained in both ways. The use of the scissor-shaped perforator appears to us in general to be preferable. There is scarcely a more simple operation than to push such a sharp instrument through a suture into the

14

skull, and by a very little caution there need be no fear of lacerating the maternal soft parts. We will only remark here that several times the sacrum projecting into the pelvic cavity has been pierced by the trephine instead of the head. In face presentation the forehead is perforated, either in the frontal suture by the scissor-shaped, or else by the trephine-shaped perforator.

The after-coming head is perforated by pushing the scissor-shaped instrument into one of the lateral fontanelles. If exceptionally they cannot be reached, the instrument may be made to enter the head, according to Michaelis, between the atlas and the occiput through the foramen magnum, or the base of the cranium is perforated between the chin and the vertebral column. In the latter case a trephine-shaped perforator is more useful.

The prognosis of the operation is in every respect favorable to the mother. By a little skill she cannot be injured by the instrument, and other dangers to her are not connected with the operation. Nevertheless it cannot be denied that a great number of the women upon whom perforation is performed die. The death is not due to the operation, but to the unfavorable conditions which rendered the operation necessary, and not infrequently, also, to its late performance. Most frequently a fatal issue is observed after perforation, but not *propter*, but *post*, when it has been performed because delivery could not be terminated by a long and forcible forceps operation. On the whole, perforation is an operation decidedly favorable to the mother. Its performance does not any way injure her, and the pressure upon the maternal soft parts is immediately, and to a great extent, moderated by it.

If the pelvic contraction is not very considerable perforation alone, with emptying the contents of the cranial cavity, and when the pains are defective, extraction, in the way about to be described, sufficed to overcome the existing disproportion.

But if the contraction is very great the bulk of the head is to be still further reduced in another way; for that purpose various methods have been recommended. In Germany and France cephalotripsy is generally employed.

The cephalotribe is a very powerful and long forceps, with very small head and pelvic curves, which by various contrivances can be so screwed together that the tips of the blades approach, and the greatest distance between them does not amount to more than four centimètres. One of the most practical instruments is that of Breisky.

The cephalotribe is applied in the same way as the forceps, only its pelvic curve being small, and the head usually high up, the handles have to be greatly pressed against the perinæum.

If both blades are well *in sitû* the instrument is locked, and the compression apparatus is put on and screwed up. This has to be done very slowly and cautiously, as otherwise the head may easily slip out of the instrument.

By a powerfully constructed cephalotribe the head is compressed with so much force that the bones of the cranial vault as, well as part of the base of the skull, are crushed. If a more complete reduction of the head is intended, *cephalotripsie repetée sans tractions*, as recommended by Pajot, may be made; that is, the instrument is taken off, applied a second and, if necessary, a third time, in different diameters, and thus compression repeated.

The cephalotribe is less frequently used by English obstetricians; they generally employ other methods of reducing the bulk of the head.

Simpson designed a new instrument, similar to the old bone pliers, but of stronger construction, which he called "cranioclast;" one blade of the pliers is introduced into the cavity of the cranium and the other applied externally. The bone now grasped is broken up, and the cranial vault thus destroyed.

A very similar method, which certainly deserves recommendation in very high degree of pelvic contraction, is adopted by Braxton Hicks and R. Barnes. The latter introduces the smaller blade of his craniotomy forceps into the interior of the cranium, and the other blade between the bone to be removed and the scalp. The bone is broken up and removed under the guidance of the fingers. If the pelvic contraction is not very considerable it suffices to break off two or three pieces of bone, after the removal of which all the other bones composing the cranial vault collapse upon the base of the skull. If the contraction is still greater, the conjugate measuring only 6½ or 5½ cent., 2·5—2·2 inches, or less, the frontal, parietal, temporal, and occipital bones are broken up by means of the forceps, and of the whole skull only the base remains.

To sufficiently reduce the bulk of the head in pelvic contractions up to one inch, Barnes recommends the use of a strong steel wire. After perforation of the head, and, if possible, also after the destruction of its vault by means of the craniotomy forceps, the skull is fixed by the crotchet, and the loop of the wire of an écraseur compressed by the fingers is introduced into the uterus above the head. The compression being removed, the loop springs open, and is brought above the head. By tightening the screw of the écraseur the head is evenly divided. The separated portion of the head is removed by the craniotomy forceps, and the rest of the head again divided until sufficient reduction is obtained. By the craniotomy forceps the rest of the head is drawn through the pelvis. Traction is now made to one side, so as to bring a shoulder into the brim. The operator then hooks the crotchet into the axilla, draws it down, and with strong scissors amputates the arm at the shoulder. This proceeding is then repeated on the other arm. Then the thorax is perforated and the ribs cut through with a pair of scissors. The thorax and abdomen are then eviscerated, and the trunk, now in a condition of complete collapse, can be delivered by moderate traction.

Barnes lays especial stress upon the fact that by his proceeding the maternal soft parts are very little exposed to pressure and contusion; but he himself admits that the operation is much more difficult and requires greater skill than the Cæsarean section.

In order to effectually reduce the bulk of the base of the cranium, which by cephalotripsy is sometimes not crushed at all, or only insufficiently, it has recently been proposed to trephine the base of the skull. Thus the "forcipe perforatore dii Fratelli Lollini" of Bologna consists of strong pliers, to which a borer is movably attached. By means of it the vault and then the base of the skull are pierced in one or more places, so that the forceps can easily crush the skull.

The "cephalotripse intra-cranienne of Guyon" appears to be more practical. Guyon pushes the piercer contained in the centre of the crown of a trephine into the cranial vault and trephines. Through the opening thus made the piercer is brought to the base of the skull, and this is perforated by means of a somewhat smaller trephining crown.

Hubert has lately recommended a very complicated proceeding—sphenotresie or transforation—to perforate repeatedly the base of the skull.

As regards the applicability and the advantages of the various methods, there is no doubt that on the whole cephalotripsy is not quite free from danger, and the explanation of it is easy. The cephalotribe can only be applied in a transverse or in an oblique direction. The head being compressed in one of these directions, the other cranial diameters, and also the one in the conjugate, must necessarily be increased, or, if this be prevented, press upon the impediment. The already greatly compressed soft parts at the anterior and posterior walls of the pelvis are again exposed to a new pressure by the application of the cephalotribe. Of course the compression is much less when the brain has been previously removed, and for this reason it is necessary always to perforate previously. By the English method the increase of pressure upon the already bruised soft parts is avoided. No doubt the gradual removal of the cranal bones by means of bone pliers is a long and tedious operation. But with a skilful hand it must succeed without any injury to the mother, and the compression of the soft parts is at no time increased, but with each removal of a bone continually lessened. Extraction also, as will soon be seen, is relatively easy under the most difficult conditions. It cannot, therefore, be denied that the English method has decided advantages over cephalotripsy in spite of its long and tedious performance. Future time will have to decide on the usefulness of the other methods.

By the reduction of the bulk of the head the principal indication, viz. the diminution and abolition respectively of the disproportion between head and pelvis, is met, and in all such cases it must be a principle, unless an immediate delivery is urgently demanded, to allow nature time to effect expulsion.

It very frequently occurs in practice that after reduction of the bulk of the head delivery cannot be delayed any longer, and, therefore, means must be employed to effect extraction artificially.

Manual extraction is possible only in the milder cases. At times it suffices to introduce two fingers into the hole in the skull, at others the hand has to be passed over the collapsed head in order to draw upon the base of the cranium. After the removal of the bones of the cranial vault the twisted scalp sometimes forms a useful handle. Turning also, which is possible after perforation, may sometimes advantageously be performed. This must be performed with caution, lest the inner surface of the uterus be injured by spicula of bones.

In the majority of cases instruments are required to effect extraction. The common forceps, by which a normal head alone can firmly be taken hold of, is not applicable for the emptied skull; and since Baudelocque the nephew has invented the cephalotribe, this instrument is pretty generally used for that purpose in France and Germany. Having crushed the skull, the instrument is allowed to remain *in situ*, or it is applied anew, so that the blades together with the portion of the skull which they compress come to lie within the contracted conjugate diameter, and are afterwards cautiously extracted.

Extraction by means of Simpson's cranioclast or a good craniotomy forceps is less injurious to the maternal soft parts and applicable in very considerable pelvic contractions. With the instrument designed by Carl Braun excellent results are obtained in Vienna, One blade of Braun's instrument is placed in the cranial cavity, the other over the face, and thus extraction performed.

If the English method is adopted and the whole cranial vault removed the base of the skull may conveniently be passed through a very highly contracted pelvis. Braxton Hicks recommends bringing the face into the pelvis and extracting by means of a crotchet applied to the lower jaw. Robert Barnes applies his craniotomy forceps over the lower jaw and the orbits and drags the base of the cranium edgewise through the narrow space. This proceeding may be adopted in a very high degree of pelvic contraction, since the grasped diameter does not usually measure more than $2\frac{1}{4}$ cm., or 1 inch.

There is no doubt that the methods of operating with the craniotomy forceps are more rational, and if performed with a little skill are, in difficult cases, less dangerous to the mother than cephalotripsy. But, on the other hand, the cephalotribe is more easily handled and is, therefore, especially to be recommended to less practised operators.

After perforation in face presentations the crotchet is useful for extraction, because it can easily be fixed. The cephalotribe applied over chin and forehead, also renders good service, as a very small diameter is thereby opposed to the pelvis.

The after-coming head can generally be easily extracted after the removal of the brain. This proceeding is, at any rate, much milder than extraction by means of the cephalotribe without previous perforation. If the pelvis is so greatly contracted that the base of the skull is too large, a crotchet may be put into the side of the skull in order to place the base obliquely and to bring it edgewise through the narrow opening.

OPENING OF A NEW PASSAGE BY THE CÆSAREAN SECTION

The Cæsarean section consists in opening a new passage by dividing the abdominal and uterine walls for the purpose of extracting the child from the womb.

This operation is attended by the greatest risk to the life of the mother, and as long as she is alive it has to be limited to the following cases:

1. When the delivery of the mother *per vias naturales*, the fœtus being alive or dead, is not practicable at all, or only so under greater dangers than those attending the Cæsarean section.

Independently of those very rare cases where large tumours, which can neither be diminished in bulk, nor displaced, fill up the pelvis, this occurs almost exclusively in cases of pelvic deformities, and principally in the rickety and osteomalacic pelvis. With regard to the contractions of the soft parts only greatly advanced carcinomatous degeneration of the cervix necessitates the Cæsarean section. Here, the birth of a mutilated child *per vias naturales* is always very highly dangerous to the mother, and since by the sacrifice of the child the life of the mother already doomed can only be protracted for a short time, this indication for performing the Cæsarean section deserves full attention when the child is alive.

2. When a living child cannot be delivered by the natural passages without sacrifice of its life and the mother desires the performance of the operation. Pelvic deformity is also the most frequent determining cause of this. It is extremely rare that the operation is performed on this account, and mothers almost always prefer the perforation of the child, and justly so.

The most suitable time for operating is when the membranes are still intact, the os rather dilated, and the pains strong. If the operation is performed earlier the very important contractions of the uterus, after the operation, are defective; if later, after the escape of the liquor amnii, the performance of the operation is more difficult and the prognosis more unfavorable for the child. The woman lies either on her bed or, more conveniently, on a mattress placed upon a table. The room is to be large and well ventilated, and the air in it pretty warm (72½° F.). Vessels containing warm water are also to be placed in the room in order to render the air moist.

The instruments required are several curved bistouries, a grooved director, a knife with a blunt end, several new sponges, ligatures, and sutures.

After chloroform has been administered, unless local anæsthesia is preferred, any coils of the intestine lying in front of the uterus are pushed aside, and the uterus so placed as to be bisected in the linea alba (in greatly contracted pelvis the uterus lies often very obliquely, and this requires attention lest the incision pass through the edge of the uterus). It is fixed in that position by an assistant, and the abdominal walls are stretched over it. Previously, however, the bladder is to be emptied by means of a catheter. The incision is made in the linea alba, commencing at the umbilicus and continued so far downwards as not to injure the bladder. The abdominal walls are divided in layers, on reaching the peritoneum the abdominal cavity is opened at one small spot; the peritoneum is divided then to the necessary extent on a director. Then the uterus appears in the wound, and an assistant must grasp it with both hands; any portions of the omentum lying upon the uterus are to be pushed back beneath the abdominal walls. Intestinal coils are not now visible.

The steps of the operation have been performed until now slowly and gradually; but as soon as the uterus is to be incised the remainder must be continued with all possible speed, yet with calm consideration. The wound of the uterus will give rise to profuse and sometimes, indeed, to enormous bleeding, so that here the operator has an opportunity of showing his presence of mind, for, to stop the hæmorrhage, there is only one effectual remedy, and that is emptying the uterus. The more profusely the blood pours out the more rapidly is the uterus to be incised.

A free incision has now to be made into the uterus, the knife is to pierce through the parenchyma of the organ at about the internal os, so deeply that the membranes or, when the waters have escaped, fœtal parts can be seen. Through the incision one finger is introduced into the uterine cavity and the wound extended. Whilst an assistant hooks one finger into the upper angle of the uterine wound, and fixes it in contact with the corresponding angle of the abdominal wound, the hand is passed into the uterus, grasps the feet, ruptures the membranes, and extracts the child by the feet. If the head appears in the opening, the child is extracted by it. If the uterine wound is too small it must be enlarged upwards by the blunt-pointed knife. Then the cord is divided and tied and the child is handed over to the nurse. The placenta is detached by the contractions of the uterus, and can easily be removed through the wound. If adherent to any spot it is to be carefully separated. Whilst the waters are escaping and the child is being extracted the uterus contracts considerably. The assistant, who fixes the uterus, must accurately follow the contracting organs with his hand, in order to prevent prolapse of the omentum or of the intestinal coils.

When the uterus contracts well, so as to close the wound and arrest the bleeding, the organ is replaced in the abdominal cavity, and the abdominal wound is closed by sutures, which have to include the peritoneum throughout the whole extent of the wound. If liquor amnii or blood in any large quantity have escaped into the abdominal cavity, it must first be removed by new sponges.

Sometimes the uterus does not contract sufficiently to close the wound in it; hæmorrhage, therefore, occurs from time to time, and this may be very considerable when the uterus feebly contracts. By friction of the uterus complete apposition of the edges of the wound may be obtained. Failing this sutures may become necessary; but, if it is possible, their use is to be avoided.

When the operation is completed the woman is put to bed. Several long strips of adhesive plaster are then placed around the pelvis, and brought over the front of the abdomen, crossing the wound; a dry napkin is placed over the wound. At first no other dressing is required. Later, when the wound begins to suppurate, it is dressed with lint.

The after treatment is merely symptomatic. Irritation on the part of the intestinal tract (as hiccough, nausea, vomiting) is suitably combated by small pieces of ice. Subcutaneous injections of morphia are also useful. The bowels are to be evacuated as soon as possible by enemata, castor oil (if this does not produce vomiting), or calomel.

It would lead us too far to indicate the treatment of all possible sequelæ. If the woman is weak, or if much blood has been lost during the operation, our first attention is to be paid to good nutrition.

We have already spoken of the Cæsarean section on the dead. The operation is performed in the same way as upon the living.

Prognosis.—Although the statistics of the mortality after the Cæsarean section vary greatly, this much is certain that the chances of the mother are very doubtful. According to the recent statistics of Mayer the results in different countries are the following:

Of 480	operations in	England	236	recovered,	244	died	= 50	per cent.
" 712	"	Germany	332	"	380	"	= 53	"
" 344	"	France	153	"	191	"	= 55	"
" 11	"	Belgium	4	"	7	"	= 63	"
" 46	"	Italy	5	"	41	"	= 87	"
" 12	"	America	8	"	4	"	= 33	"
Of 1605	operations		738	"	867	"	= 54	"

Michaelis has arrived at the same conclusion. His statistics show of 258 cases 118 recovered, 140 died = 54%. Kayser found of 338 operations 210 bad and only 128 good results, that is, death occurred in 62%.

It can scarcely be doubted that even these results appear somewhat too favorable. The great majority of operations with favorable issue are published, whilst those which end fatally are rarely brought under the notice of the profession at large.

A good result may be expected when the operation is performed at the right time and under favorable external circumstances; moreover, when there has been no great loss of blood, and no unpleasant accidents have occurred during the operation. Not infrequently, however, under such favorable circumstances, death occurs, whilst, on the other hand, cases are known where, under the most unfavorable external conditions, in a greatly osteomalacic pelvis, and the woman being in a very weak state, the operation has unexpectedly been followed by a favorable result. Nay, even in some cases of far advanced carcinomatous degeneration of the cervix the women have recovered to succumb, at last, to the progress of the new growth.

When the hæmorrhage has not been very considerable, and the woman has got over the shock of the operation, and the uterine wound has well contracted, the principal danger arises from a secondary peritonitis. If no symptoms of this appear, the bowels have acted, and the woman feels herself better at the end of the first week, there is reason to hope for a good result.

The repetition of this operation upon the same woman by no means gives unfavorable results, and it even appears that, in such cases, the prognosis becomes better; according to Kayser the mortality is then only 29%. Cæsarean section has been performed relatively frequently twice, very rarely three times, upon the same woman, and a Mrs Adametz has even four times been successfully operated upon by Michaelis. The last time the abdominal cavity was not opened at all on account of old adhesions. Oettler, in Greiz, communicates a case where he successfully operated four times upon a rickety woman, and in the 'Revue Therap.' of the 15th Sept., 1870, a similar case is recorded.

All the children who at the commencement of the operation are in perfect health are born alive. Until the uterus is incised the child can in no possible way be endangered by the operation. After the incision the child is exposed to asphyxia, when the bleeding is very considerable, and most easily when the incision passes through the placenta. But always in such circumstances, on account of the hæmorrhage the child is to be extracted so rapidly that the asphyxia cannot assume a very high degree. Statistics of the result to the child (according to Kayser and Michaelis 30% died) are valueless, since the state of the children before the operation has not been duly considered. According to Michaelis all the children have been born alive when the operation has been performed before or immediately after the rupture of the membranes, and this shows, to a great extent, that the prognosis for the children is favorable, when previous to the operation they are perfectly healthy.

2.—Malposition of the Child

A. *Correction of the position by turning*

Version or turning is an operation by which the position of the child is changed, so that instead of the presenting fœtal part another part of the fœtus, the cephalic or pelvic extremity, is made to present.

The object of turning is a double one. It either serves to convert a position absolutely, or in the actual case only, unfavorable into a better one in order to obtain better chances of a safe delivery, or it is intended to change the position of the child from one in which it cannot be extracted into one which allows of immediate extraction. In both cases the fœtal position is changed in order to facilitate delivery; in the one a natural, in the other an artificial delivery. In the first case the operation is completed with the turning, in the other it is to be followed by extraction.

Version is made either by the head or the pelvic extremity.

a. Turning by the head—cephalic version.—Cephalic version can only be made for the purpose of changing the presentation and never for facilitating extraction, for the head situated above the brim is not suitable for immediate extraction. We have, therefore, here a very important restriction of the applicability of cephalic version; it is out of place whenever circumstances demand instantaneous delivery.

A great many circumstances have been mentioned as contra-indications. Of all these, provided the child be alive, only one merits serious attention. It is prolapse of the funis. If this has occurred whilst the os is slightly dilated it is preferable to let the cord and child remain as they are, because the deviated head does not press upon the funis. A sufficiently dilated os, however, allows an immediate termination of labour by podalic version and extraction, which, of course, is preferable to reposition of the cord followed by cephalic version. All the other contra-indications mentioned as such by some obstetricians cannot be accepted unconditionally. Thus, for example, pelvic contraction. This may frequently make cephalic version impracticable. Yet, at the commencement of labour and in only slight pelvic contraction, the operation is often successfully performed. That the head be close to the os is useful, but not necessary. Also, the intact membranes facilitate the operation, but their rupture does not render it impossible. Regular uterine action is by no means required, especially when the os is only a little dilated. Cephalic version is most easily performed just in the absence of pains, and the delay in the delivery is less injurious after the head has been made to present. The operation certainly does not require a complete dilatation of the os uteri. We shall show that the operation is then most beneficial when the os is only slightly dilated.

When in a given case cephalic version is indicated it remains to be determined at what period of the labour this should be performed.

Mattei, Esterle, C. Braun, Hecker, and Hegar, have recommended turning by the head to be performed even during pregnancy. This proposal may be adopted when there has been an opportunity of examining during pregnancy, and the more so since the change of position is then generally very easily effected. Only no great advantage can be expected from it, for just in those cases where, at term, the head does not present, the tendency of the child to change its position is usually so very considerable that the chances of retaining the turned child in the head position are very slight.

At the commencement of labour turning by the head deserves much greater consideration. If the os is closed or, at the most, able to admit one or two fingers, turning by the head, except in some few cases, is preferable to that by the feet. Even if the pelvis is moderately contracted, the operation is not absolutely contraindicated, for the head will have time enough to accommodate itself to the brim. In placenta prævia it is always better to turn by the feet, for the slightly dilated os is more effectually closed by the feet than by the head.

Mattei and Hegar have also proposed to convert breech into head positions towards the end of pregnancy or at the commencement of labour. Provided the turning can be effected without difficulty, no serious objections can be made to that proposal because, on the whole, head positions are more favorable to the child than breech positions.

The case is different when the os is almost or entirely dilated. Moreover, when the waters have escaped, turning by the head is usually impossible, or at least so difficult that protracted attempts are certainly not justifiable. Before the membranes have burst the child may easily and successfully be turned by the head, but whether this be advisable is quite another question. The contracted pelvis forms, for reasons to be mentioned as we proceed, a contra-indication. But even in the absence of any disproportion it must always be borne in mind that under such favorable conditions turning by the feet is an operation wholly indifferent to the mother, and the child must almost necessarily derive benefit from it, when extraction follows in a suitable time. Besides, after podalic version labour can be terminated at any time, whilst after turning by the head strong pains are necessary, else delivery is delayed for a long time. All that has been above said as contra-indicating cephalic version applies more forcibly still to those cases where the os is dilated; then only may this operation be performed, when the head is not too far from the brim, the pelvis normal, the membranes intact, and the uterine action such as to allow a reasonable prospect of a speedy termination of labour.

With regard to the prognosis of cephalic version, it can in general be said that it is equally favorable to mother and child. It is because the operation itself is harmless and the head position is, of all presentations, the most favorable.

The methods of operating are the following:

1. *Presentation of the head by the mere posture of the woman.*—The mildest and the least hazardous proceeding is to effect a head presentation by means of a suitable posture of the woman. Success is, as a rule, certain when the head has deviated to one side and the membranes are intact; for instance, if the head is situated somewhat to the right, the woman is placed on her right side. The fundus uteri with the breech falls over to the right, and the head engages in the brim. A cushion placed under the right side of the mother, so as to press upon the side of the abdomen, may facilitate the engagement of the head.

In suitable cases this proceeding is successful; but unless the head is quickly fixed by the pains the previous faulty presentation is easily assumed again when the woman changes her posture, or if she remains in that posture the head is liable to deviate to the opposite side. From time to time the position of the head is to be ascertained. As soon as the head presents the woman assumes the dorsal posture, whilst the uterus is prevented from deviating too far to one side. By the rupture of the membranes the head can be fixed with great certainty; but as long as the os is undilated it cannot be done, since prolapse of the funis may occur, which will place the life of the child in the greatest danger. External pressure upon the uterus also serves to fix the head, and gives very good results when the os is dilated; it is, however, uncertain when the os is not at all or scarcely dilated.

2. *Turning by external manipulations alone.*—Before internal manipulations are to be thought of turning should always be tried by the more *gentle* external manipulations. For that purpose the woman is to lie on her back, and the operator sits on the edge of the bed (at about her thighs), or he stands at the bedside at a place corresponding to the woman's chest, and manipulates the abdomen from above. One hand pushes the breech upwards with some force, and the other directs the head towards the brim. The manipulations are made during an interval in the pains, and during the pain the fœtus is fixed in a position approaching as much as possible the one desired.

3. *Turning by combined external and internal manipulations.*—If external manipulations alone fail in engaging the head in the pelvic inlet, this may be tried by combined external and internal manipulations; not only when the os admits the whole hand, but as soon as one or two fingers can be passed internal manipulations may be practised. According to the teachings of Braxton Hicks, one hand manipulating inside pushes the shoulder away, whilst the other, acting externally upon the head, approaches it to the fingers within the os, and the head thus plays between both hands. If now the breech does not completely ascend to the fundus, the hand within the vagina is withdrawn, and whilst one hand continually presses upon the head the other pushes the breech upwards.

We have already stated above the reasons why this operation is very exceptionally performed when the os is only slightly dilated.

If it is to be done it is advisable either to manipulate externally or to act simultaneously from within and without. The os being sufficiently dilated to admit one hand, the internal manipulations of Busch or d'Outrepont may be adopted when Braxton Hicks's method fails. It is always useful to assist the hand manipulating internally with the other hand externally (for instance, by pressing upon the head or pushing the breech upwards).

Busch's method is especially applicable when the membranes are intact, and contain a large quantity of liquor amnii. It consists in the following manipulations:—When the head of the fœtus is to the left, for example, the right hand is passed through the os without rupturing the membranes. Arrived at the head, the membranes are ruptured, the head grasped by four fingers and guided into the brim. Two fingers are left there to control the position of the head, and somewhat feeble pains are strengthened by friction. The fingers are removed when the head is firmly engaged.

D'Outrepont's method is often successful in cases where favorable results can no longer be expected from that of Busch, especially when there is a small quantity of liquor amnii, or after its escape. If the head lies to the left the left hand is passed up into the interior, and the shoulder is pushed into the right side of the mother. Here also the external pressure upon the head with the right hand can never be dispensed with.

β. *Turning by the breech.*—In general, turning by the breech is then indicated when, the os being slightly dilated, cephalic version was originally intended, and this has become impracticable, because the head is too far away from the inlet. Under such conditions a transverse or oblique position is usefully converted into a breech presentation.

If alteration of the posture of the woman or external manipulations alone suffice, turning by the breech may be tried. But if internal manipulations are necessary, it is always better to bring one foot down into the vagina. The breech position has hardly any advantages over an incomplete foot position, and when the breech has entered the pelvis a speedy extraction may become impossible, whilst this can easily be done by one foot. At the time when the breech is made to engage in the brim, it cannot be said with certainty whether speedy extraction will be necessary or not, and since after the descent of the breech a foot cannot possibly be brought down, it is certainly always for the interest of the child to turn by one foot.

In exceptional cases, where the hand cannot reach the feet at all, or only with the greatest difficulty, as when a transverse position has persisted too long, it then becomes a necessity to turn by the breech. Betschler says that in the live child, even in very difficult cases, the breech can be brought to the brim by a finger hooked on the perineum, and Schmitt recommends bringing the breech downwards by leverage movements. In the dead child, and when the shoulder is not too low down, the breech can be brought down by hooking one finger in the anus (Meissner thinks it also applicable to the live fœtus), or by a blunt or sharp hook, and thus embryotomy is at times avoided.

The operation of turning by the breech is the same, with

the exception of the last-mentioned cases, as that of cephalic version.

γ. *Turning by the feet.*—Under turning by the feet (podalic version) is understood the artificial conversion of a head or transverse position into one with the feet presenting.

The operation is indicated in the following cases:

1. In transverse or oblique positions, for the purpose of correcting those positions, when cephalic version is either quite impossible or only practicable under the greatest difficulties, or when turning by the feet, for reasons above mentioned, deserves the preference.

2. In head presentations, when there is reason to suspect that, in the given case, the natural course of the labour will be more unfavorable for the mother, the child, or for both, than the course of an artificially produced footling presentation. This may occur in faulty presentation of the head or face, in the descent of extremities or of the funis in front of the head, in placenta prævia, in malformations of the fœtus, and in the contracted pelvis. The conditions under which podalic version, in these cases, is successfully made, will be given in the special pathology of parturition.

3. When labour must be immediately terminated and the presenting part cannot be drawn upon; in such a case podalic version is performed with the sole object of extracting the child.

Version alone does not completely meet this last indication. It has to be followed by extraction also. The other indications only requre turning, and the termination of labour is left to nature. However, it often occurs that subsequently symptoms appear which render extraction necessary.

The practicability of version is always dependent upon two conditions, viz. (1) the pelvis must not be absolutely too narrow, and (2) the presenting part must not so occupy the brim that the hand cannot in any way pass by it.

It is not particularly necessary that the os uteri be so dilated as to allow the hand to pass through it; on the contrary, the chances of version are the greater the earlier it is performed. The operation is certainly most easy when, in the absence of any deformity of the pelvis, the os is dilated, the child freely movable, and the membranes intact.

The prognosis is, under the circumstances before mentioned, in every way favorable for the mother and the child. The operation itself, if dexterously performed, is neither dangerous to the mother nor to the child. Pelvic positions are, on the whole, more unfavorable to the child than head-positions; but by the intervention of art, at a suitable moment, the chances are greatly increased, so that turning, if performed under favorable conditions, certainly promises excellent results.

The results, however, are different when the operation is performed under unfavorable conditions, as in a contracted pelvis after the waters have escaped for a long time and the uterus has firmly contracted. Even in these cases the act of turning is very rarely indeed hazardous to the child (it is different as regards the subsequent extraction); but the mother is far more exposed to dangers, which, however, are not due so much to the turning as to the unfavorable circumstances under which this has to be performed. For the introduction of the hand into the uterus, and the version of the child, may produce a severe inflammation of that organ, or, if a predisposition exists, it may be the determining cause of a rupture.

It is of the greatest importance to make an accurate external and internal examination before performing the operation. Sufficient information as to the position of the child is obtained by the external, and the condition of the os and the position of the presenting fœtal part is ascertained by an internal examination.

The dilatation of the os must also be taken into consideration, whether it admits the whole hand or not. There is no need to put off the operation till the os is completely dilated; but if it will admit one or two fingers the proceedings recommended by Braxton Hicks and R. Barnes are to be adopted. If, for instance, the child is in the first head position, and the membranes are intact, two fingers of the left hand are passed through the os and applied to the head. If the os is high up, the introduction of the half or the whole hand into the vagina is necessary, and the operation should then be performed under chloroform. Whilst the head is pushed by the fingers towards the left iliac bone the right hand externally presses the breech greatly to the right. If in this way the head has been made to deviate the fingers are applied to the shoulder. The child being in a transverse position and the membranes intact, the breech is often remarkably quickly brought into the os by the transverse tension of the uterus, or the knee can be reached with one finger, and is then dragged down into the os. If this, however, does not succeed, strong pressure upon the breech will usually bring the foot downwards, and allow of its being drawn into the os. The smooth yet firm membranes greatly hinder a firm grasp of the foot, and often this is not obtained until after the membranes have been ruptured. Occasionally, when the head has deviated from the brim, external pressure is usefully applied, at one time to the breech, and at another to the head. In transverse positions one often succeeds more easily in bringing a knee or a foot within reach of the internal os. If the turning of the child in this way fails, we must wait until the os is more fully dilated.

Though the os be widely open, so as to admit one hand, turning is, nevertheless, when a suitable occasion presents, to be attempted in the way just described. If there is any difficulty about it the whole hand may be introduced, and the feet seized at the place where they are situated.

For the operation the woman must be in a suitable posture. This applies, not only to cases when version itself is surrounded by difficulties, but to all cases where turning is to be followed by extraction, and where this is not uncommonly easy, as, for instance, in the second twin.

The lateral posture offers material advantages for the act of turning; but extraction in this position is certainly more difficult than when the woman lies upon the cross-bed. However much for reasons of convenience and usefulness the lateral posture deserves to be recommended, if turning alone is intended, or if extraction will apparently be free from impediments, we yet prefer, in cases where extraction is necessary, and where it promises to be difficult, especially in the narrow pelvis, the use of the cross-bed. However, during the operation itself the lateral posture can easily be converted into the oblique one above mentioned, in which extraction is so very easy. The knee-elbow posture, which in certain cases may be very useful, can almost always be replaced by the lateral posture.

Previous to the operation everything must be prepared for it and for the reception of the child. A sufficient number of napkins must be at hand, also warm water to bathe the child, and an elastic catheter for the treatment of asphyxia. It is also desirable to have a sling and a sling carrier at hand. Bladder and rectum are always to be emptied before the operation.

In cases where the delay of a few minutes is of no great consequence, and where no weighty contra-indications exist, turning is best made under chloroform; for not only does it spare the woman the pain produced by the introduction of the hand into the vagina, which is almost always very great and sometimes even very intense, but it also greatly facilitates the various steps of the operation itself.

The hand which most naturally corresponds to the position of the child's feet is used in turning. In head presentation, the feet being, for instance, in the right side of the uterus—first position—the left hand is used; and, on the contrary, the right when the child is in the second position. In transverse position, also, that hand is taken which corresponds to the side of the mother towards which the feet lie—that is, the right hand when the feet are to the left, and the left hand when the feet are to the right.

In transverse positions, when the os is sufficiently dilated and the membranes intact, the operation is, under ordinary circumstances, performed in the following way:

The shirt sleeves are turned up, and the back of the hand is well smeared with oil, and the fingers brought together in the shape of a cone. Whilst the other hand guards the genital fissure, so as to prevent any dragging upon the labia minora and the pudendal hair, this is introduced into the vagina. When the hand has passed through the os, it is pushed onwards with the dorsal surface applied to the inner surface of the uterus, between it and the ovum, straight upwards to the place where the previous examination has shown the feet to lie. If the feet, for example, lie behind and to the right the hand is passed upwards to the right sacro-iliac articulation, and if they are in front and to the left, behind the left ramus of the pubis. The other hand must always be placed upon the uterus externally, and push the small fœtal parts contained within the uterus downwards, so as to act with the hand manipulating internally. By pressure also upon the head or breech the turning of the child is greatly facilitated. If a foot is felt through the membranes it is seized, and as the membranes thereby are ruptured, it can be brought down into the os and the vagina. The breech having descended so far as to be engaged in the brim, the operation of turning is completed. Sometimes, however, the membranes are so tough that the seizure of the foot does not rupture them; then Hüter's proposal of turning within the intact membranes may be adopted; but the foot, still covered by the membranes, is sometimes so slippery that, in order to obtain a firm hold, the membranes must be ruptured. Besides, it matters little, when the forearm is within the vagina, when or where the membranes are ruptured, for the sudden escape of the waters is prevented by the arm acting as a plug.

In head presentations the operation is performed in the same way. The foot in front is usually close to the head, and can easily be grasped. Sometimes turning fails, because the head is fixed too firmly in the brim, and by drawing upon the foot, it descends lower down together with the foot. The head may then be pushed back with the thumb, whilst the foot is held between the index and middle finger and drawn upon. If this also gives no satisfactory result, the double manipulation, first recommended by Justine Siegemund, may be adopted. It consists in putting a sling round the foot and drawing upon it, whilst the other hand pushes the head back. If the foot cannot be brought into the vagina, and we do not wish to let it escape, a sling is put round it by means of a sling carrier. Braun's instrument for replacing the funis is very useful for the purpose. It is brought to the foot, the sling put round it, and then held.

If the waters have only shortly before escaped, and the child is still freely movable in the uterus, turning, though not quite so easy as when the membranes are intact, is yet practicable without any great difficulty. Prolapse of an arm in transverse position by no means impedes turning. It is generally left there because it remains then extended along the abdomen of the child, and its artificial liberation is not required. If there is any fear that in turning it may recede into the uterus it may be put into a sling; at any rate there can be no question of replacing it.

When the waters, however, have escaped a long time previously, the operation may present considerable difficulties.

As the energy of the uterine action increases with the amount of the resistance, it follows that in neglected transverse positions the pains set in in rapid succession, and force the presenting shoulder continually deeper into the true pelvis. The intervals between the pains become shorter and shorter until, finally, the uterus has reached the climax of its activity, and persists in uninterrupted tetanic tension. Such forcible contraction, uninterrupted by even a short interval, is easily followed by inflammation of the parenchyma of the uterus, and of its serous covering.

Since this rigid contraction of the uterus drives the presenting shoulder with great force into the brim, and keeps it fixed there, the introduction of the hand will meet with considerable opposition.

Chloroform is the sovereign remedy in such cases, and should always be employed.

In deep narcosis the tension of the uterus subsides, and the operator is able, quietly and gradually, to insinuate his hand past the presenting part without using any great force.

When chloroform is not at hand large doses of opium may be given, or by a warm bath relaxation of the contracted uterus can be brought about. Venesection, also, in the erect posture of the woman, until she faints, has a decided influence upon the relaxation of the uterus. Such a treatment, however, must be restricted to very extreme cases, for women, under such conditions, cannot bear the loss of much blood.

In these difficult cases of turning the woman is always to lie on her side; for it greatly facilitates the introduction of the hand and the seizure of the feet of the child, particularly when they lie more in front above the symphysis. Sometimes the knee-elbow posture would offer advantages, but then the use of anæsthetics must be given up.

If the space of the brim is very limited, Levret, Stein senior, Deleurge, and Birnbaum propose to bring the other arm down, so as to enable the hand to pass.

When the hand has passed the presenting part there is often great difficulty in seizing the foot, while the knee is close at hand. The advice of Simpson, Simon Thomas, and R. Barnes, not to turn in such cases by the foot, but by the knee, deserves all consideration in such difficult cases. The knee is nearer than the foot, and to grasp the latter the whole hand is required, whilst the knee can be brought down by hooking the index finger over it.

Sometimes when the lower extremities are too far from the brim, they may be made to approach, the hand manipulating internally by rotation of the trunk round its longitudinal axis. Von Deutsch expects great advantage from this proceeding.

Or it may be comparatively easy to engage the breech in the brim if the shoulder does not lie too deep, when there are insurmountable difficulties in seizing one of the feet or a knee.

Even if one foot has been seized turning the child by it in such protracted cases may be almost impossible; double manipulation will then be advantageous, or the other foot must also be brought down, and the child turned by them.

B.—*Embryotomy as a means of Extracting the Child*

Embryotomy has already been mentioned as a means of delivering the woman in cases of malformation or diseases of the fœtus; in the highest degrees of pelvic contraction, the breech presenting; or when, after craniotomy, the strongly developed fœtal trunk cannot pass through the pelvis without reduction of its bulk. It is also indicated when a dead child, presenting by the shoulder, cannot be turned, or when this at least is more dangerous to the mother than embryotomy. If the pelvic deformity is not too great it may be advisable to defer embryotomy for some time, in the hope that delivery may be effected in the natural way, by what is termed spontaneous evolution.

In a given case it is, of course, a matter of some difficulty to decide which method of delivery is the least hazardous to the mother. Embryotomy is not an operation entirely indifferent to the mother, although no immediate dangers can accrue to her if performed with some skill. All that can be said is that a gentle attempt at version can always previously be made. But by no means is force to be used in turning. If this is too difficult, and spontaneous evolution is out of the question, recourse must be had to that destructive instrument, and at the sight of the mutilated child the accoucheur must console himself with the consciousness of having preserved the life of the mother.

Embryotomy can be performed after two methods. Either the thoracic and abdominal cavities are eviscerated, in order to allow of version and extraction in the faulty position, or the head is separated from the trunk, and they are each of them separately extracted.

The first method is, then, only to be recommended when the neck is so high up that it cannot be reached without great difficulty, and when, consequently, the breech and the feet are nearer to the brim.

The operation is performed in the following way :—Guided by the left hand, a scissor-shaped perforator is introduced, and pushed into the thorax, so as to open an intercostal space. If a prolapsed and greatly swollen arm materially impedes the operation, it is to be amputated at the shoulder-joint by strong scissors ; but if this can possibly be avoided it is better for this reason alone that it forms an excellent handle for extraction. The finger will then be able to separate the two contiguous ribs so that several fingers can be passed into the thoracic cavity in order to remove its contents. From this perforation is continued through the diaphragm into the abdominal cavity, or a new incision is made into the abdominal walls in order to empty the abdominal cavity also. The proceedings following evisceration are different. Podalic version is, as a rule, very difficult, even at this time, and is, moreover, unnecessary. Where the shoulder is high up it is best to bring the pelvic extremity down, either by the finger, or, if necessary, by the crotchet or the craniotomy forceps, and thus to imitate the process of spontaneous version. If the shoulder is in the pelvis and the arm prolapsed, the process of spontaneous evolution may be imitated by dragging the shoulder low down, in a direction opposite to that where the breech lies, and the pelvic extremity is then drawn past the thorax. Occasionally it may be useful to adopt the process of Michaelis. He proposes to break the vertebral column and to extract the child doubled up, that is, thorax and abdomen and head and pelvis together. Simpson recommended spondylotomy, that is, complete division of the child at the most prominent point of the vertebral column.

If the neck is easily accessible decapitation is preferable, for even after evisceration delivery may still offer very great difficulties. If the arm has prolapsed this is strongly drawn upon in order to make the neck descend deeper ; then the index finger, or, in difficult cases, a blunt-hook is put round the neck, and firm traction made upon it. Guided by one or two fingers placed around the neck the soft parts and the vertebral column are then divided by strong and bent scissors.

C. Braun has constructed an instrument (the decollator) to wrench off the head from the trunk. It consists of a steel rod fitted to a transverse handle, and terminates in a sharply-bent hook provided with a knob. When the neck has been made accessible in the way above described the hook is introduced under the guidance of the hand, put around the neck, and fixed by firm traction. By repeated rotatory movements in one direction, with simultaneous traction, the head is separated from the trunk. Experience, extending until now over thirty cases, seems to be in favour of this instrument.

A number of other instruments have been designed for decapitation. Of these the écraseur of Stiebel junior, deserves most to be recommended.

After the head has been separated from the trunk, the extraction of the latter can easily be effected either by the arm or by means of a hook fixed in it. The extraction of the head also is not difficult when the pelvis is not too much contracted. Under normal conditions, if the head is not expelled by the pains, it may be removed from the uterus by external pressure as the placenta is, or one hand is introduced and the head is extracted, either by the lower jaw or by grasping it by the orbits and the base of the skull. In a greatly contracted pelvis the extraction of the head may meet with serious obstacles; but it is usually successful by drawing upon the lower jaw, assisted by powerful external pressure, the more so since the brain can escape through the foramen magnum. In other cases the head, fixed externally, is perforated and extraction again tried. If it then also fails the head is to be extracted by the craniotomy forceps, one blade being introduced into the cranial cavity through the opening made by the perforator, or the cephalotribe is applied, and after having grasped the head it is so turned that its blades come to lie in the smallest diameter of the pelvis. Thus extraction is accomplished.

Literature.—Denman, Introduction to the Practice of Midwifery. London, 1795, p. 395.—F. A. Mai, Progr. de necess. part. quand. praem., &c. Heidelb., 1799.—Wenzel, Allg. geb. Betr. und über die künstl. Frühgeb. Mainz, 1818.—Reisinger, Die künstl. Frühgeb., &c. Augsb. u. Leipz., 1820.—Ritgen, Die Anzeigen der mech. Hülfen, &c. Giessen, 1820, and Gem. d. Zeitsch. f. Geb., i, p. 281.—Burckhardt, Essai sur l'acc. prém. art. Strassb., 1830.—Stoltz, Mém. et observ. sur la prov. de l'acc. prém. &c. Strassb., 1835, and Gaz. Méd. de Strassb., 1842, No. 14; and 1843, No. 1.—Hofmann, Neue Zeitsch. f. Geb., B. 15, p. 321, B. 16, p. 18, and B. 23, p. 161.—Krause, Die künstl. Frühgeb. Breslau, 1855.—Germann, M. f. G., B. 12, p. 81, 191, 271, 361, and B. 13, p. 209.—Elliot, Obstetric Clinical. New York, 1868, p. 157.—Thomas, American Journal of Obstetrics, vol. ii, p. 732.—Spiegelberg, Arch. f. Gyn., B. I, p. 1.—Stadfeldt, s. Virchow-Hirsch'scher Jahresbericht über, 1870, p. 540.—Litzmann, Arch. f. Gyn., B. II, p. 169.—Roederer, De non damn. usu Perfor., &c. Gotting., 1758.—Osborn, Essays on the Practice of Midwifery, &c. London, 1792.—Boër, Natürliche Geburtshülfe, 3 Aufl., 3 Bd. Wien, 1817, p. 199.—W. J. Schmidt, Heidelberger klinische Annalen, I, p. 63.—Wigand, Die Geburt des Menschen. Berlin, 1820, B. II, p. 52.—K. Chr. Hüter, Die Embryothlasis, &c. Leipzig, 1844, and C. Hüter, M. f. G., B. 14, p. 297 and p. 334.—Credé, Verh. d. geb. Ges. in Berlin, 3, 1848, p. 1.—Kiwisch, Beiträge zur Geb., II Abth. Würzburg, 1848, p. 43.—Hennig, Perf. und Cephalothrypsis. Leipzig, 1855, and M. f. G., B. 13, p. 40.—C. Braun, Zeitschr. d. Ges. d. Wiener Aerzte, 1859, p. 33.—Lauth, De l'Embryothlasie, &c. Thèse, Strasbourg, 1863.—Barnes, Obstetrical Transactions, VI, p. 227; and Obstetrical Operations, 2 ed., 1871, p. 289.—Rabe, Deutsche Klinik, 1869. No. 47—51.—M. Duncan, Transactions of the Edinburgh Obstetrical Society, 1870, p. 1.—Rokitansky, Wiener Med. Presse, 1871, No. 8, &c.—François Rousset, Traité nouveau de l'Hysterotomotokie ou enfantement Caesarien, &c. Paris, 1581, lateinisch von Caspar Bauhin, 1586.—Simon, Mémoires de l'Acad. de Chir., T. I, Paris, 1743 p. 623, and T. II, 1753, p. 308.—Levret, Suite des Observ., &c. Paris, 1751, p. 237.—Lauverjat, Nouv. Méth. de Prat. l'Opér. César., &c. Paris, 1788, deutsch von Eysold. Leipzig, 1790.—G. W. Stein, Kleine Schriften. Marburg, 1798, p. 205.—Graefe, Graefe und Walther's J. für Chir., 1826, B. IX, p. 1.—Schenk, Siebold's J., 1826, B. V., p. 461.—Michaelis, Geb. Abhandlungen, Kiel, 1863, p. 34.—Winkel, M f. G., B. 22, p. 40. Justine Siegemund, Die Kgl. Preuss. und Chur.-Brand. Hof-Wehe-Mutter, &c. Berlin, 1752, pp. 37, 40, 43, 62, and 64.—H. Deventer, Neues Hebammenlicht, &c. Jena, 1717, pp. 302, 307 *seq.*—W. Smellie, A Treatise, &c., vol. i, 3 ed. London, 1756, p. 352, *seq.*—Aitken, Principles of Midwifery, &c. London, 1786.—Osiander, Neue Denkw., I, Band 2. Göttingen, 1799, p. 36. Grundr. d. Entb., 2 Th., 1802, p. 35; and Handb. der Entb., 2 Aufl., B. II. Tüb., 1830, p. 321.—Labbé, De la version du Fœtus, Strasb., 1803; Eckard, Parallèle des acc. Nat., &c., Strasb., 1804; and Flamant, Journ. Compl. des Sc. M., T. XXX, Cah. 17, p. 3.—Wigand, Hamburger

Mag., 1807, 1, B. 1, St., p. 52, und drei Abhandl., &c. Hamburg, 1812, p. 35 (s. Wittlinger's Analecten, I, 2, p. 362).—D'Outrepont, Progr. von der Selbstwendung u. d. Wend. auf d. Kopf. Würzburg, 1817. Abh. u. Beiträge. Th. I, p. 69, und in dem neuen Chiron., B. I, H. 3, p. 511.—Busch, Geb. Abh., 1826, p. 27.—Ritgen, Anzeigen d. mech. Hülfen, &c., p. 411, Gem. deutsche Z. f. G., B. II, p. 213, and B. IV, p. 261.—Mattei, Gaz. des Hôp., 1856, Nr. 55.—Velpeau, Traité élém. de l'Art des Acc., T. II, Paris, 1829, p. 703.—Nivert, De la version cephalique, &c. Paris, 1862.—v. Franque, Würzb. Med., Z., 1865, B. VI.—Hegar, Deutsche Klinik, 1866, Nr. 33.—Betschler, Rust's Magazin, &c., B. XVII, 1824, p. 262.—W. J. Schmitt, Heidelberger klinische Annalen., B. II, 1826, p. 142.—A. Paré, Briefve Collection de l'Administration Anatomique, &c. Paris, 1550, et les œuvres, &c., 6 éd. Paris, 1607. De la gener. chap. xxxiii.—Guillemeau, De l'heureux Accouch. des Femmes, &c. Paris, 1609.—Mauriceau, Traité des Mal. des Femmes grosses, &c., 6 éd. Paris, 1721.—De la Motte, Traité compl. des acc., &c. Paris, 1722.—J. v. Hoorn, Die zwo, &c. Weh-Mütter, Siphra u. Pua, &c. Stockh. and Leipz., 1726, p. 125.—N. Puzos, Traité des acc., &c. Paris, 1759.—Levret, L'art des acc., 2 éd. Paris, 1761.—Deleurye, Traité des acc., &c. Paris, 1770, übersetzt von Flemming. Breslau, 1778.—F. B. Osiander, Neue Denkwürdigkeiten, I, 2. Göttingen, 1799, p. 108 seq., und Handb. der Entbindungskunst, 2 Aufl., 2 Band. Tübingen, 1830, p. 320, seq.—J. H. Wigand, Die Geb. d. Menschen. Berlin, 1820, Bd. II, p. 442.—Oehler, Gem. d. Leitschr. f. Geb., B. 7, p. 105, and Neue Z. f. G., B. 3, p. 201.—Michaelis, Neue Z. f. Geb., B. 6, p. 50.

PART VII

SPECIAL PATHOLOGY OF PARTURITION

Labour is said to be healthy when the mechanism of the expulsion of the ovum is subject to no disturbance. This has already been shown to depend upon the normal condition of the expelling forces, on the one hand, and upon the obstacles they meet with, on the other. Now, as the latter consist of the relations of the child to be expelled to the passages through which it has to pass, deviations from the normal mechanism may be caused by—

1. Anomalies of the expelling forces.
2. Contraction of the maternal passages.
3. Anomalies (in shape or position) of the fœtus which obstruct its passage through the normal parturient canal.

We shall now consider, in their order, all these causes of the abnormal mechanism of labour.

CHAPTER I

ANOMALIES OF THE EXPELLING FORCES

As the resistance which the expelling forces meet with varies greatly, so also does the intensity which they must reach to terminate labour vary equally greatly.

In one case so moderate a force suffices to expel the child that the woman scarcely heeds it, and is completely taken by surprise by the commencing labour. In another case the greatest efforts of the contracting muscular fibres of the uterus and the assistance of the auxiliary forces are necessary to overcome the obstacles to delivery. It is evident that in the first case, although the pains are unusually feeble, we are not entitled to speak of pathological weakness of the expelling forces, since they accomplish the expulsion of the child; and in the other case it is equally evident, although the pains have reached an extraordinary degree of intensity—so great, perhaps, as to awaken the fear of a rupture of the uterus—that there is no reason to call it an anomaly of the expelling forces. Here the anomaly lies in the abnormal obstruction; the too great increase of the expelling forces is only a consequence of that, and in general a very salutary one, without which labour could not have been terminated naturally. It would tend to the worst practical consequences if such pains were considered pathological, and therapeutic measures accordingly taken against them. The test by which alone we can estimate the pains is the effect they produce, and thus even the most powerful pains may still be too feeble for a given case.

Therefore no general measure can be given of too strong or too feeble pains. We have seen very feeble pains suffice in one case, and the strongest possible even insufficient in another to terminate labour. We can only speak of relatively too strong and relatively too feeble pains.

Let us add that, in general, the strength of the pains accurately increases with the amount of the obstacles, and when this relation between the expelling force and the obstacle is disturbed, the obstacle is, in the great majority of instances, the primary active agent, and the anomaly of the expelling forces is the secondary, the consequence.

These considerations will serve to justify the course we adopt in regard to the anomalies of the expelling forces. As they are rarely of primary origin, we shall here speak of them shortly, and shall treat of them in connection with the abnormal obstacles as caused and produced by them.

Too feeble pains are very frequent at the commencement of the first stage, so that the first part of labour passes on very slowly, and there may even be a complete cessation of the pains for some time. At this period the weakness of the pains has no practical significance. Since with such feeble pains the woman is able to walk about and attend to her lighter duties, as long as there is no urgent indication to hasten on the labour, all that is required is the necessary patience of the attendant and the woman. But if she and her relatives become very uneasy on account of the retardation of stronger pains, or if a longer delay also is to be feared from a medical point of view, which exceptionally occurs through premature rupture of the membranes, labour may be made to advance by the application of the douche, or by the introduction of an elastic catheter.

In the further progress of the first stage and during the stage of expulsion, primary defective pains are rarely seen.

Sometimes, but by no means always, this occurs in consequence of a general debility, or after exhausting diseases. More frequently it depends upon congenitally feeble development of the uterine muscular fibres.

The condition of the latter is, above all, of the greatest importance. Their contractility may have been diminished by too great distension of the uterus during pregnancy, as by twins and by hydramnios, and, without doubt, also by endometritis, metritis, and their consequences. A very frequent cause of debility of the uterine muscles, and one practically of very great interest, is frequent and rapidly succeeding labours, especially when they have been very difficult. In the latter case there is almost always an anomaly of the pelvis, which, in conjunction with the defective pains, acts very unfavorably. Here, as well as in the cases above mentioned, treatment directed to the strengthening of the pains cannot do much, and artificial means must replace them as soon as it appears possible and necessary. Only when the weakness of the expelling force is due to an enormous distension of the uterus do we possess, in suitable cases, a very effective remedy for increasing the uterine action in the artificial rupture of the membranes.

When the shape and size of the uterus are altered by neighbouring tumours its action is very easily disturbed. Independently of new growths in the abdomen, the accumulation of fæcal masses in the rectum, or of the urine in the bladder, deserves to be noticed in this respect. The latter is quite a common cause of feeble pains, and, after emptying the bladder by means of the catheter, the pains almost always become stronger.

It is also a well-known fact that mental emotions influence the strength of the uterine action, and this is worthy of the attention of the practitioner. In such cases opium is the most effective remedy for increasing the pains.

It rarely happens that the pains are unable to dilate the os and to push the head through the normal pelvis without there being a known cause for this bad effect of the uterine action. If in such cases the bag of waters be still intact, though the ovum be not over-distended, artificial rupture of the membranes is the best means to advance labour. Even where only the small quantity of liquor amnii in front of the head has drained off the os is seen rapidly to dilate and the head to descend. If there are objections to the artificial rupture of the membranes the introduction of an elastic catheter will have a very precise effect. After rupture of the membranes the defective uterine action may be increased by gentle friction of the fundus uteri. The action of internal remedies is always very uncertain, and the most efficacious of them, ergot of rye, produces a general, uniform tension of the uterus, rather than regular and powerful pains. After delivery this acts very beneficially upon the closure of the maternal vessels of the placenta, but it has very slight if any influence whatever upon the progress of the child.

Too strong pains are not injurious when the labour is properly managed. When the resistance is very slight and the pains very strong, the os is very rapidly dilated, and the head is somewhat impetuously pushed through the genital fissure—*Partus præcipitatus*. If no support is given to the perinæum, especially when the woman is in an unsuitable position, considerable injury may be inflicted upon it. There is no necessity for moderating the powerful pains. They are either required to overcome a mechanical obstacle, or, in its absence, the child is rapidly born, and the strong pains produce a good contraction of the uterus.

The case is somewhat different with painful uterine action. This term is also a strictly relative one. There are women who, even during uninterrupted convulsive uterine action, do not complain of pain, whilst others, even at the commencement of labour, make a great outcry at the slightest contraction of the uterus.

So much, however, is certain, that the suffering produced by the uterine action may reach such a degree of intensity as to become unbearable, and may lead to momentary mental disturbance. Since we possess means to remove the pain with certainty and without danger, it would be cruel on our part not to use them in cases of great suffering.

The general tonic contraction of the uterus—tetanus uteri—is always due to extraneous causes, which, as a rule, are very evident and close at hand. This disease is never primary.

Convulsive contractions of some part of the uterus, the so-called strictures, which are rather frequent during the third stage of delivery, are very rare indeed before the child is completely expelled. They are chiefly due to repeated and less careful examinations of the external os. The previously flaccid os feels very tense and firm in the intervals between the pains, and slightly contracts during a pain. This is easily and safely removed by two or three not very superficial incisions into the os. Convulsive strictures of the internal os, happily very rare, may be a serious and important impediment to delivery. Warm baths and narcotics, as well as abstaining from continual trials at delivery, may here prove very beneficial.

The accessory force of the pains, the abdominal pressure, which acts upon the progress and expulsion of the child, may, at the commencement of labour, be regulated by the will, but towards the termination of labour it sets in involuntarily. It is seldom subject to any anomalies. In the rare cases, where it is completely absent or almost so, labour, nevertheless, proceeds under otherwise normal conditions quite undisturbed.

CHAPTER II

ANOMALIES OF THE MATERNAL PASSAGES

I.—Of the Soft Parturient Canal

A.—*Malformations of the Genital Canal*

The one-horned uterus, with or without an atrophied accessory cornu, has not been observed to disturb the process of labour. In the different forms of double uterus, labour, although frequently quite normal, has more often been prolonged than in the single uterus. The anomalous labour, independently of the mechanical impediment, which several times the septum of the vagina offered, appeared to be due chiefly to the oblique position of the impregnated half of the uterus, which exerted an unfavorable influence upon the pains and upon the position of the child, which was frequently a shoulder presentation. Several times also the uterus has been ruptured. The action of the one-horned or double uterus does not seem to be defective when the organ is in its normal position. If only one half of a double uterus contains a fœtus, the os of the other half sometimes remains entirely closed during labour, at other times it also dilates. If both halves are impregnated each may expel its fœtus independently of the other, and there may be a long interval between the expulsions. Even in simultaneous labour the contractions of the halves are quite independent, so that one may contract whilst the other dilates. In the third stage the double uterus easily produces abundant flooding, and, on account of its unequal contractions, this may be particularly dangerous when the placenta is inserted into the septum.

B.—*Occlusion and Narrowing*

1. *Of the Uterus.*—As complete atresia of the os prevents conception, it follows that an occlusion of the os observed in labour must have taken place during pregnancy.

Very frequently there is a superficial and easily separable agglutination of the external os. It is due to an inflammatory process of the lips of the os from a previous blenorrhœa. During labour the advancing head is seen to push the lower uterine segment forwards to the outlet, and to thin it more and more. This thinning may be so great that the head appears to be covered only by the membranes. By an accurate examination the os feels like a small and soft dimple, directed greatly backwards. If during a pain the finger or the uterine sound be forcibly pressed against the dimple, the agglutination of the os will suddenly give way. The os itself now very rapidly and completely dilates, and labour proceeds without impediment. Often the pains themselves succeed in breaking down the adhesions of the os. It very rarely occurs that the os only partially dilates after the agglutination has been torn through, and remains rigid so as later to require incisions.

There is very seldom so firm an adhesion between the maternal and fœtal membranes in the immediate vicinity of the internal os that the lower uterine segment cannot retract over the ovum. Separation by the finger or rupture of the membranes renders possible the dilatation of the os.

A firm cicatricial adhesion of the os forms during gestation still more rarely than agglutination. It follows inflammation of the cervix, and occasionally also cauterization. The os is distinctly felt to be occluded by cicatricial tissue. This may cause rupture of the anterior vault of the vagina unless assistance be given. To prevent this the occlusion must be cut through by means of a bistoury and scissors.

The cicatricial closure of the os is very frequently incomplete. More or less fine openings remain pervious, and conception is, therefore, possible, though difficult. Most frequently this stenosis is due to an ulcerative inflammation in the lying-in state. The treatment is the same as has been given above.

The narrowing or rather deficient dilatability of the os may also be due to an abnormal condition of its tissue. Induration produced by heterologous new formations will be considered as we proceed. Here we shall only speak of simple rigidity of the os. This is the consequence of chronic inflammatory processes of the uterus, and particularly of the cervix. It is most frequently met with in older pluriparæ, in whom, during pregnancy, the cervix has already been found greatly enlarged and thickened; this is noticed in a high degree when a previous prolapsus of the uterus has disappeared with the commencement of pregnancy. The internal os is, for the most part, comparatively easily forced open, but the thickened external os dilates only to a certain extent, in spite of strong pains. A few free incisions with a probe-pointed bistoury or curved scissors almost always removes the impediment. Spontaneous dilatation of the indurated os, especially when observed early in the commencement of labour, may very suitably be promoted by warm injections.

A hypertrophied cervix very easily leads to an effusion of blood into the diseased tissue. It may acquire an enormous size and, as the so-called "thrombus of the cervix," form an impediment to delivery.

2. *Of the Vagina and Vulva.*—Narrowness of the vagina is, in the majority of cases, congenital. The whole vagina is either narrowed uniformly, or ring-like strictures are met with. Labour is somewhat delayed by a very narrow but otherwise normal vagina. But the pains always succeed in dilating the softened vagina sufficiently for

the passage of the head. Longitudinal rents of the vagina are an exception. However, there are almost always solutions of the continuity of the mucous membrane, but these do not easily extend into the connective tissue of the pelvis.

The vagina is not so very seldom the seat of bridge-like folds, stretching from one side to the other; frequently they require no operation, for on account of their thinness they are easily torn by the advancing head.

The vagina is liable to be partially narrowed by all kinds of ulcerative processes, which heal with abundant formation of cicatricial tissue. The gradually advancing head often overcomes these obstructions, but occasionally superficial incisions will be necessary.

Abnormal rigidity of the hymen, which has not been ruptured in coitus, usually only slightly delays delivery, because the advancing head greatly stretches and tears it. Rigid and unyielding cicatrices, following previous destructive processes of the vulva, may urgently require operative assistance.

A too narrow genital fissure, with slightly yielding margins, may greatly delay the expulsion of the child; strong pains may produce considerable laceration. Lateral incisions into the tense margins may here also be requisite.

C.—*Changes in the Position of the Uterus*

A deviation of the uterus to one side, most frequently the right, is normal, and even if it be rather considerable does not, as a rule, interfere with the progress of labour. A great lateral deviation, accompanied by considerable flaccidity of the uterine walls, as frequently occurs in pluriparæ, may determine a transverse position of the child.

The inclination forwards of the uterus is almost always rectified at the commencement of labour by the woman lying on her back. The sole consequence it gives rise to, when the pelvis is normal, is a more backward direction of the os and a slower obliteration of the anterior lip.

A descent of the lower uterine segment, together with the presenting fœtal part, even to the pelvic outlet, very frequently occurs in primiparæ. As this position is only possible in a wide pelvis, so that the os can easily recede at the commencement of the pains, it rather facilitates delivery on account of the fœtal parts being so low down.

Considerable hypertrophy of the cervix may simulate prolapsus of the uterus. Here, as well as in the extremely rare cases of complete prolapsus of the gravid uterus at term, the impediment to delivery is chiefly due to pathological changes in the os. This is the seat of a chronic inflammation and induration, allowing only a slow dilatation, which renders incisions necessary.

D.—*Tumours of the Soft Parturient Passages*

1. *Fibroids.*—The influence of fibroids on the process of parturition varies greatly, according to their situation. The submucous and the larger interstitial fibroids only very exceptionally allow of conception, and when this has taken place absorption very frequently

follows. During labour they may act as a predisposing cause of rupture of the uterus, for the submucous tumours, by continually pressing upon one spot, extenuate the contractile tissue of their matrix, and the very existence of interstitial fibroid tumours shows that the uterine tissue is pressed asunder. In the third stage of labour they may give rise to profuse flooding, when they are so situated that the placental insertion cannot sufficiently contract.

Submucous fibroids with short pedicles—polypi—may be pushed forwards in front of the presenting fœtal parts, and by a careless examination may be mistaken for them. Assistance is often necessary, for the child has either to be extracted or the polypus has to be removed first. The removal of a polypus by the scissors is, if possible, advisable even in regard to the lying-in state. Sometimes the pedicle tears spontaneously. The non-pedunculated submucous fibroids can also be removed when they are situated in the lower portion of the cervix. Danyau and Braxton Hicks have enucleated very large tumours of this kind, and in this way have allowed delivery to take place by the natural passages.

Subperitoneal fibroids, in many cases, do not at all interfere with the progress of labour. If they are very large, and not situated at the fundus, but lower down, they may become a very serious impediment to delivery by obstructing the brim. If they do not completely obstruct the brim they may be so greatly softened that quite unexpectedly the child is expelled past the tumour, or at least can be extracted, though sometimes only after reduction of the bulk of the head.

In these cases extraction by the feet is always easier than by the forceps applied to the presenting head. An attempt should always be made first to push the tumour back into the false pelvis. Sometimes it recedes spontaneously from the true pelvis, drawn upwards by the uterus, which gradually retracts from above the ovum. This often succeeds by steady perseverance, even if the tumour appears to be very large, and at first completely immovable. If it fails we must have recourse to the Cæsarean section, the result of which, under such conditions, is almost always fatal to the mother and most usually so to the child. According to Lambert, two only of fifteen mothers on whom the operation was performed were saved.

2. *Carcinoma.*—The cervix is almost exclusively the seat of the carcinomatous degeneration; on this pregnancy has the most disastrous influence. Very profuse hæmorrhage may occur even during gestation, and during labour the fœtal part by pressing upon the os, often produces extensive ruptures of the cervix and of the lower uterine segment, for the cervix, whose extensile tissue is replaced by the rigid product of the new formation cannot dilate, or the pressure causes mortification of the new growth, and death ensues in childbed from the consequences of the gangrene.

Treatment must be chiefly directed to procure a sufficient dilatation of the os, and this is principally obtained by free incision into the degenerated masses. It is in the interest of the mother as well as of the child to terminate labour artificially as soon as it is practicable. When the child is alive the practicability of the Cæsarean section will have to be considered, and it appears that, for the sake of the child, this is too seldom resorted to, since we know from experience that the mothers extremely often die during labour or in the puerperal state.

Cauliflower excrescences give rise to profuse hæmorrhage, and render the prognosis equally unfavorable to the mother, because the new growth mortifies after delivery. A radical operation, therefore, is to be recommended, and should be performed before labour has set in. We must be on our guard lest we mistake these growths for placenta prævia; in one case the papillary proliferations were supposed to be the fingers of the fœtus.

3. *Tumours of the Vagina and the Vulva.*—Neoplasms very rarely form in the vagina and vulva. Fibroids, polypi, or carcinomatous tumours, if their size is at all considerable, must be extirpated during labour. Cysts occur most frequently, but they are almost always so small as not to form an impediment to delivery. In the case reported by Peters the cyst was so large that it had to be punctured before the child could be extracted. Abscesses of the connective tissue or of Bartholini's glands should always be opened.

Thrombus or hæmatoma of the vagina or vulva forms, as a rule, only after the expulsion of the child, and for this reason it is very rarely an impediment to delivery; if it, however, happens to be so, the tumour must be opened, and the child, unless it be quickly expelled spontaneously, is to be extracted.

Great œdema of the vulva may also give rise to a very painful impediment to delivery. It can, however, be lessened to a great extent by scarifications.

4. *Tumours of the neighbouring Organs.*—Occasionally great danger arises from a hernia which contains a coil of intestine, and is so situated in the pelvis that it is compressed between the bony pelvis and the advancing head. This may be the case in simple vaginal hernia, that is, when by the prolapse of the posterior wall of the vagina, Douglas's pouch has been enlarged below, or when the intestines have, together with the peritoneum, passed through the pelvic fascia and the muscles of the floor of the pelvis in front of, or more frequently behind the broad ligament, forming a projection of the skin of the perinæum—perineal hernia. If the hernia is situated in front of the broad ligament its contents descend into the posterior portion of one labium.

The hernia is always to be reduced during labour. Delivery should be terminated as quickly as possible, in order to prevent strangulation and contusion of the intestine.

The rectum may be over-distended by fæcal masses, so as actually to impede the progress of labour. The sensation produced by an accumulation of fæces is characteristic. The impressions of the fingers through the vagina remain in it as in soft clay. The rectum should be emptied by enemata, or, if necessary, by means of the finger.

Retention of urine very often occurs in parturient women. If the urine has accumulated in large quantity the uterine action is thereby impeded, and the presenting part is prevented from entering the pelvis. Externally a smaller tumour is seen beside the uterus, and situated somewhat more to the opposite side. It is below the umbilicus, and separated from the uterus by a distinct furrow.

If a cystocele has previously existed the distended bladder forms a large tumour projecting into the vagina, and situated greatly to one side. The prolapsed portion of the vaginal wall, together with a more or less large diverticulum of the bladder, is pushed downwards by the advancing head, and this may not only cause delay in the labour, but also produce considerable contusion, and even rupture of the walls of the bladder and vagina.

Great difficulty is often met with in emptying a distended bladder by means of the catheter. The urethral opening may sometimes be so greatly drawn into the vagina as to be quite out of sight, and the urethra itself almost completely compressed by the head. A male catheter is always to be used for parturient women; but previously the position of the bladder must be made out by palpation, and the end of the instrument held in the hand is to be greatly depressed in a direction opposite to the position of the bladder, so that the end of the catheter can be made to enter it. If this method fails, an attempt may be made to pass the catheter in the knee-elbow posture. A cystocele can best be prevented by pushing the anterior vaginal wall upwards as soon as the head enters the pelvis.

Vesical calculi may also obstruct labour to a serious extent. If discovered before the head is engaged, they should be pushed into the false pelvis. There they are innocuous. But if they are impacted between the head and the symphysis, so that the head cannot pass by them at all, or without great danger, they are to be removed by opening the bladder through the vagina, the wound being immediately closed by sutures.

Ovarian tumours, which not infrequently cause abortion, may obstruct labour in the same way as subperitoneal fibroids. Reposition, if practicable, is here also the best treatment, or the bulk of the tumour may be reduced by puncture. The contents of the cyst may be so thick that a very large trocar is required. By the pressure of the pains the tumour, although a cyst, may become so tense as to feel actually as hard as wood, and give no trace of fluctuation. A solid tumour, unless ovariotomy is to be performed during labour, must be treated like subperitoneal fibroids.

Tumours, such as carcinoma, sarcoma, and fibroids, may also arise from the pelvic connective tissue and from the periosteum of the pelvic bones. Attempts to displace them are almost always unsuccessful; they must either be extirpated or the Cæsarean section must be performed, unless they soften during labour, as in the case related by D'Outrepont.

II.—Of the Bony Pelvis

A.—*Examination of the Pelvis*

To form a just opinion of the capacity of a given pelvis from an obstetrical examination is a most difficult task, and it will readily be understood that every means are to be employed which can assist in elucidating that point. It is true that an accurate manual examination is the sole and exclusive way of obtaining approximately sure results; nevertheless all the other circumstances, and especially the history of the case, which may excite the suspicion of a probable pelvic contraction, deserve careful consideration.

In the first place, the history will show whether diseases have been or are still existing which tend to produce changes in the shape of the pelvis. Those diseases are, with very rare exceptions, rickets and osteomalacia. Since not only these but also other deformities of the pelvis appear to be inherited just as is the case with normal and roomy pelves, the inquiry ought to be extended to the labours of the nearest relatives. In the case of a pluripara it is of the greatest importance to inquire into the course of previous labours. Yet one thing is to be remembered, that even in considerable congenital pelvic contraction the first labours have almost always a more favorable course, so that one will frequently hear from women with no slight contraction of the pelvis that their first labours were normal. But if previous labours required operative procedures, such as perforation and cephalotripsy, indicating the presence of a mechanical impediment, or if the child's head was distorted or showed distinct traces of pressure upon the scalp, there is reason to diagnose—when an abnormally large head can be excluded—a contraction of the bony pelvis. Here also it has to be remembered that unsuitable assistance (by means of the forceps), when the pelvis is normal, may also leave traces of pressure upon the bones of a normal head.

The stature and growth of the body will first have to be considered in the inquiry.

Whilst in very unusually short women there is a strong presumption that the pelvis is narrow, a medium or rather more than a medium height is by no means an absolute guarantee against pelvic contraction, although, in general, a fine stature promises a normal pelvis. In well-built women with broad hips and a large sacral surface, there will rarely be abnormalities of the pelvis, whilst very narrow hips, and still more so distortions of the lower extremities, are very suspicious; the latter for this reason especially, because they most frequently are the consequences of a previously existing rickets. Attention is also to be paid, as regards that disease, to an unusual curvature of the clavicles, and to the so-called rachitic rosary along the costal cartilages, and to a distortion of the superior extremities. Scoliosis, being frequently of rickety origin, also points to that affection. Moreover, it is also to be remembered that pendulous belly is more frequently met with in the contracted pelvis, and a great forward inclination of the uterus in a first pregnancy is always very suspicious.

However valuable all these conditions are, they only allow an inference to be drawn as to the probability of the existence of a pelvic deformity, whilst as to the degree of deformity they indicate nothing even approaching to the facts.

Sufficient information, as far as this is possible in the living woman, can only be obtained by an accurate examination of the pelvis itself.

This is done either by the hand or by means of instruments, externally and internally.

To measure the individual distances by an external examination Baudelocque's pelvimeter is used. The woman must lie on the bed, and, according to the direction of the distances to be measured, either in the dorsal, lateral, or knee-elbow posture. When the knobs of the compasses have been placed upon the points to be measured the distance is read off from the scale attached to the instrument.

A number of external measurements of the pelvis have been given. Some of these are either quite useless or only of importance in recognised and very rare abnormalities of the pelvis.

We shall confine ourselves, in this place, to mentioning only the most important of these measurements.

In the false pelvis the most important of these measurements are—

1. The distance between the anterior superior spines of the ilia (Sp. J.).

2. That between the crests of the ilia (Cr. J.).

To obtain constant results, and such as may be used for mutual comparison, it is advisable always to adopt the same method of mensuration.

In measuring the distance between the iliac spines we must not try to make out their actual distance, as would be done in the dried pelvis. It is sufficient to place the knobs of the compasses outside, and close to the origin of the tendon of the Sartorius muscle; whilst as regards the distance between the iliac crests this is taken between the most distant points of their external edges. The first measures, on an average, 26 centimètres, or 9·6 inches; the latter, 29 centimètres, or 10·75 inches.

The significance of these measurements depends much less upon their absolute size than upon their relation to each other. Thus, the rickety pelvis is distinguished by an increase of the length of Sp. J. in relation to that of Cr. J., so that either the difference between the two distances is less or their length is equal, or even the length of Sp. J. is more than that of Cr. J. The latter being the case, it is useless to measure Cr. J. in the way above mentioned, because the distance between the crest continually diminishes from the spines backwards. Under such conditions it is sufficient to say that Sp. J. is greater than Cr. J.; or the distance of Cr. J. is measured two and a half inches behind the iliac spines.

The greatest distance between the trochanters (Tr.) is of little significance. Only its very remarkably small extent is so far suspicious as the transverse diameters of the true pelvis must, in accordance with it, also be contracted. Slight deviations of this measurement from the normal do not permit any conclusion as to the transverse diameters of the true pelvis. That measurement is easily taken, and is on an average 31½ centimètres or 11·6 inches.

By far the most important measurement of the whole pelvis is the conjugate, for in the great majority of pelvic deformities the contraction is either exclusively, or at least chiefly, in that direction. The so-called diameter of Baudelocque (D. B.) or the external conjugate is not without value in estimating its size.

This measurement is taken by means of the pelvimeter, the woman lying on her side. The depression below the spinous process of the last lumbar vertebra is taken as the posterior point. As a rule this is distinctly marked and easily found, and sometimes it can readily be seen. At the sides of the posterior surface of the sacrum two depressions are seen where the skin is more firmly attached to the subjacent bone, namely, the posterior superior spines of the ilium. If these two depressions are connected by a line, the point sought for lies in the normal pelvis from one to two inches above the middle point of that line, so that lines drawn from the fossa below the spinous process of the last lumbar vertebra to the posterior superior iliac spines, and thence to a point lower down, where the gluteal muscles unite, map out a quadrangular space, which, in the normal pelvis, has well-nigh the shape of a lozenge (Michaelis).

In the deformed pelvis, and especially in the rickety, the spinous process of the last lumbar vertebra is much lower down; therefore the upper angle becomes more obtuse, and the above-described depression may be either within or even below the line connecting the two posterior superior iliac spines. The space thus becomes triangular.

As the anterior point that portion of the symphysis is taken where the distance is the greatest. This is always situated at the upper margin of the pubic bones.

To ensure accuracy in measuring, the posterior point is first looked for. It can almost always be felt without difficulty, since the spinous processes of the lumbar vertebræ are far more prominent than those of the sacrum, and the fossa also is usually very distinct. One knob of the compasses having been placed into that fossa and fixed by one hand, the other knob is placed upon the above-mentioned spot of the symphysis, and, after moderate compression of the instrument, the measure is read off the compasses *in situ*.

In rare cases the fossa is not so distinct, and in order to find the last lumbar vertebra all the vertebræ have to be counted from above downwards.

In the normal pelvis this measurement is on an average 20¼ centimètres (7·8 inches).

It is supposed by some that by subtracting 9 centimètres from D. B. the length of the true conjugate is obtained with approximate accuracy. This supposition is, however, unfounded; the conjugate can never be calculated from D. B. with certainty. It is, of course, of assistance in recognising, to a certain extent, the contraction of the conjugate in general, but is almost entirely valueless as regards the degree of contraction.

On the whole, it may be said that an external conjugate of less than 18 or 19 centimètres (7—7·4 inches) ought greatly to arouse the suspicion of a pelvic contraction, whilst one of 21 or more centimètres (8·2 inches) will certainly be only in very rare instances associated with slight contraction.

From the measurements of the diagonal conjugate (C. D.) more certain conclusions can be drawn as to the size of the conjugate.

The diagonal conjugate is the line which connects the lower border of the symphysis with the nearest point of the promontory. The anterior point of this diameter is, therefore, the sharp edge of the inferior pubic ligament, and the posterior is in most cases situated in the middle of the promontory. Only in very asymmetrical pelves the most prominent point of the promontory is displaced to one side. In deformed pelves it may also occur that at the place where the first and second sacral vertebræ unite a shorter diameter is obtained. In such a case this is taken as the posterior point, for practically it is alone of importance to determine where the antero posterior diameter of the pelvis is the shortest.

This measurement is taken in the following way:—The index and middle finger of the left hand, the other fingers being doubled into the palm, are introduced into the vagina, the perinæum being slowly but strongly raised. By lowering the forearm, in order not to come too far backwards into the sacral excavation, the promontory is sought. Then the ulnar side of the tip of the left middle finger is placed against the promontory, and the radial side of the same hand is brought closely to the pubic arch. With the tip of the right index finger the place is felt where the inferior pubic ligament rests upon the left index finger or its metacarpus. On this spot a distinctly visible mark is made. Now the left arm and hand are withdrawn in the same position they occupied during the measurement, and by means of the compasses the distance is measured from the ulnar margin of the tip of the left middle finger to the mark just mentioned.

In the normal pelvis the promontory can, under favorable circumstances, be reached, and consequently C. D. measured.

It is easier in pluriparæ with a wide genital fissure and yielding perinæum, more difficult in primiparæ with a short and narrow vagina and a high and rigid perinæum, also in considerable sensitiveness or strictures of the vagina. Great descent of the presenting fœtal part and tumours of the vagina render it impossible.

To facilitate the measurement the following rules must be observed:

The woman must be in a suitable posture. A good gynæcological examination chair is the best; if this is not at hand, lying either on a table or across the bed with the nates well raised will do.

The left elbow must be supported on the knee in order not to fatigue it during the examination.

In introducing the left hand the labia must be held apart by the right hand, so as to prevent the hairs being dragged upon, and so causing pain.

The perinæum must never be suddenly pushed upwards, but very gradually and slowly raised with some force.

The bladder, and especially the rectum, is to be emptied before the measurement is taken.

It is only possible by practice to obtain sufficiently reliable results. At first other measurements taken as a check will frequently show very different figures, whilst after some practice the results frequently agree completely or differ only by a quarter or not more than half a centimètre.

By the assistance of all the above-mentioned precautions the promontory can in most cases be reached, or it can at least be seen that C. D. has a minimum length, which excludes a pelvic contraction. Under favorable conditions a length of 13 centimètres or more is obtained.

On an average $1\frac{3}{4}$ centimètre (·68 inch) are to be deducted from the diagonal conjugate in order to obtain the length of the true conjugate. The amount to be deducted varies somewhat, but only inconsiderably. The more obtuse the angle made by the conjugata vera and the symphysis and the higher the latter itself is, the greater is the amount to be deducted.

By making the following deductions in the various kinds of pelves only a slight error can possibly occur.

1. In the normal and the generally contracted pelvis, $1\frac{3}{4}$ centimètre.

2. In the flat, non-rickety pelves, somewhat more than $1\frac{3}{4}$ centimètre.

3. In the rickety pelves, 2 centimètres or more.

Manual measurement of the diagonal conjugate will suffice in all cases where, during labour, the treatment has to be decided upon; but in some cases, and especially where a time has to be fixed for the induction of premature labour, it is important to know as accurately as possible the length of the true conjugate, and this knowledge must be obtained by direct measurement.

Vanhuevel's pelvimeter is the most useful for this purpose. It consists of a longer and a shorter arm, movable upon each other. The tip of the longer arm is placed against the projecting margin of the promontory, and the knob terminating the shorter arm is pressed upon a definite point of the anterior surface of the symphysis. The measurement thus obtained is the true conjugate diameter plus the thickness of the anterior wall of the pelvis. If now the longer arm is applied to the posterior surface of the symphysis, and the knob of the shorter arm again to the same point as before, the thickness of the symphysis will be known. This being deducted from the first measurement, the difference represents the diameter of the true conjugate.

Instrumental mensuration has many disadvantages. Not only is skilled assistance necessary, but the mensuration itself is a little painful. Where the question as to the induction of premature labour has to be decided the pelvimeter may be very conveniently used, and, if necessary, under chloroform. It is especially useful for the comparison of its results with those obtained from measuring the diagonal conjugate. (As the longer arm cannot at times be well applied to the posterior surface of the symphysis, the thickness of the anterior wall of the pelvis is made too large, and the conjugate consequently too small.)

Irregularities of lesser degree can scarcely be recognised in the living woman with any certainty. The external oblique diameters (from the anterior superior spine of the ilium of the one side to the posterior superior spine of the other) have, indeed, very little value. An accurate internal examination gives the most trustworthy results, for this allows the position of the promontory to be ascertained and both sides of the pelvis to be felt by the finger. Both hands should be used in the examination at short intervals, for if only one hand is employed an impression may be conveyed that there is some irregularity.

The measurements of the pelvic outlet have hitherto been greatly neglected. This is due to the difficulty met with in the operation, and also to the rarity of contraction of the outlet. Breisky has lately proposed to take the measurements in the following way:

The antero-posterior diameter of the outlet, from the apex of the sacrum to the vertex of the pubic arch, the coccyx not being included, cannot accurately be measured internally. External measurements are, therefore, preferable. In the lateral posture of the woman the posterior point can easily be seen. It corresponds to the upper edge of the anal opening, and forms, in lean persons, a slight projection. The lower cornu of the sacrum can usually be felt without difficulty. To be quite sure, the index finger may be introduced into the rectum, and by moving the coccyx between the index finger and thumb the articulation is easily felt. The anterior point of the diameter is the sharp edge of the inferior pubic ligament within the vertex of the pubic arch. Whilst one hand applies the knob of the compasses on the posterior point, the thumb, introduced into the vagina, presses the other knob upon the inferior pubic ligament, and the distance between this point is read off on the compasses. This external measurement somewhat exceeds the proper antero-posterior diameter. It is not sufficiently known how much is to be deducted in order to obtain the latter—perhaps 1 to 1½ centimètre.

Breisky measures the transverse diameter of the outlet in the following way:—The woman lies on her back with the sacrum raised, the thighs flexed and slightly separated. He then feels for the internal margins of the ischial tuberosities, and measures the distance between them by means of Osiander's compasses; since between the knobs of the compasses and the bony points soft parts are situated, to the measurement thus obtained 1½ centimètre is on an average to be added.

We have examined Breisky's statement, and are convinced that the measurement of the antero-posterior diameter of the outlet can be taken without too great difficulty; yet it is somewhat tedious, and in a sensitive vagina also painful. There is a more practicable, and at least an equally reliable, way of measuring the transverse diameter. The woman being in the lithotomy position, the situations of the ischial tuberosities are marked on the skin of the nates with a piece of blue chalk, and the distance between measured by means of Osiander's compasses. There need be no addition on account of the soft parts.

B.—*Anomalies of no Obstetric Importance*

A too wide pelvis in no way interferes with the process of parturition. Labour is here precipitated by the same conditions as in the normal pelvis. From an obstetrical point of view it cannot be considered pathological.

A too high pelvis, if not contracted, offers no disadvantage.

Anomalies in the pelvic inclination, if not observed at all, may lead to some inconvenience. Too great inclination may prevent the head entering the brim, and too slight inclination the passage of the head through the outlet. The former can be diminished by raising the upper half of the trunk and the nates, so that the lumbar region is the lowest, the woman then being in a half-sitting posture; the latter can be increased by supporting the lumbar region, the nates being low down.

With the exception of the last-mentioned cases, the contracted pelvis alone causes any such impediment to delivery.

c.—*The Contracted Pelvis*

1. *Definition and varieties.*—Whether in a given case the pelvis will allow the undisturbed passage of the child through it depends, not only upon the diameters of the pelvis, but also upon the size of the child, and especially of its head. Since enlargements of the head which interfere with the mechanism of labour are of extremely rare occurrence, and since, on the other hand, the size of the head of a mature and healthy child is never much below the average measurements, the custom has arisen of regarding the head as of a definite magnitude. This is the more excusable as we do not possess any accurate method of estimating the size of the head still contained in the uterus. It is, however, of some importance not to lose sight of the fact that in an individual case a very small head may make a narrow pelvis appear a wide one, and *vice versâ.*

In our subsequent considerations we shall take the head to be of a definite size, as known from average measurements, and shall estimate the contractions of the pelvis in relation to such a head of medium size.

The pelvis may be narrowed at any spot, or in any or all of its diameters, and it is, therefore, impossible to give a set of rules which would serve to recognise all kinds of pelvic contraction.

The study of the narrow pelvis, so very important to the practitioner, is greatly facilitated by the circumstance that in the vast majority of instances the contraction is exclusively, or at least chiefly, situated at the brim.

Several of the very rare forms of pelvic contraction will for the present be neglected. The other forms, where the contraction is exclusively and chiefly situated at the brim, will be considered from a common point of view. Such a consideration is of vast importance, for these are the forms which almost alone come under our notice in practice.

We omit at present the consideration of the following very rare forms of narrow pelvis, viz. the spondylolisthetic, the synostotic and kyphotic transversely contracted, the synostotic obliquely contracted, the one contracted exclusively or chiefly at the outlet, the one narrowed by bony tumours, and the osteomalacia, which occurs relatively frequently in some parts only. Each of these will be separately described afterwards. Here we shall speak simply of the contracted pelvis without any addition; we mean those forms of contracted pelves, where the brim alone is the seat of the contraction.

By far the most frequent contraction of the brim is in the direction of the conjugate, and the narrow pelvis is usually estimated according to the size of that diameter. This is true in the great majority of cases; only the obstetrical importance of a narrow pelvis varies greatly according to whether, with the contraction of the conjugate, the transverse diameter has preserved its normal width or has also been contracted.

The narrow pelvis is, therefore, divided into—

1. *The generally and uniformly contracted pelvis.* The contraction of all the diameters at the brim is pretty uniform, and is also continued into the pelvic cavity and outlet.

2. *The flat pelvis.* The conjugate is either alone or chiefly contracted. According to whether the transverse diameter of the brim is of normal length or also contracted, it is thus distinguished—

a. The simple flat pelvis.

b. The generally contracted flat or the generally irregularly contracted pelvis.

But there is an important question to be decided, viz. where does the limit between a normal and a contracted pelvis begin?

In practice a contraction will be recognised when it is so considerable as to impede the passage of the head through the pelvis. The question is, what degree of impediment is sufficient to say the pelvis is contracted?

Whilst formerly the actual mechanical impediment to labour was attributed to the narrow pelvis, Michaelis has shown, first, that the influence of pelvic contraction has a far wider range, and is shown, also, by abnormalities in the fœtal position and presentation, as well as in the uterine action. Moreover, he has shown that one who is acquainted with the deviations in the positions of the head in the various kinds of pelvic contraction will be able to diagnose from the position of the head the kind of contraction, even in those cases where a less experienced observer would only see the normal course of labour.

Michaelis's experience has led him to consider a pelvis contracted when its conjugate measures less than 9½ centimètres, or 3·7 inches. This, however, by no means gives a complete definition of a contracted pelvis; for, on the one hand, numerous very serious disturbances during parturition may occur, as Michaelis himself admits, in the generally contracted pelvis with a conjugate of somewhat more than 9½ centimètres; and, on the other hand, in the flat pelvis, with a conjugate of somewhat more than 9½ centimètres, there are scarcely any real impediments to the passage of the head, although deviations in the presentation of the head, characteristic of the flat pelvis, can be observed, and these are just the cases which most frequently explain to us the mechanism of labour in the contracted pelvis.

For these reasons it is not advisable to leave quite out of consideration, as being perfectly like the normal, pelves with a conjugate of somewhat more than 9½ centimètres. It is better to reckon them amongst the contracted pelves; only so much must be added that they do not offer any mechanical impediment to the passage of a normal head, but their influence is shown solely by the deviation in the mechanism of labour.

We shall therefore classify all the various kinds of pelvic contractions under three great heads, viz.—

1. Pelves absolutely too narrow, which under no conditions admit the birth of a living and mature child. To this class belong those pelves the shortest diameter of which measures 6¼ centimètres (2·5 inches).

2. The narrow pelvis, which under favorable circumstances permits the birth of a live child, but where the danger of a fatal termination continually threatens both mother and child. The limits

of this kind are in the flat pelvis when the conjugate is only contracted from 6½—9½ centimètres (2·5—3·7 inches), and in the generally contracted up to 9¾ centimètres (3·8 inches). Pelves of this kind, where the contraction is the least, usually give rise to another serious disturbance in frequently recurring labours.

3. The narrow pelves, which do not give rise to any considerable impediment, but which only influence the position of the head so as to cause it to deviate from the normal. They form a gradual connecting link with the normal pelvis.

Usually the first two kinds alone are considered as contracted pelves; and this is so far correct, as they only seriously impede the process of labour. In order to fully appreciate the influence of a pelvic contraction upon labour and upon the position of the head, it is of the greatest importance to be acquainted with the mechanism of labour in the slightly contracted pelvis. The more roomy pelves of the second order are distinguished from those of the third order only in this, that of many labours one or more have a fatal issue, whilst the remainder have just the same course as in the pelves of the third order.

2. *The usual forms of Pelvic Contraction, the Brim being exclusively or chiefly Contracted*

(*a*) *The generally uniformly Contracted Pelvis*

The generally uniformly contracted pelvis shows on the whole the normal female shape, only all its diameters are uniformly smaller than the normal. There are two varieties.

1. Pelves which are distinguished from the normal only by the

FIG. 10.

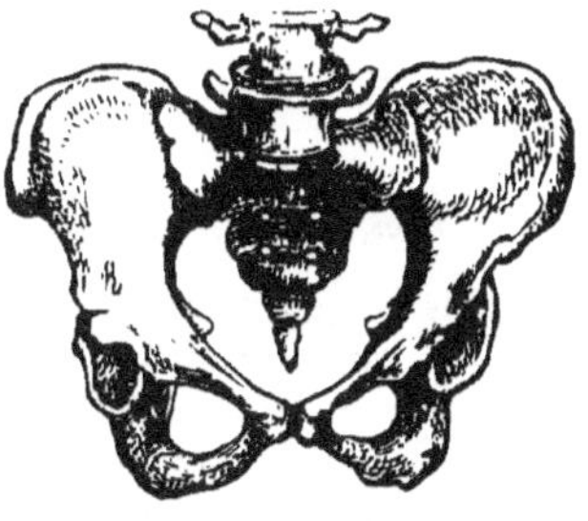

smallness of their bones. Their thickness, structure, and joints in no way differ from those of the normal pelvis. The bones are either proportionately thick and strong, or they are somewhat thin and slender, often showing a small normal pelvis of exquisite beauty. This pelvis is chiefly found in persons of small stature; but it is met with also in well-built, thin, and portly women of middle or even more than medium height.

2. Pelves the shape of which show the regular female type, but the bones more resemble in size and thickness, and mostly also in the joints, the juvenile pelvis—the so-called dwarf-pelves. This

variety is very rare, and only occurs in very small and symmetrically built women—actual dwarfs.

Sometimes these two varieties of pelves show in every respect the normal type of a female pelvis, and are distinguished only by the smaller capacity of the canal (particularly in the second variety), due to an originally small development; more frequently, however, they somewhat retain the shape of the juvenile pelvis (arrest of development). This is especially seen in the sacrum. It is narrower across, particularly as regards its lateral portions; it is also more straight, and is less depressed in the pelvis. The pelvis, therefore, has remained in an earlier stage of development, and the growth of the single bones as well as their joints has been prematurely arrested. Perhaps in this kind of generally contracted pelvis the primary affection consisted in a premature union of the individual bones, which was followed as a secondary consequence by an arrest of growth in a direction perpendicular to the (premature) ossification.

The degree of contraction varies greatly in all these cases. It is, of course, most considerable when originally small development coincides with premature arrest of growth. The contraction is here always less than 3 centimètres, or 1·17 inch, in all the diameters. The antero-posterior diameters are most frequently narrower than the others. As a rule, the brim is the most contracted, and exceptionally only the outlet.

The generally uniformly contracted pelvis is much less often met with than that with the conjugate alone contracted.

Fig. 10 shows a pelvis of this kind, from the collection in the Museum of Bonn. Its diameters are—

	Ant.-post.	Transv.
Brim	9 cm.	11¼ cm.
Cavity	11 „	11½ „
Outlet	10¼ „	10 „

A child of eight months had been delivered by means of the forceps, but the woman died of eclampsia.

(b) *The Pelvis with the Conjugate contracted, or the Flat Pelvis.*

The characteristic of this large group of narrow pelves (the other kinds of pelvic contractions are very rarely met with by the practitioner) consists in a flattening from behind forwards. These are grouped, according to whether the transverse diameter is of normal size or not, into the simple flat pelvis and a generally contracted flat pelvis; a further subdivision is made, according to whether the contraction is due to rickets or not.

1.—*The simple Flat Pelvis*

(*a*) *Flat Non-rickety Pelvis.*—The simple flat non-rickety pelvis is by far the most frequent in those countries where rickets do not very often occur, but even where this disease is more often observed it is of still more frequent occurrence than the rickety.

At first sight such a pelvis at times appears normal and well formed, but by instrumental mensuration the flattening of its antero-posterior diameter can be easily discovered. This flattening

FIG. 11.

is due to a greater descent of the sacrum into the pelvis without a concurrent rotation around its transverse axis. All the antero-posterior diameters of the pelvis are therefore contracted, although this is more so at the brim, so that the relations of the antero-posterior diameters of the cavity and of the outlet to the conjugate remain approximately the same as in the normal pelvis. The size of all the pelvic bones, and especially of the sacrum, appears to be pretty uniformly diminished, so that the transverse diameters would also be lessened if the descent of the sacrum did not oppose it. The sacrum being connected with the posterior spines of the iliac bones by very strong unyielding ligaments, it would, when pushed forwards, exert a powerful traction upon the posterior iliac spines. If the pelvic ring were not closed in front by the symphysis, the pubic bones would be separated by this. As, however, the firm connection by means of the symphysis opposes it, the still relatively flexible bones of the pelvis are elongated in their transverse diameter by the traction upon the posterior iliac spines, and the symphysis is somewhat approached to the sacrum. Supposing a pelvis to be generally uniformly contracted in a slight degree, the great descent of the sacrum into the pelvic cavity in early youth would have the following effect: the antero-posterior diameters are somewhat reduced by the descent itself; the transverse diameters are somewhat elongated, from their greater transverse traction, and this elongation is aided by the approach of the symphysis to the sacrum. Consequently two causes act in producing the elongation of the transverse and the reduction of the antero-posterior diameters. As the pelvis was supposed to be somewhat contracted, the transverse diameters will have the normal length, whilst the antero-posterior will be considerably smaller. (Very frequently, however, the transverse diameters remain in such pelves somewhat below the average measurements.) Thus, we have seen that the flat pelvis is produced by the same mechanism by which the pelvis of the new-born child is converted into that of the adult woman, and the difference is that in the originally small pelvis all these changes are too excessive.

The causes of this pelvic deformity may probably be due to too early and too long sitting and running about of little girls, or a too great increase of the pressure of the trunk by raising and carrying weights before puberty.

The individual bones are normal as regards texture and strength. The iliac bones do not deviate in position or shape from the normal. Their curves are usually very distinct.

It has already been said that the simple flat non-rickety pelvis is the most frequent of all narrow pelves, and, doubtless, more frequent than all the other varieties taken together, and is, therefore, a very frequent cause of dystocia. However, it does not show the highest degree of contraction, for its conjugate rarely measures less than 8 centimètres, or 3·14 inches.

Occasionally flat pelves are met with where the transverse diameter of the outlet is also contracted. This is a very important complication, and is of great practical consequence. Sometimes a so-called double promontory is formed at the junction of the first with the second sacral vertebra (which do not ossify). The two vertebræ meet at an obtuse angle, and project into the pelvic cavity, virtually contracting it. If the distance between this double promontory and the symphysis is as small or even smaller than the true conjugate, such an anomaly decidedly influences the course of labour, and requires to be noticed in measuring the pelvis.

Fig. 11 is a beautiful flat pelvis, from the collection of the Munich Lying-in Hospital. The woman had been delivered by means of the forceps, and died of peritonitis. It measures—

	Ant.-post.	Trans.
Brim . . .	$8\frac{3}{4}$ cm.	. . . 13 cm.
Cavity . . .	$10\frac{1}{2}$,,	. . . $12\frac{1}{2}$,,
Outlet . . .	10 ,,	. . . $11\frac{1}{2}$,,

(*b*) *The Flat Rickety Pelvis.*—The flat rickety pelvis has the following characteristics:

The texture of the bones is usually normal; they are rarely greatly atrophied, and sometimes even tense and thick.

FIG. 12.

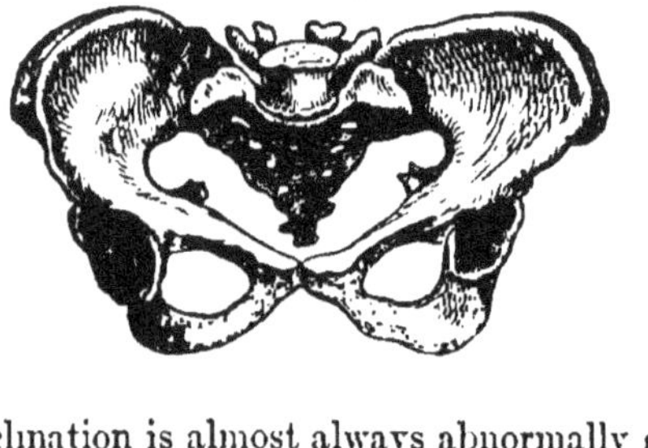

The pelvic inclination is almost always abnormally great.

All the bones of the pelvis are, like those of the flat non-rickety kind, abnormally small. This is most pronounced in the iliac bones, and is there most easily recognised. The sacrum also is short and narrow; but as the lateral concavity of the sacrum is defaced by the descent of its vertebræ, the transverse diameters may reach the normal length.

The principal changes in the pelvis are due to the greater descent of the sacrum into the pelvic cavity, and to its simultaneous rotation around its transverse axis. This altered position of the sacrum is attended by a double change in its shape. First, the bodies of the sacral vertebræ are forced down between their auricular surfaces, consequently the concavity of the sacrum in the transverse direction is lost, and is converted into a straight line, or the bodies are even forced out from between the auricular surfaces, and then the sacrum projects into the pelvis, not with a transverse concavity, but with a convexity. Secondly, the curve of the sacrum from the promontory to the apex is also increased. The upper portion of the sacrum is greatly bent backwards, and the lower portion forms a sharp, hook-like projection forwards, which usually commences with the fourth, rarely with the third or fifth, vertebral body. In consequence of the descent of the upper part of the sacrum the posterior spines of the iliac bones project far more than normally beyond the posterior surface of the sacrum. The bodies of the vertebræ, and especially that of the first vertebra, are greatly compressed along their posterior surfaces.

The superior expanded portion of the iliac bones are small, greatly inclined horizontally, slightly curved, and widely separated in front. The length of Sp. J. increases in proportion to that of Cr. J., and is sometimes even greater than the latter.

The pubic arch is very wide, the symphysis makes a greater angle with the conjugate, the ischial tuberosities are wider apart, and the acetabula are directed more forwards than in the normal pelvis.

The cavity of the pelvis is influenced by these changes in the following way :—By the descent of the promontory the pelvis is flattened from before backwards, and this flattening is here, as well as in the non-rickety flat pelvis, increased by the transverse extension of the pelvis. The extension may be so considerable that, in spite of the original smallness of all the bones, the transverse diameter of the brim is abnormally large, and in the ilio-pectineal line not far from the auricular surfaces of the sacrum a kind of angular flexure appears. As regards the brim, the transverse diameters are of normal or even more than normal size, and the oblique diameters are inconsiderably shortened, whilst the sacro-cotyloid distance is greatly contracted, and the conjugate the shortest of all.

All the diameters enlarge towards the pelvic cavity. Here the antero-posterior diameter sometimes acquires its normal size on account of the rotation of the sacrum around its transverse axis, and in the outlet of the rickety pelvis the transverse diameter especially is often abnormally large.

We shall now explain the way in which rickets act in producing the changes just described.

Children in the first years of life are most liable to be attacked by the peculiar disturbance of nutrition called rickets. If the disease is of moderate duration and intensity the pelvis consists of pieces of firm bone, covered on all sides by soft osteoid plates, which pieces are connected together by cartilage of normal texture. The union of each two pieces of bone is, therefore, effected by the just-mentioned soft osteoid substance, which normally would have become osseous tissue, and by the harder cartilage interposed. The bones themselves are thinner, and more easily subject to curvatures and fractures.

Independently of the pelvic deformities which are due to the altered pressure, as in scoliosis, &c., it is the compression of that osteoid substance upon the epiphyses of the bones and the shifting of the bones along each other which produce the changes observed in the shape of the pelvis.

The smallness of the bones is partly due to the compression of the osteoid substance, partly to the long-continued inactivity of the muscles of the thighs and pelvis; but the shape of the rickety pelvis is chiefly determined by the shifting of the bones along each other.

Rickety children have either never learned to walk or have forgotten it again. They are forced to sit in bed, and the pressure of the trunk produces the changes which are known as belonging to that disease, for the trunk presses the upper portion of the sacrum down into the pelvic canal. The promontory thus comes lower down and more forwards, the sacrum having turned around its transverse axis. The frequent half-sitting and half-lying posture in bed produces the crook-like curvature of the lower portion of the sacrum, which is fixed by the sacro-spinous and ischial ligaments. Since the individual sacral vertebræ have no firm connection, either amongst themselves or with their lateral portions (auricular surface), they are pushed out forwards from between the auricular surfaces, which are fixed, and their posterior surface is compressed. The descending sacrum again draws by means of the firm ilio-sacral ligaments upon the posterior spines of the iliac bones, the consequence of which is the greater transverse extension of the pelvic ring. The separation of the iliac bones is partly due to that, but partly also to the greater pressure of the tympanitic and distended intestinal canal.

The ischial tuberosities are pressed outwards, partly by the traction of the femoral muscles arising from them, and partly by sitting on the soft bones. This also causes the enlargement of the pubic arch and the obliquity of the symphysis.

Rickets occurs almost everywhere; in some countries, however, more frequently than in others. It is by far the most frequent cause of a very considerable pelvic contraction, and this may be to such an extent that a pelvic cavity scarcely exists.

The other forms of rickety pelvis deviating from the ones described, as well as those of greater asymmetry, will be mentioned as we proceed.

Fig. 12 represents a rickety pelvis from the collection of Bonn. It measures—

	Ant.-post. D.	Trans. D.
Brim . . .	6·3 cm. . . .	13·8 cm.
Cavity . . .	8·1 „ . . .	14·5 „
Outlet . . .	8·5 „ . . .	12·0 „

2. *The generally Contracted Flat Pelvis, or the generally Unequally Contracted Pelvis*

(*a*) *The generally Unequally Contracted Non-rickety Pelvis.*—In the common flat pelvis the transverse diameter is, as we have stated above, of a size which still lies within the physiological limit. Very exceptionally this is less than 12½ centimètres, 4·9 inches, so that the pelvis is generally contracted; the contraction of the conjugate, however, still predominates. This kind, to which Michaelis first

FIG. 13.

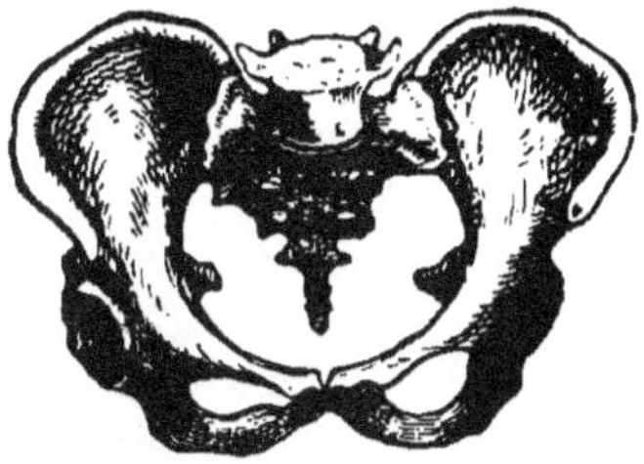

called attention, probably always depends upon a smallness of all the constituent parts, connected with a moderate degree of flattening. The smallness of the parts determines the very considerable contraction of the conjugate, and the transverse extension is insufficient to make the transverse diameter reach the normal length.

Fig. 13 shows a pelvis of this kind, to which, on the whole, little attention has been paid. It belongs to the Bonn Collection. Putrid twins had been delivered without much difficulty. The measurements of this pelvis are—

	Ant.-post. D.	Transv. D.
Brim . . .	8·6 cm. . . .	12·2 cm.
Cavity . . .	9·8 „ . . .	12.3 „
Outlet . . .	9·8 „ . . .	13·0 „

(*b*) *The generally Contracted Rickety Pelvis.*—The more unusual forms of rickety pelvis, of which there are three varieties, will be described together.

FIG. 14.

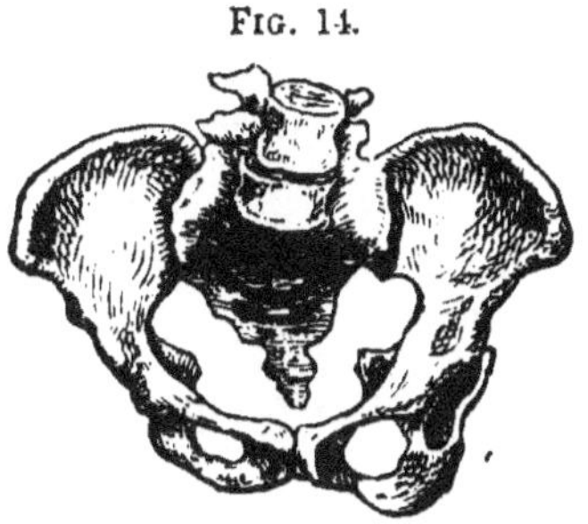

1. The generally contracted flat rickety, or the generally unequally contracted rickety pelvis. These pelves show very distinctly

all the characteristic peculiarities of the rickety pelvis. The bones are greatly atrophied, and their growth has been arrested, so that besides the predominating contraction of the conjugate the transverse diameter of the brim, at least, is also narrowed to a great extent. The smallness of the pelvic bones is especially apparent in the superior expansions of the ilia. It is not improbable that this kind of pelvis develops when rickets appears very early and with great intensity, but is soon cured with complete osseous union of all the bones.

The pelvis of Fig. 14 belongs to a very rickety distorted skeleton in the collection at Bonn. It has a double promontory. The conjugate from the upper promontory measures 8·25 centimètres, or 3·25 inches, and from the lower 7·25 centimètres, or 2·8 inches. The other measurements are—

	Ant.-post. D.	Transv. D.
Brim . . .	7·25 cm. . . .	11·75 cm.
Cavity . . .	8·5 „ . . .	9·75 „
Outlet . . .	9·25 „ . . .	7·75 „

2. The generally uniformly contracted rickety pelvis. Pelves of this kind are very rare. They have not, with the exception of the superior expansions of the ilia, the characteristic marks of rickets at all, or only in a slight degree. The brim may be quite uniformly

Fig. 15.

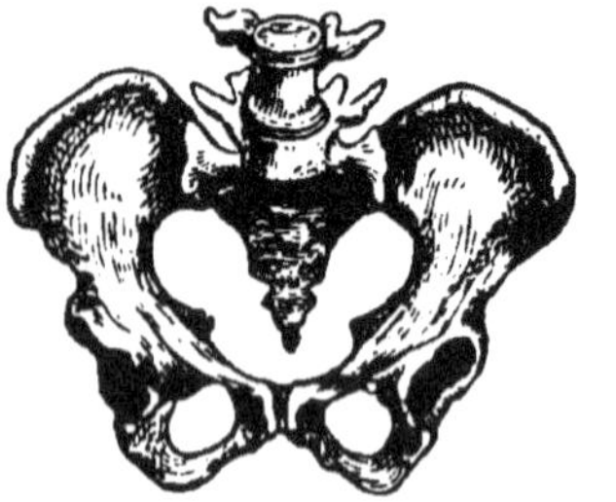

contracted, but towards the pelvic outlet the contraction of the antero-posterior diameter diminishes, whilst that of the transverse diameter often increases. The pubic bones are sometimes of a beaklike shape, approaching in that respect the osteomalacic form. This pelvic shape is produced by the pressure which the continued use of the lower extremities exerts through the acetabula; the pelvis being uniformly compressed from all sides. Thus the osteoid substance situated between the bones is chiefly affected by that compression, the result of which is that the growth of the bones is greatly interfered with, whilst the changes in the shape of the bones are slight or quite wanting.

The pelvis represented in Fig. 15 belongs to a skeleton also in the Bonn collection. Its diameters are—

	Ant.-post.	Transv.
Brim . . .	8 cm. . . .	10¼ cm.
Cavity . . .	10¼ „ . . .	9¼ „
Outlet . . .	9⅛ „ . . .	7 „

3. The pseudo-osteomalacic (Michaelis), or the distorted rickety pelvis (Litzmann). This has a striking resemblance to the osteomalacic pelvis. The promontory is low down, and the lower end of the sacrum approaches it on account of a curve in the third or even in the second sacral vertebra. The acetabula also come closer together, since the symphysis is pushed forwards, so that the brim assumes a triangular shape. The difference between this kind of pelvis and the real osteomalacic one most frequently consists in the smallness of the bones, and especially in that of the upper portions of the ilia. They often preserve their rickety shape and position without the peculiar deformity of osteomalacia; yet in some cases this also is distinctly pronounced.

This kind of pelvis is due to too great use of the lower extremities, whilst the bones are still considerably softened. Under such circumstances the upper portion of the sacrum sinks downwards under the pressure of the trunk, and the acetabula are pressed into the true pelvis by the heads of the femora. Such a high degree of softening of all the bones may be produced by rickets persisting for a long time, when osteoid plates are deposited externally upon the bones, whilst the bone already formed undergoes in the course of time a physiological reabsorption by the development of the medullary cavity. O. Weber has shown that very intense rickets also produces an osteoporosis, that is, a softening of the fully formed bone. The disease producing the pelvis just described may be considered to be rickets complicated by osteomalacia.

Fig. 16 represents such a pelvis, which was first described by Naegele.

FIG. 16.

The shape of the rickety pelvis is greatly modified by the curvature of the vertebral column, which to some extent usually accompanies rickets. Scoliosis then only influences the shape of the pelvis when the sacrum participates (as in fact it commonly does) in the compensatory curvature. In general, the characteristic peculiarities of rickets are distinctly pronounced in these pelves. The transverse concavity of the sacrum is absent, and the bodies of the sacral vertebræ are forced out from between their auricular surfaces. The promontory is pushed somewhat towards the side of the curvature, and the auricular portion of that side is often considerably smaller than that of the other. The contiguous hip-bone is directed upwards, inwards, and backwards; the acetabular portion is higher up, and the superior expansion quite perpendicular. The ischial tuberosity is as a rule bent outwards as usual, and the pubic arch is wide. The symphysis is pushed over to the opposite side, and the ileo-pectineal line of the contracted side is straighter. The transverse diameter of the contracted side is greater than that of the opposite side, whilst the sacro-cotyloid distance is considerably smaller. The origin of these asymmetries can be easily explained by the greater pressure on one of the acetabula. In very considerable scoliosis the acetabulum may be so approached to the promontory that the shape of the pelvis resulting from it is very much like that in the pseudo-osteomalacic form. The sacro-cotyloid distance may thereby become so narrowed that no portion of the head can enter it, and thus the whole of this half of the pelvis is lost for the mechanism of parturition. The pelvis, or rather the wider half of it—if there can be at all the question of an entrance of the head—would then be like a generally contracted pelvis, where the conjugate diameter would be represented by the sacro-cotyloid distance, and the transverse by the oblique diameter, in the way as shown diagramatically in the accompanying Fig. 17. To this category probably belongs the pelvis described by G. Braun, as well as that mentioned by Lambl, which is in the Pathological Museum of Florence.

FIG. 17.

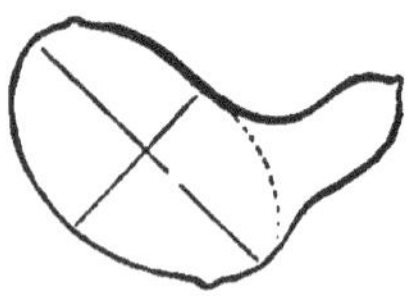

The peculiar shape of the rickety pelvis is still more modified, when rickets is followed by marked kyphosis (for the peculiarities of the rickety pelvis are in almost all points just the reverse of those of the kyphotic pelvis). Very few of the characteristics of rickets remain, and such a pelvis can often only with difficulty be recognised as being rickety; and the more so since in the kyphotic pelvis the proportion of Sp. J. to Cr. J. is somewhat in favour of Sp. J. The complications increase still more when, as most often occurs, the rickety kyphosis is accompanied by scoliosis. As a rule the sacrum then shows most of the distinctive peculiarities of kyphosis. It is twisted round its transverse axis. The conjugate is either not narrowed at all or only slightly so, and sometimes even a little enlarged, whilst the antero-posterior diameter of the outlet is more contracted. The promontory is very high up; and as the transverse surface of the sacrum is either straight or even convex as in rickets, the line drawn from the symphysis to the lower edge of the first or second sacral vertebra represents the normal conjugate. Exceptionally the bodies of the sacral vertebra may be pushed backwards from between their auricular surfaces, as is the case in the kyphotic pelvis. The conjugate diameter is then frequently diminished; sometimes, however, it is normal or even larger; almost always it is smaller than the transverse. The hip-bones usually retain the peculiarities of rickets; they are atrophied, with a very decided S-shaped curvature, and widely separated in front. The pubic arch also is usually wide, the tubera ischii are greatly separated and bent outwards; sometimes, however, the outlet is shaped as in the kyphotic pelvis.

It is not difficult to explain the origin of these changes. The rickety peculiarities only remain in part, and are most clearly seen in the absence of the transverse concavity of the sacrum and in the small, flattened, and widely separated hip-bones. Kyphosis, coming on at the commencement of rickets, causes the ischial tuberosities to turn greatly inwards, and the subsequent sitting posture still more increases the narrowness of the outlet; but if the rickets had previously drawn them outwards their position remains unaltered. If the kyphosis is higher up, so that it is compensated for by a marked lordosis of the lumbar vertebræ, the pelvis retains all the peculiarities of rickets.

c.—*Position and Attitude of the Fœtus in the Contracted Pelvis*

The position of the fœtus during labour is greatly influenced by a narrow pelvis, so that malpositions are at least four times as frequent in the contracted as in the normal pelvis. In primiparæ this proportion is even more conspicuous.

It has already been said that in primiparæ the firm ovoid uterus forces the head, which, on account of the specific gravity of the child, is directed downwards, almost straight into the inlet, and that when this is normal the unyielding abdominal walls force the lower segment of the uterus containing the head into the pelvis, in the latter part of pregnancy; consequently, in primiparæ, the head, at the commencement of labour, is already placed more or less deeply within the pelvis. This condition is altered, even in primiparæ, when the pelvis is too narrow to permit the easy entrance of the head. In the majority of instances the firm uterus, nevertheless, determines the position of the child, so that the head lies upon the brim. But frequently the head, since it does not fit into the brim, is deviated to one side or forwards, so that it rests then either upon the margin of the hip-bone or upon the symphysis. Sometimes, and especially on account of an anomaly in the position of the uterus, the presentation of the fœtus is still more irregular. The narrow pelvis predisposes to pendulous belly, and to a considerable mobility of the uterus; consequently, in the erect posture of the woman the uterus is no longer at an angle of about 35° with the horizon, but lies almost horizontally, or the fundus is even lower down than the lower uterine segment. From what has previously been said of the causes of the normal position of the child, it will be intelligible that, under such circumstances, transverse and pelvic positions easily develop. The great mobility of the uterus also, which allows the fundus to fall from one side to the other, thereby frequently changing the child's position, easily predisposes to malpositions.

If at the commencement of labour the head is deviated to one side, the pains which contract the uterus transversely often succeed in re-establishing the longitudinal position. The head is again made to present. Sometimes, however, the transverse position continues during labour, or even the originally presenting head is pushed away from the inlet, which is too narrow to engage the head, and the transverse position develops secondarily.

The attitude also of the child easily becomes irregular from such changes in the position. The chin is separated from the chest and the face is made to present. In pelvic position, where the breech can less easily enter the brim, the feet pass into the vagina, or in transverse and in head positions an upper extremity and very frequently the funis presents in front of the head. In primiparæ prolapse of the funis is almost exclusively due to pelvic contraction; for at term, in primiparæ, if the pelvis is normal, the head has either entered or at least is fixed immovably upon the brim, and as the opening os retracts, closely applied to the head, prolapse of the funis is impossible; but if the head, on account of a pelvic contraction, remains high up, there is almost always room for the cord when the os dilates; and, in fact, prolapse would be far more frequent if the shortness and the coilings of the cord around the fœtus were not sufficient to prevent it.

An abnormal presentation is sufficiently often met with in primiparæ under these conditions; this frequency, however, still more increases with each subsequent labour.

The flaccid uterus allows transverse positions more easily to take place during pregnancy; and as the pelvis is too narrow to receive the head, the pains are often unable to bring about a head presentation. Even when the head originally presented it is often made to deviate, for whenever the woman assumes the lateral posture the easily movable uterus also falls to that side. The more flaccid the uterus, and the more exhausted by previous labour, the more frequent and complicated are the positions of the fœtus. The normal attitude also of the child is lost in the flaccid sac now formed by the uterus; and therefore, in pluriparæ with narrow pelvis, we meet with positions and attitudes which do not occur under any other circumstances.

D.—*The Presentation and Mechanism of Labour in the Contracted Pelvis*

Even in cases of pelvic contraction where labour terminates apparently quite normally, that is, where the forces of nature are alone sufficient to deliver the woman in not too long a time and without harm to herself and the child, an experienced practitioner can with great probability diagnose from his examination without measurement of the pelvis, not only that there is a contraction, but also the degree of it. It is because "the position of the advancing fœtal part and the way in which it passes through the contracted place are characteristic of the various kinds of narrow pelvis."

When speaking of the normal mechanism of labour it was stated that when the resistances to the anterior and posterior portions of the head were equal, the latter, since it is nearer to the direction of the expelling force, must precede, because the same resistance, acting upon the longer arm of the lever, becomes greater than that acting upon the shorter. Now, in a pelvis partially contracted at the conjugate the normal entrance of the head with the biparietal diameter in the conjugate is, of course, impeded when this is the contracted part of the pelvis. The impediment is now nearer to the occiput than to the forehead. On account of the increased resistance to the occiput the forehead will now descend lower down, in spite of its being the longer arm of the lever; but, when the chin is somewhat removed from the chest, the direction of the expelling forces, acting through the vertebral column, now comes nearer to the forehead. Thus the anterior arm of the lever becomes shorter, the posterior longer. Since, then, not only the resistance to the occiput has increased, but this also acts upon a relatively or absolutely longer lever arm, it follows that the forehead must descend still lower. Accordingly, in the simple flat pelvis, the head enters the brim with the forehead low down, and as soon as the pains have fixed the head in the inlet the sagittal suture is found running almost transversely or a little obliquely close to the sacrum, and the large fontanelle is situated not far from the promontory, in the first head position to the right, in the second to the left of it. Instead of the parietal, the smaller and more compressible bitemporal diameter of the child's head has entered the narrow conjugate. In this position, viz. the sagittal suture running transversely and the large fontanelle lower down, the head is pushed into the pelvis. There is not, however, room enough in one half of the pelvis for the whole of that portion of the head which is situated behind the coronal suture. The head, therefore, passes through the contracted conjugate with a greater diameter than the bitemporal, although usually its greatest diameter, the biparietal, cannot pass through such a conjugate. When the head has passed through the narrow place the forehead then meets with greater resistance from the pelvis, and accordingly the occiput descends lower. The small fontanelle can now be easily reached; soon it comes lower down than the larger fontanelle. If the influence of the posterior pelvic wall preponderates over that of the anterior whilst the large fontanelle is low down, this at first comes forwards, and later only returns backwards. Again, if both fontanelles are approximately on the same level, the head remains for a long time in the transverse position, and the small fontanelle only turns forwards after it has previously come lower down. Since the outlet is normal, the further course of the labour is as usual.

If the contraction of the conjugate is greater the forehead descends still lower down, for the reasons above stated, and the promontory presses upon the upper angle of the frontal bone situated behind. It is very difficult for the head to pass in this direction, and it only exceptionally takes place.

If in a very narrow and greatly inclined pelvis the liquor amnii escapes suddenly, when the head is resting upon the symphysis, it may be arrested in that position. In such a case the sagittal suture does not run close to the sacrum, but immediately behind the symphysis, and the parietal bone situated behind lies flat upon the brim. The head cannot enter the pelvis in this position, but remains movable above the brim, and only when the forces of nature or art have produced a change in the position, so that the sagittal suture comes nearer to the sacrum, does the head enter the pelvis.

In the generally uniformly contracted pelvis the mechanism of labour is the same as in the normal pelvis; only its details are more pronounced. The head may enter the brim in any position, but when the pains are strong, the small fontanelle always comes very low down. This is on account of the relatively greater increase of resistance at the longer anterior lever arm than at the posterior, so that in the greatly contracted pelvis the small fontanelle is distinctly felt in the pelvic axis. In that position the head enters the pelvis and the small fontanelle turns forwards, as soon as the influence of the posterior wall predominates. If the cavity and outlet are free from contractions the unusually deep descent of the small fontanelle ceases, and the mechanism of labour now becomes normal; but if the uniform contraction extends to the outlet the small fontanelle continues the lower, and first becomes visible above the frenulum instead of below the pubic arch.

In the generally but unequally contracted pelvis the mechanism of labour is made up of that peculiar to the flat and to the generally uniformly contracted pelvis. The narrower the conjugate is in proportion to the transverse diameter, the more the head presentation resembles that in the flat pelvis; and the narrower the transverse diameter, the more the head presentation resembles that of the uniformly contracted pelvis. There is, as a rule, the constant transverse position of the head peculiar to the partially contracted, and the great descent of the occiput peculiar to the uniformly contracted pelvis.

In individual cases the small and large fontanelle alternately descend whilst the head advances, so that through this the transverse position may continue down to the outlet.

Face presentations are more frequent in the narrow than in the normal pelvis. The course of labour is aggravated, the face retains for a long time the transverse position, and usually the rotation forwards of the chin occurs rather late.

In pelvic positions in the narrow pelvis the feet are more frequently found to present than the breech. This is readily understood; for the narrow pelvis does not allow the breech to enter the brim early, and the feet therefore sink into the lower uterine segment; and secondly, the obliquity and want of form of the uterus, which are especially frequent in pluriparæ with narrow pelves, determine the deviation of the breech when this is low down from the brim, and thus the feet come to lie within the dilating os. For the rest the mechanism is normal. The expelling force acts upon the breech so far backwards that the child's sacrum is always the lowest, and consequently, even in the narrow pelvis, the rotation with the back forwards does not fail to take place. In the partially contracted pelvis the aftercoming head always enters the brim transversely, and the bitemporal diameter comes into the conjugate, so that therefore the promontory is close to the ear, and in the further progress of the head it presses against the parietal eminence. If the pelvic contraction is somewhat considerable, the passage of the aftercoming head is almost always delayed so long that the child's life is lost unless assistance be given. The passage through the narrow pelvis is still more difficult when the chin is separated from the chest, so that the head has to be passed through the pelvis with the occiput first. The head passes through the generally uniformly contracted pelvis in the ordinary way with the chin flexed upon the chest.

In transverse positions the child usually presents with the shoulder, as in the normal pelvis. Frequently the transverse position is primary; sometimes it develops in the course of labour from a previous head presentation. The back of the child is here more frequently towards the maternal spine than in the normal pelvis. More rarely the chest presents, whilst complications, such as prolapse of an arm or of the funis, are very frequent. It is very rare in the normal pelvis that the labour is terminated by nature, and in the contracted there is still greater difficulty. Yet it is not impossible in the flat pelvis, when the transverse diameter is sufficiently large.

E.—*The Course of Labour in the Contracted Pelvis*

Labour is on an average much longer in the contracted pelvis than in the normal. This protraction is partly, but not exclusively, due to the narrowness of the parturient canal. Even the first stage, that of the dilatation of the os, is greatly protracted, because the presenting part being high up cannot press upon the os with the same degree of force as under normal conditions, and the bag of waters is less efficient in dilating the os. The stage of expulsion is delayed by the mechanical disproportion. But when the head has once passed through the narrow place it is quickly born, and only exceptionally, when the pains have thereby been greatly exhausted, the expulsion from the pelvic canal requires a longer time.

Delivery may also be delayed by a too early escape of the liquor amnii. This is far more frequent in the narrow than in the normal pelvis, and often the waters drain off before uterine action has set in, and whilst the os is completely closed.

It is chiefly, however, the character of the uterine action which determines the duration of labour, and its favorable or unfavorable issue.

It is a general rule, to which there are, however, many exceptions, that the expelling forces augment in the same ratio as the resistance increases. In the great majority of instances of pelvic contraction the energy of the pains reaches such a degree of intensity as is very rarely seen in the normal pelvis. The pains are most powerful in the flat pelvis, and they follow each other in such rapid succession that they threaten to rupture the uterus. But this uterine action, increased as it is beyond the ordinary limits, is by no means pathological in the narrow pelvis, for it is necessary to overcome the abnormally great resistance.

The first stage at its commencement is generally slow and tardy, but when the os has begun to dilate the force of the uterine action increases. The head is now fixed in the inlet, gradually adapted to its shape, and pushed through the contracted part. Just at that period the extraordinary energy of the pains is of the utmost importance, and nothing could be more injurious than feeble pains, which happily are very rare in such cases. The woman is to be advised to bear down, and all the auxiliary forces must be called into play in order to assist the strenuous efforts of the uterus. The more powerful the expelling forces at that time the more favorable is the labour for the mother and the child. In these cases it is highly desirable that the abdominal pressure should begin to act whilst the head is still high up.

The timely dilatation of the os is also of very great importance; if it does not occur serious danger may threaten in two directions. The os may very quickly dilate and retract over the presenting fœtal part before it has entered the pelvis. The consequence is that the vagina is enormously stretched, and may then be ruptured, or even completely torn from the lower uterine segment. This occurs most frequently in neglected transverse positions, but also when the head is delayed entering the pelvis. The greatest strain is put upon the vagina when the head is very high up, and the uterine action little assisted by the abdominal pressure. The consequences of tardy dilatation are not less grave. In the pendulous belly, which is often considerable, and the position of the head with the anterior parietal bone upon the symphysis, the os is placed far backwards, and its anterior lip is very slowly obliterated. Whilst the pains now firmly press the head against the promontory and the symphysis, the pressure upon the anterior lip between the head and the symphysis impedes its complete obliteration, and inflicts severe contusion. The lip either enormously swells or is bruised and torn.

More rarely there is a primary feebleness of pains. This may correctly be considered an anomaly, and it is no doubt one of the most unfortunate complications of pelvic contraction. It almost always requires the assistance of art; for to overcome a greater obstacle a greater force is necessary, and where this fails nature is unable to terminate labour. In the flat pelvis feeble pains are only secondary in consequence of previous very difficult births, but in the generally contracted pelvis they are sometimes primary. The partial pressure to which in the flat pelvis the lower uterine segment is subjected on the part of the promontory and the symphysis seems to arouse the greater energy of the uterine action, whilst the pressure exerted on all sides in the uniformly contracted pelvis has, apparently, not the same effect. It is probable that here the uterine muscles are less completely developed (an arrest of development of all the organs of generation).

Secondary feebleness of pains is oftener met with, and is of greater significance; this is sometimes observed in pluriparæ after frequently repeated labours. Unusual efforts during previous labours, especially when they have followed each other in short intervals, greatly impair the muscular development of the uterus. In pluriparæ with narrow pelves the uterus is often during pregnancy like a flaccid sac, predisposing particularly to transverse positions, and during labour such an organ cannot produce those powerful contractions which are so necessary in the contracted pelvis. Hence the great danger to mother and child, which becomes more imminent with each successive labour, and also the frequent necessity for operations in the later births.

If the obstacle opposed to the advance of the head is very great, the uterine activity also increases correspondingly. The intervals between the pains not only become shorter, but in them the uterus remains in a state of considerable tension, which at last becomes a permanent rigidity. If the head presents, and its entry into the pelvis is not absolutely impossible, it is often effected by this utmost exertion of the uterus, which in that respect proves salutary after all. It is different, however, with transverse positions. Here the uterine contractions meet with obstacles which cannot be overcome. The uterus narrowly contracts around the child, and presses the presenting shoulder violently against the brim. This fruitless activity of the uterus, the child remaining in the unfavorable position, has been called tetanus of the uterus. If operative procedures are urgently indicated, we may meet with the most dangerous obstacles on the part of the rigid uterus. Yet this condition of the uterus is in itself not abnormal, but rendered necessary by the malposition of the child; and to this also are to be ascribed the fruitless, though powerful uterine contractions, and the difficulty or impossibility of operative procedures.

F.—*Diagnosis of the Narrow Pelvis*

The diagnostic points in pelvic contraction have already been mentioned, whilst speaking of the measurements of the pelvis. We shall here only briefly recapitulate the most important.

The following conditions observed during pregnancy would point to the existence of a pelvic contraction:—a remarkably small stature, this may be due to arrest of development (a real dwarf), or in consequence of rickets; distorted bones, also due to rickets; a very considerable degree of pendulous belly (especially in primiparæ); a peculiar mobility of the uterus, which is able to be pushed from one side to the other with great ease: transverse positions, and the situation high up of the presenting part in the last months of pregnancy in pluriparæ. Certainty as to the existence and a correct knowledge of the degree of contraction are only obtained by an accurate measurement of the pelvis.

The size of the true conjugate is calculated with some certainty from that of the diagonal conjugate. Baudelocque's diameter gives a valuable indication of the presence of contraction, but nothing as to the degree of it. Such a position of the spinous process of the last lumbar vertebra, that it does not lie at least one inch above the line connecting the posterior-superior spinous processes of the ilium, is found exclusively with the narrow pelvis. It occurs almost exclusively in the rickety pelvis that the distance between the spines of the ilia (Sp. J.) is proportionately greater than that between the crests of the ilia (Cr. J.). A remarkably short distance between the spines and the crests of the ilia must excite the suspicion that the transverse diameter is also contracted. There is, however, no accurate method by which the transverse diameter of the pelvis can be measured; it is therefore advisable always to feel the lateral walls of the pelvis, and with some experience a distinct contraction of the transverse diameter will generally be detected by the touch. The peculiar character of the mechanism of labour gives a sufficient idea of the size of the transverse diameter.

During labour every deviation from the normal process must call attention to the probability of a pelvic contraction. Thus the premature escape of the liquor amnii is a rather frequent occurrence in the narrow pelvis; abnormal positions of the child, and particularly prolapse of the funis are very suspicious of pelvic deformity, and this is especially the case in primiparæ. A high position of the head, whilst the pains are strong and the uterus dilating, is, in primiparæ, a pretty sure sign of an existing deformity. All these circumstances are mostly so far valuable that they call for an accurate examination of the pelvis. The degree of contraction is only learnt from the measurement of the diagonal conjugate. It is far more difficult to ascertain the size of the transverse diameter during labour; but although it cannot be measured with accuracy, its size can be approximately known from the peculiar mechanism of the labour. Thus, for instance, if in a pelvis, the conjugate of which is known from measurements to be contracted, the head is situated at the brim with the small fontanelle lower down this position would indicate, with great probability, that the pelvis was generally contracted. If the large fontanelle is lower down, it would show that the transverse diameter has, at least approximately, the normal size. The easy rotation forwards of the small fontanelle is, in the case supposed, indicative of a pretty uniform contraction of the pelvis; and the contraction of the conjugate would preponderate when the head remained a long time transversely.

G.—*Consequences and Termination of Labour in the Narrow Pelvis*

The prognosis of labour in the narrow pelvis chiefly depends upon the degree of contraction, yet to a great extent it is modified by the treatment. The prognosis for the mother differs from that for the child.

In general it may be said that moderate degrees of contraction are less unfavorable for the mother than for the child. The proportionate dangers to both keep pace with the increasing contraction. In absolute narrowness the prognosis is very unfavorable for the mother, whilst suitable operative procedures, as the Cæsarean section, afford better chances for the child, which otherwise is certainly lost.

Let us first consider the consequences to the mother. The greatest danger accrues to her from bruising of the soft structures. Here it is of importance to know that a rapid and temporary compression need not necessarily cause mortification of the soft parts, whilst a pressure much less in intensity, but continually acting, may have the most disastrous consequences. This is also the reason why, with suitable treatment, pelvic and transverse positions do not easily prove fatal to the mother, whilst head position may inflict severe injuries upon her. When the disproportion is slight, and, at the same time, the pains are powerful, the head may pass so rapidly through the pelvis that the maternal soft parts are not at all or only slightly injured; but the liability to injury is, of course, greater when the contraction is more considerable. The head, pressed in between the promontory and the symphysis, bruises the maternal soft parts situated over them to a greater or less degree.

The bruising may be followed, either by an innocent local inflammation, or it may be so severe as to cause gangrene and reabsorption of the decomposed organic matter, producing the symptoms known as puerperal fever. If the os has not yet retracted over the head the anterior lip may be lacerated by the pressure; or the uterine parenchyma, together with the peritoneum, are crushed, and perforation into the abdominal cavity follows. Bruising of a part of the posterior wall of the bladder gives rise to obstinate retention of urine, catarrhal or diphtheritic cystitis, or even to a vesical fistula. In rare instances the passage of the head through a narrow pelvis causes rupture of one or more pelvic joints—by far the most frequently the symphysis.

Besides these injuries, the mother is exposed to all the accidents which follow protracted labour. Great delay in labour after the complete evacuation of the waters causes metritis and endometritis, attended by febrile symptoms, which may terminate in death from exhaustion. Rupture of the uterus also may be the result of very powerful pains where the obstacle is insurmountable.

The operations necessitated by the degree of the pelvic deformity are also more or less dangerous to the mother. The Cæsarean section, which must be had recourse to in cases of absolute pelvic contraction, is most so of all. Perforation, again, is free from danger to the mother, and so is turning when performed early enough; but if the transverse position has persisted for a long time the inflammation of the uterus may be intensified by the then necessary manipulations, or rupture of the uterus, which had until then threatened, now takes place. The unskilful use of the forceps may be attended by the most serious consequences, yet in the hands of the experienced physician, who knows the limits of its applicability, that instrument can scarcely do harm.

Pelvic and transverse positions, if suitable aid be rendered and the pelvis be not too narrow, are not injurious to the mother; for the soft breech does not bruise the maternal soft parts, and the after-coming head is, unless soon expelled by the pains, so rapidly extracted that the duration of the pressure is too short to inflict severe injuries. Transverse positions, if early recognised and suitably treated, are as harmless as they are hazardous when allowed to continue a long time.

The consequences of a contracted pelvis are of no slight importance to the child. The long duration of labour, which is almost always caused by deformity of the pelvis, imperils the life of the child. In every case of labour unforeseen circumstances, which are not to be looked upon as pathological, as, for instance, prolapse of the funis, at a place where it is exposed to pressure, may bring the child into danger. The longer a labour lasts the longer a child is exposed to dangers, and consequently the probability of a fatal issue will certainly be very great, when in the narrow pelvis the birth of the child is delayed a long time after the rupture of the membranes. The waters, however, never escape completely, although the head may not be quite closely surrounded by the lower uterine segment, unless by incautious examination a considerable quantity of air has been allowed to enter the uterine cavity, for the pressure of the atmosphere causes a certain quantity of liquor amnii to remain within the uterus, which helps to maintain the ovoid shape of the fœtus by filling up the vacant space between the individual parts of the child; still, such powerful pains as are required to propel the head through a narrow pelvis, after the rupture of the membranes and escape of the waters, are dangerous to the child in two respects. First, because the blood is expelled from the maternal vessels of the uterine walls, as they are under a more considerable pressure, on account of the strong contractions, followed only by short relaxations; and as the uterine contents are under the same pressure as the uterine walls, the blood does not enter the placenta, but the abdominal vessels of the mother. This impedes the exchange of gases between the blood of the mother and that of the fœtus. The venous condition of the fœtal blood thus produced causes irritation of the vagus, which, in its turn, gives rise to a diminution of the frequency of the pulse of the fœtus, and the original want of oxygen in the fœtal blood is thereby increased. The consequence of such powerful, rapidly succeeding pains is a very intense venous condition of the fœtal blood and asphyxia, or even death of the fœtus in utero. Secondly, the placenta is either partially or completely detached by the powerful contractions of the uterus. Such premature detachment is more likely to occur after the escape of the waters, since by it the uterus, and consequently the placental insertion, is very considerably diminished. These accidents, to which the child is exposed, may also occur in the normal pelvis, but they are far more frequent in pelvic contractions, because the deformity delays labour and decreases the uterine activity.

Prolapse of the funis is another very serious accident in the narrow pelvis. In the normal pelvis the dilated os usually surrounds the head so closely that the presentation of the funis, and after the rupture of the membranes its prolapse, is impossible. In the narrow pelvis, however, the head remains a long time high up, and the os is chiefly dilated by the protruding bag of waters, and in the space between the head and the os the funis can easily get into the bag of waters. If the membranes rupture when the head is still high up, the funis must almost of necessity fall into the space between the head and the os, unless it be absolutely too short or coiled round the child's body. Prolapse of the funis is a very grave accident for the child; but in pelvic deformity complicated by prolapse of the cord, the child is always lost unless suitable aid be rendered, and even then the child's life is, nevertheless, greatly endangered.

But, apart from the grave complications just mentioned, the head position even, which is known to be otherwise the one most favorable to the child, is in the narrow pelvis attended by many dangers. This is chiefly due to the mechanical injury which may be inflicted upon the head in passing through the contracted part. We shall briefly consider the changes to which the head is liable within the narrow pelvis, and their influence upon the life of the child.

The shape of the head may be greatly changed by the caput succedaneum. This is not only not injurious to the child, even if it be very considerable, but is in many respects quite the contrary. When the whole fœtal body is exposed to a uniform pressure that part which is free from it and lies within the os becomes the seat of the caput succedaneum. A tumour of some size only then forms, when the os tightly grasps the head on all sides, and the powerful pains force the head into the inlet; that is, when on the whole the disproportion of the pelvis is not too great. Moreover, a large caput succedaneum assists in fixing the head within the brim, and its wedge-shaped form allows the head to pass more easily through the narrow part. Therefore neither the formation nor the existence of the tumour is in any way dangerous to the child, and can only be considered as favorable.

In the contracted pelvis the head undergoes different changes. Since the cranial bones are movable over each other within their sutures, and the head is not yet completely unyielding, it can change its form when pressed by the pains against the pelvic bones of the mother, and thus accommodates itself to the shape of the inlet. The first effect of a narrower pelvis is, therefore, an overlapping of the bones along their sutures. As the head enters the contracted conjugate with the transverse diameter one parietal bone is, as a rule, pushed under the other, in order to effect the diminution of that transverse diameter. Usually it is the posterior parietal bone which is pressed against the promontory, and is pushed beneath the anterior. This is almost exclusively the case in the generally contracted pelvis, whilst in the flat pelvis the anterior often lies beneath the other. In the frontal suture the displacement is the reverse of that in the sagittal suture, whilst the occiput is almost always placed under the parietal bones.

In general this overlapping of the bones along their sutures is certainly a favorable change in the form of the head, and if not too excessive is free from danger. But if the cranial cavity is too much compressed rupture of the sinuses situated under the sutures may take place, and cause hæmorrhage, either externally under the scalp or internally upon the surface of the brain.

Another change in the shape of the head is possible on account of the flexibility of the individual bones. In the flat pelvis the parietal bone situated behind is, as a rule, flattened by the pressure of the promontory, and the convexity of the anterior parietal bone is increased as soon as the skull has partly entered the pelvis. In extreme cases the head has at first sight a remarkably asymmetrical shape. The posterior frontal bone, which very rarely passes by the promontory, is seldom, therefore, flattened; and if it occurs it is chiefly in the generally contracted pelvis.

Slight solutions of continuity are very frequently associated with this flattening of the bones. They consist in fissures running from the periphery of the bone towards the centre of ossification, but in the majority of instances they give rise to no serious consequences. These fissures are, in point of fact, injuries, which are often inflicted by the narrow pelvis upon the bones as well as upon the soft parts.

Injury to the soft parts is alone produced by the persistent pressure of the promontory or of the symphysis upon them. This never takes place when the after-coming head rapidly passes through the pelvis, although the pressure may have been very considerable. The promontory is the most projecting portion of the bony pelvis, and, therefore, most frequently produces the so-called pressure marks. They are very rarely met with in the generally contracted pelvis, but are most frequently found in the flat pelvis, situated on that parietal bone which during labour had been directed backwards. They usually extend from the larger fontanelle along the coronal suture; at times, when the transverse diameter of the pelvis is somewhat contracted, their direction is more towards the ear. They are seldom found in the vicinity of the parietal eminence or upon the frontal bone close to the coronal suture.

In the majority of cases only reddened streaks are found, which, within a few hours after birth, have already turned pale. In other instances there are blue sugillations; the epidermis is detached and a dermatitis then results which lasts for some time; sometimes the pressure causes mortification of the scalp to a varying depth. Then a circumscribed black spot is seen, surrounded by an area of inflammation. In the course of a few days the slough reaching often down to the bone is thrown off and the wound heals by suppuration. They are very rarely dangerous to the child, and they are only so far important as they help to establish the diagnosis of the mechanism of labour. If there is only one pressure mark, it is almost always caused by the promontory; if two, they are produced by the promontory and the symphysis, or, at least, by a spot in the anterior wall of the pelvis close to the symphysis.

Injuries to the cranial bones are of a far greater significance as regards the prognosis. Those most frequently met with are the furrow-shaped incurvations of the edges of the parietal bones close to the coronal suture. They are often met with in a slight degree, so that the edge of the parietal bone projects somewhat above the coronal suture, whilst a distinct impression is seen immediately adjoining the margin; more rarely furrows are found into which a finger can be put. Usually there is only a simple incurvation of the bone, more rarely a distinct fracture. The incurvation is not dangerous; it becomes more serious when complicated, as is frequently the case, by a separation of the squamous suture.

The so-called spoon-shaped depressions of the cranial bones are of more serious consequence to the life of the child. They occur on the frontal and parietal bones between the parietal eminence, the large fontanelle, and the coronal suture. They are seen as deep depressions of the bone, and the elevation of the parietal eminence forms the steep descending wall of the fossa. Cephalhæmatomata (that is, effusion of blood between the bones and the epicranium) which are very common in all injuries to the head, regularly develop within the fossa at the seat of the infraction. The angle of the injured bone close to the large fontanelle is thereby greatly elevated, and one or more fissures are, as a rule, to be found around the circumference near the suture. On the frontal bones these depressions occur very seldom spontaneously; they are most frequently produced there by the wrong application and the violent use of the forceps, when the frontal bone is forcibly dragged past the promontory. They may sometimes be produced on the parietal bones by the uterine action alone, but more frequently they also are caused by the forceps or by forcible traction upon the after-coming head, especially when the latter is firm and unyielding. The prognosis of these depressions is very grave, though by no means absolutely fatal. Of 65 cases of spoon-shaped depressions, which we collected, 22 of the children (34 per cent.) were stillborn or dying; 10 (15·4 per cent.) died in consequence of the injuries received; 33 (50·8 per cent.) remained alive, and, as far as their subsequent history can be traced, their health does not appear except in a few instances to have been seriously damaged. The depression may gradually disappear; more frequently a mark remains on the forehead as an indication of the danger connected with the birth.

Separation of the temporal and parietal bones along the squamous suture is of greater practical importance. This may be met with when the head presents, but is far more frequent in the after-coming head. The separation rarely results in an overlapping of the parietal upon the temporal bone; usually they lie in the same plain, and a large interval between them is occupied by the dura mater. By laceration of the sinus and hæmorrhage this injury very frequently proves fatal, because the bleeding spot is too near the base of the skull.

A fissure between the condyles and the squamous portion of the occipital bone is far more serious on account of the situation of the injury. It is very seldom seen in the presenting, more frequently in the after-coming, head. By the compression of the squamous portion of the occiput on the sides the condyles are torn off. In the great majority of instances this injury is attended by hæmorrhages into the cranial cavity, and proves fatal on account of its immediate vicinity to the medulla oblongata.

We have thus mentioned a variety of injuries to the head which are followed by more or less dangerous consequences. We shall only remark that not too abundant effusions of blood upon the surface of the brain are usually free from danger, whilst those at the base of the skull are always fatal; sometimes, however, the death of the child occurs during labour from simple cerebral compression without any palpable changes.

The result to the child of the various kinds of treatment will be mentioned as we proceed.

The danger which accrues to the child from a narrow pelvis is least in the first labour, and increases with every succeeding one. Whilst in the first three labours living children are twice as numerous as still-born, this proportion becomes five times more unfavorable in the next three labours (Michaelis). This is due to the frequent functional inability of the uterus, which leads to abnormal positions of the child and feeble pains, and to the fact that with the age of the mother and the increasing number of her labours the transverse diameter of the fœtal head becomes not inconsiderably larger.

H.—*The Treatment in the Contracted Pelvis*

Prophylactic treatment, that is, the prevention of the origin of the narrow pelvis, has until now been almost entirely neglected; yet this deserves the full attention of the medical attendant, for there is no doubt that where little girls have suffered from rickets a suitable regimen can do a great deal, either to entirely prevent the changes in the pelvis or to limit them considerably. As the contraction of the conjugate is due to the pressure of the trunk, which forces the promontory into the pelvis, this agent may be rendered ineffectual by prohibiting sitting upright, standing, and walking.

Perhaps the development of the common flat pelvis could also be somewhat restricted if little girls were not allowed to sit at too early an age or too long at a time, and if they were prevented from lifting weights, &c., or in any way burdening the trunk before the pelvis was fully developed.

The treatment of labour in the narrow pelvis essentially depends upon the degree of the disproportion. Unfortunately this can never be made out during labour with absolute certainty. For although the pelvic measurements may give very accurate results, the size of the head about to be born cannot be determined with anything like accuracy. Even when the pelvic measurements are well known, labour must be carefully watched, in order to estimate correctly the impediments which may present in a given case.

If the contraction is very slight—that is, when in the flat pelvis the conjugate measures more than 9½ centimètres, and in the generally contracted more than 9¾ centimètres—treatment is as little required as in the normal pelvis, nor is this, if necessary for other abnormalities, such as malposition of the child, for instance, interfered with by that pelvic contraction. As a rule, such pelves are only of interest on account of the peculiar mechanism of labour; however, it is always to be borne in mind that an unusually large head may make the slight contraction a considerable impediment to labour.

The case is different where the conjugate measures only from 6½ to 9½ centimètres, 2·5 to 3·7 inch, and we shall first consider the treatment which may be necessary in head presentations.

When the conjugate measures not less than 9½ centimètres the head presents at the brim in the way before described. Whilst the os is very slightly dilated, the cranial bones are made to overlap each other, and a caput succedaneum is formed. Thus the head enters the brim, and is pushed through the pelvis, the os gradually dilating. In such cases the mother, as a rule, does not suffer at all, and the child also is usually born in perfect health. Exceptionally, and even when labour progresses rapidly, fatal injuries may be inflicted upon the head by the great compression. Yet, as a rule, labour in such cases is equally favorable to both mother and child, and treatment is not called for. All the accoucheur has to do is to study the way in which the head passes through the pelvis, and to watch carefully over the labour.

In other cases the head enters the brim when the os is as yet little dilated, and whilst the dilatation continues it is fixed there. The progress, however, of the head through the pelvis is retarded for a long time.

Even then no active interference is required unless the state of the mother demands it; but if the sounds of the fœtal heart are diminished in frequency, showing the danger that threatens the child, turning may be attempted, provided that the os is sufficiently dilated; we have, however, supposed that the head is fixed in the brim, and turning will, therefore, be impossible without the use of chloroform. The child being already asphyxiated, not much benefit will be derived from this procedure, because too much time is required for the administration of the anæsthesia. The forceps also proves in such cases of no avail, as will be shown later on, and nothing, therefore, remains but to let the child die, and to facilitate the delivery for the mother by perforation of the dead child. If from the long duration of labour fears are entertained for the safety of the mother, it is necessary to abbreviate labour by artificial means. When the os is sufficiently dilated, and chloroform has been administered, the head may be pushed a little to one side for the purpose of turning and extracting. A gentle attempt at turning is always harmless, and often succeeds, when the head is apparently impacted, thus affording the best chance for the mother and the child. Such an attempt, however, must not be delayed until the life of the mother is evidently in danger. Let the operation be proceeded with when from the long duration of labour contusion of the maternal soft parts may be suspected, although there are no distinct symptoms of it. If turning cannot be performed, the head is then usually so firmly pressed into the narrow place that powerful pains soon propel it so far as to enable it to be grasped by the forceps and extracted. It will, however, be advisable to await the effect of the pains when the head is so firmly impacted that turning is impossible. As a rule, in such cases the head will be expelled by the natural forces in not too long a time.

Sometimes the head remains impacted at the narrow spot and symptoms arise which, in the interest of the mother, urgently demand the termination of the labour. These symptoms are due either to a contusion of the maternal soft parts between the head and the pelvis, and to the inflammation resulting from it, or to the exhaustion of the woman from her continual straining efforts. Rupture of the uterus is unfortunately seldom suspected, since, as a rule, it occurs suddenly, and only takes place when the uterine contractions are very powerful. When, therefore, the vagina becomes dry and the abdomen sensitive to pressure, when there is fever and great exhaustion, all other considerations must yield to that for the life of the woman. For if assistance is withheld until the maternal pulse becomes very frequent and small, and her strength completely exhausted, the woman is almost sure to die, and the child can scarcely ever be saved. As a rule, perforation is in such cases the sole means of terminating labour.

Under these circumstances it is usually recommended to apply the forceps in order to save the mother and the child. It is certainly extremely seldom that this expectation will be realised; on the contrary, both mother and child are often sacrificed by it.

The reasons why we do not recommend the forceps in these cases are the following:—The head enters with the sagittal suture in the transverse diameter of the contracted pelvis. But when the head is high up the forceps can only be applied in the transverse diameter of the pelvis, and consequently the blades must grasp the forehead and the occiput. This position of the blades is faulty, for the tips of the blades as well as the handles are too far apart, so that the maternal soft parts are greatly exposed to injury and the blades also can easily slip. The forceps applied in this way would also compress the longitudinal diameter of the head, and this compression would increase the pressure to which the maternal soft parts are already exposed from the head lying between the promontory and the symphysis. It is evident, also, that this enlargement of the head in the transverse diameter would render its passage through the contracted part still more difficult. The favorable action of the pains, which accommodate the head to the brim by elongating the longitudinal and contracting the transverse diameter of the head, would be directly counter-balanced by the forceps. In spite of our objections to the use of the forceps in these cases, we do not doubt, when the pains are strong and the disproportion not too great, that the head may be dragged through the contracted pelvis by forcible traction with the forceps.

Unless delivery is effected almost exclusively by the natural forces, it will always terminate fatally to the child on account of the very considerable compression to which the head is subjected on all sides, and the bruising also of the maternal soft parts will be much greater.

We therefore distinctly caution the young practitioner not to rely upon the forceps when the head is high up in the contracted pelvis. When, however, the head is not yet so firmly impacted but that the hand can pass by it under chloroform version is indicated. If this is impossible, one who is very skilful in obstetric operations may apply the forceps, and by moderate, yet strong, tractions, try to drag the head through the contracted part. He will know when to remove the forceps in case his attempt does not succeed. But the tyro begins to draw with greater force when the head does not follow, and, covered with perspiration from the anxiety he undergoes, and ashamed at the failure of his operation, he drags so violently that the otherwise innocent forceps becomes one of the most dangerous of instruments. Only when the impediment has been completely or almost overcome by the for ces of nature should the forceps be applied in the narrow pelvis. In most cases the natural expelling powers are able to push the head rapidly through, but it may occur that they have previously been greatly exhausted, and that the head remains within the pelvic cavity. Here the forceps is in its right place, here it again becomes the innocent and saving instrument, because it is used, not on account of the pelvic contraction, but of the feeble pains.

If the head has not yet passed through the brim and the state of the mother requires the immediate termination of labour, nothing remains, when turning is impracticable, but the sad alternative of either allowing the mother and child to die or of shortening the life of the fœtus, which in either case is always lost, by an operation, viz. by perforating the living child, and thus to save the life of the mother. The operation will not benefit the mother if it is performed too late, when her condition is such as to exclude the hope of rescuing her. The time for operating is when the injuries inflicted upon the mother as yet allow a favorable prognosis. It is certainly one of the most difficult problems in obstetrics to determine correctly the time when there are still sufficient chances for the mother and when the life of the child must be destroyed, which might, perhaps, be saved. The sole means of evading perforation of the living child in such cases is the Cæsarean section. Mother and child may be saved by it, but not even the child is saved with certainty. In practice the question will comparatively rarely be raised, since the great majority of the mothers willingly assent to embryotomy. After perforation we wait a short time in order to allow the natural forces to expel the head. If this is delayed, and the woman must be quickly delivered, the cephalotribe is applied.

There are cases where the os is dilated by the presenting head, yet the head does not tend to enter the brim. The treatment is essentially different, according to whether the woman is a primipara or a pluripara.

To sufficiently dilate the os of a primipara such powerful uterine action is necessary that if it fails in pressing the head at least partly into the brim it can by no means be foreseen how long labour may last, and what dangers may arise from it if left to itself. Since the head is movable above the brim, and the os dilated, it is evident that the pains have been strong enough to effect dilatation of the os, but are unable to fix the head within the brim. The conclusion, therefore, is correct, that in the further progress of the labour, even in the most favorable event, the head will be pushed through the narrow part by such strong and continued pains as to place the child in the greatest danger, and also that this passage will be so slow that serious contusions of the maternal soft parts will be unavoidable. Such cases very urgently call for operative interference, and this at a time when the dangerous symptoms have not yet occurred. As the dilated os and the movable head are both convenient for turning, and as this is entirely free from danger, the immediate performance of that operation with subsequent extraction is indicated in such cases, and, if possible, it should be done before the membranes have been ruptured. For the extraction of the aftercoming head is much more easy than the passage of the presenting head, and because the narrow place is much more quickly traversed this is the most favorable for the mother, and not more unfavorable for the child than when the head comes first. Without doubt in this way the chances of the mother are much greater than when operative proceedings are delayed, and very recent observations show with great certainty that the prognosis for the child is not worse.

The case is different with pluriparæ. From the fact that the head is still movable above the brim, whilst the os is already dilated and the membrane intact, no conclusions can be drawn as to the inability of the pains to fix the head, or as to the further progress of labour, for in pluriparæ the slightest uterine action often suffices to dilate the os. It is not so very rare in the moderately contracted pelvis to find that the head remains high up until the membranes are ruptured, and with their rupture a sharp pain forces the head through the contracted place, and rapidly effects delivery. When the measurements of the pelvis and the course of the previous labours show that the mechanical disproportion is not very considerable, active interference may be delayed, though the os be dilated, especially as long as the membranes are intact. If they rupture, and the head does not very soon afterwards enter the pelvis, it will always be possible to turn, though not as easily as before. But not even then must the operations be deferred too long, because the size of the child's head, which greatly determines the degree of the disproportion, is never accurately known; and when the head is fixed in the brim the time for turning has passed, so that it will be necessary to perforate the living child. If it be, however, known from the previous labours that moderately large children have with great difficulty and danger passed through that pelvis, version is to be performed as early as possible for the reasons above stated. Labour is rendered thereby quite free from danger to the mother, and the results as regards the child are also better, and certainly not worse than by waiting.

Simpson has very conclusively demonstrated the reasons why the aftercoming head more easily passes through a narrow pelvis. He correctly called attention to the fact that the head represents a cone, the thick and obtuse end of which enters when the head comes first, whilst the narrower portion of the cone enters when the head comes last. It is also of great importance that when the head passes last the uterine action can be assisted by powerful traction upon the trunk. This traction is the more useful because it allows the head to accommodate itself to the shape of the brim, whilst the forceps applied to the advancing head does not allow its favorable configuration, as we have already explained above.

Martin has recommended turning in cases of asymmetry of the brim, in order to bring the occiput situated within the narrower into the wider portion of the inlet. We do not doubt that such a change of position may be attended with favorable results; we only wish to point out that in the asymmetrical rickety pelvis (of the ankylosed obliquely distorted pelvis we shall speak further on) the diagnosis of a moderate distortion is extremely difficult, and we must also leave it undecided whether the intended change of position can be obtained by turning. This indication has, at any rate, very little practical significance.

We therefore recommend version in all cases of pelvic contraction, where this is not absolute. Our reasons for doing so are—(1) in the interest of the mother; (2) in very great pelvic contraction with a conjugate diameter of 7¼ centimètres (2·8 inches) we have extracted live children; and 3rd, because perforation of the aftercoming head is neither more difficult nor more dangerous to the mother than that of the presenting head.

Occasionally we may have the pleasure of extracting a live child of moderate size through a conjugate which measures less than eight centimètres; at the worst it will only be necessary to perforate the aftercoming head; sometimes perforation is avoided, which would have been necessary had the head come first, by the spoon-shaped depression which diminishes the transverse diameter of the head. It is true that only half the number of children whose heads have been thus injured escape alive, but, seeing that perforation is an absolutely destructive operation. the results of that depression of the head are always more favorable.

As version has already been shown to give such good results in simple head presentations, the operation will, of course, be necessarily resorted to where dangerous complications exist. Where the forehead or face is placed upon the brim, or an extremity or the cord prolapsed, turning is, in fact, the only proper treatment. The operation is also indicated in the narrow pelvis, for the same complications as in the normal, but especially in transverse positions, and in dangers threatening the life of the mother or the child.

Pelvic positions in the narrow pelvis are, when medical assistance is at hand, neither unfavorable to the mother nor to the child.

The death of the fœtus in utero greatly alters the line of treatment to be adopted. When the head is still movable above the brim turning and extraction are more easy and less dangerous than perforation to the mother, because it is not known beforehand whether the natural forces will be able to expel the perforated head. But if the head is firmly fixed in the brim, and the labour is greatly protracted after the death of the child, perforation is always to be performed, for the pains have much less difficulty in pushing a perforated head through the pelvis than one whose bulk is not reduced; there is no good reason to abstain from the mutilation of the dead body of the fœtus if the mother can profit by it. The pains are allowed to effect the delivery of the perforated head, and the cephalotribe is only made use of when it becomes desirable in the interest of the mother to terminate the labour.

In the generally contracted pelvis the results of the various operations are somewhat different. The extraction of the after-coming head is more difficult, so that the child will less often be delivered alive by turning. The objections to the use of the forceps in that form of pelvic deformity are not the same as in the flat pelvis, but even here it will be very difficult to extract a child uninjured, especially if the contraction continues into the pelvic cavity. It is, therefore, advisable not to make too forcible attempts with the forceps, and if the child dies in the mean time to remove the instrument immediately and to perforate. As by perforation all the diameters of the head are diminished, it appears to be the operation in the highest degree applicable to the generally contracted pelvis.

If the conjugate measures less than 6½ cm. (2·5 inches) there is no hope of delivering a living child *per vias naturales*. The Cæsarean section is then the sole means of delivery, and, the mother assenting, it will the more readily be performed, as craniotomy in such cases is by no means an operation entirely free from danger to her. In practice, as long as there is the possibility of another way of delivery, the Cæsarean section is very rarely permitted, and in the highest degrees of pelvic contraction it will, as a rule, be necessary to perform craniotomy whether the child be alive or dead. But there is certainly a limit to the pelvic deformity where craniotomy becomes a far more dangerous operation than the Cæsarean section, and where the latter is preferable, even when the fœtus is dead. It is exceedingly difficult to decide what is this limit. In Germany 5½ cm. (2·1 inches) are, as a rule, taken as the extreme limit of craniotomy. The English bring the child through a flat pelvis the conjugate of which only measures 4 cm. (1·5 inch), and Barnes entertains the hope that it will be possible to extend the extreme limit of craniotomy to a conjugate of 2¾ cm. (1 inch). In such extreme cases it is well not to have a too high opinion of one's own skill, and unless there are the necessary instruments at hand it is preferable to perform the Cæsarean section, not only because it is easier, but because under these circumstances it is the least dangerous to the mother.

Whilst speaking of the treatment of the narrow pelvis we have all the time supposed that the physician has been summoned to a case at the commencement of labour. But if he has an opportunity of detecting the deformity during pregnancy the induction of premature labour is then the treatment most beneficial to the mother, but less favorable to the child.

III.—*The rarer Forms of Contracted Pelvis*

A.—*The Spondylolisthetic Pelvis*

The spondylolisthetic pelvis is very rarely met with. Until now only eight such pelves have been the subject of obstetric observation; in seven of these this peculiar deformity has been ascertained by post-mortem examination and subsequent maceration. In the cases recorded by Hugenberger, Barnes, Lehmann, and Blake, the deformity was probably the same.

The peculiarities of this kind of pelvis are the following:—The last lumbar vertebra slides down from the sacrum into the pelvis. The lower surface either partly projects into the pelvic cavity, or partly rests upon the upper surface of the sacrum, sometimes even its posterior surface lies upon the anterior surface of the sacrum. This change of position takes place gradually, so that the intervertebral cartilage atrophies and the shape of the corresponding lumbar vertebra is greatly modified by the friction with the sacral vertebræ. Frequently synostosis takes place between the two vertebræ, and the further sliding downwards is then arrested. By the descent of the lumbar vertebræ into the pelvis the conjugate diameter is more or less diminished. As the promontory for obstetric purposes is now replaced by the lumbar vertebræ, the conjugate will have to be measured between the symphysis and that part of the lumbar vertebræ which is nearest to the upper and inner margin of the symphysis. This was in three cases the lower margin of the fourth, in one the lower margin of the third or the upper margin of the fourth, and once even the lower margin of the second lumbar vertebræ. The degree of the contraction varies greatly. The antero-posterior diameter of the brim was, in the macerated pelves, 5, 5½, 6, 7¼, 7½, 7¾, twice 8¼, and 9¼ centimètres; the contraction was, therefore, as a rule, very considerable.

FIG. 18

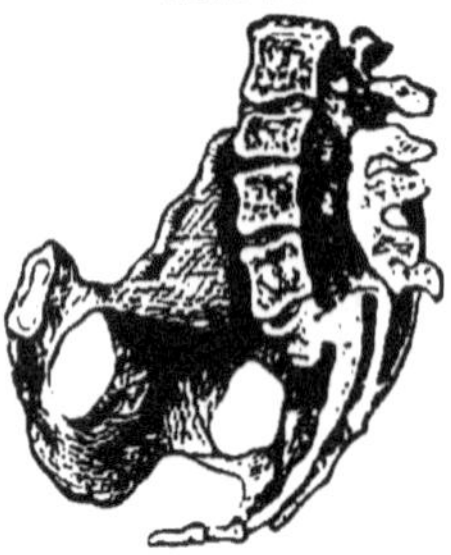

The contraction of the antero-posterior diameter of the brim is not, however, the sole change which the pelvis undergoes. Changes take place in it similar to those in the kyphotic pelvis. As the whole vertebral column—that is, the entire weight of the trunk—sinks into the pelvis, the centre of gravity would fall more forwards if this were not compensated by a diminished pelvic inclination, which, however, occurs quite constantly. The upper end of the sacrum is pressed backwards by the dislocated lumbar vertebræ, so that the base of the sacrum forces the posterior spines of the iliac bones asunder, and the apex of the sacrum projects more into the pelvis. The antero-posterior diameter of the outlet is somewhat contracted by this, whilst the proper conjugate—that is, the line between the symphysis and the first sacral vertebra—is enlarged. The transverse diameter gets gradually narrower towards the outlet, because the ischial tuberosities approach more closely on account of the separation of the iliac bones above, and of the traction of the ilio-femoral ligaments, which are greatly stretched through the diminished inclination of the pelvis. In some cases this contraction of the transverse diameter of the outlet was very considerable. As another consequence of the rotation of the sacrum, a greater mobility of the pelvic joints has been observed by Breslau.

The diagnosis is apparently easy on account of the great lordosis and the projection of the lumbar vertebræ into the small pelvis. If the last lumbar vertebra can be distinctly felt to join the sacrum at an angle, then, of course, there is no difficulty in recognising this kind of deformity; but if this is not the case it may be very difficult to distinguish between it and a rickety pelvis, for in rickets also there may be a high degree of lordosis, the last lumbar vertebra, which has descended into the pelvis, may be taken for the promontory, and its junction with the sacrum for a great curvature of the latter. Other rickety changes (in the clavicle, costal cartilage, and extremities) will, therefore, have to be looked for. There is a difference, also, between the two kinds of pelvis in the transverse diameters of the false pelvis. If the history also gives a negative result with regard to rickets, if no other rickety changes are visible, and if—as is the rule in the spondylolisthetic pelvis—the distance between the crests of the iliac bones is from $2\frac{1}{2}$ to 3 centimètres greater than that between the spines, it is absolutely certain that the pelvis is not a very deformed rickety one, for which alone it could have been mistaken.

Rickets being excluded in this way, there can scarcely be any doubt as to the existence of spondylolisthesis. Exostosis, just at that place, and of so regular a shape, would not occur simultaneously with so considerable a lordosis. If, moreover, the history shows that at an early period or immediately after puberty, there has been disease of the sacral region, the diagnosis of spondylolisthesis is certain; there is not always the peculiar waddling gait. On account of the descent of the vertebral column the place of division of the aorta is sometimes so low down that it, or at least both common iliac arteries, can be felt. Olshausen first called attention to it as a very important diagnostic sign, and Breslau in his second case could feel the aorta pulsating on the posterior wall of the pelvis. Hartmann was able to feel the place of division of the aorta as high up as the upper margin of the fourth lumbar vertebra.

The prognosis is very grave in all cases where the contraction is somewhat considerable, and certainly more unfavorable than in a rickety pelvis of the same degree of contraction; for in the spondylolisthetic pelvis the contraction is not limited to a short space, but it begins on account of the lordosis of the lumbar vertebræ already in the false pelvis, and does not cease when the narrow part has been passed, but continues into the pelvic cavity; besides, the outlet is also narrowed.

The treatment essentially depends upon the degree of contraction. On the whole it appears advisable to perform the Cæsarean section when the brim is greatly contracted.

Breslau's case, in which the conjugate measured $7\frac{3}{4}$ centimètres, is so far very instructive as the neglect of the Cæsarean section caused the death of the woman who was not delivered. In the cases of the Prague-Olshausen's and Belloc's pelvis the operation had been performed. The woman to whom the Paderborn pelvis belonged had been operated upon twice, the first time with success. In the cases of Spaeth and Ender the child had been perforated. In Hartmann's case, where the diagonal conjugate measured 11 centimètres, premature labour had been twice successfully induced. Mature children, of course, would have been stillborn.

B.—*The Kyphotic Pelvis.*

The conditions requisite for the production of this form of pelvic deformity is that the kyphosis should be situated so low down that its influence on the pelvis cannot be compensated by a lordosis still lower down. It is most characteristic when the kyphosis is situated in the lumbar vertebræ; but kyphosis in the lumbo-sacral region also impresses upon the pelvis that peculiar shape even more markedly, only here the shape of the sacrum itself is altered. If the lower dorsal vertebræ are kyphotic the changes in the pelvis are still very distinct, though not quite so marked; but if the kyphosis is higher up in the dorsal vertebræ the characters of the pelvis are less defined, and from a considerable compensating lordosis of the lumbar vertebræ the pelvis may undergo changes of quite a different kind. It is only when the curvature is due to caries of the vertebral bodies that the changes are most characteristic of kyphosis, for that produced by rickets can only alter a pelvis which has already undergone rickety changes, and they are just the opposite to those of kyphosis.

The most important changes in the pelvis observed in kyphosis occurring in the lumbar region are the following:

The upper portion of the sacrum is pressed backwards. The vertebral bodies project backwards from between their auricular surfaces, and the surface of the sacrum is greatly hollowed out from side to

FIG. 19.

side. The anterior surface of the sacrum is also much elongated, and the promontory is very high up and far back. The longitudinal diameter is increased, the transverse diminished. The curvature from above downwards is only marked in the lower portion, the upper is often convex, and the whole anterior surface of the sacrum has an S-shaped curvature. On account of the rotation of the sacrum the antero-posterior diameter of the pelvic cavity, and still more so that of the outlet, is considerably smaller than the true conjugate.

The superior portions of the iliac bones are flat and transversely extended and their S-shaped curvature diminished. The distances between the spines and the crests of the iliac bones (and especially the former) are greater than the normal, whilst the posterior superior spines are more approximated on account of the smallness of the sacrum. The anterior inferior spines are remarkably developed, the lateral walls of the true pelvis are unusually high, and the ilio-pectineal line has a very straight course.

The ischial bones are very near to each other, and the distance between their tuberosities and spines is much less than normal. The pubic arch is consequently small, and its shape is very much like that of the osteomalacic pelvis.

The conjugate and the oblique diameters of the brim are enlarged, the former more so than the latter. The sacro-cotyloid distance is also greater, whilst the transverse diameter is usually lessened and very frequently even absolutely smaller than the conjugate. In the pelvic cavity the antero-posterior and the oblique diameters are diminished, the former less than the latter, and this contraction considerably increases towards the outlet. The antero-posterior diameter of the outlet is of the normal size, or only slightly smaller; but compared with the unusual size of the conjugate it is contracted, and the oblique diameters are very considerably smaller.

In lumbo-sacral kyphosis the changes in the pelvis are far more marked, and the sacrum differs from the one above described. Instead of being elongated, it is smaller than normal; as a rule very small indeed, and without a proper promontory. In the Brüssels pelvis and that described by Mohr an abnormal mobility of the joints was observed.

The kyphosis may be compensated by a great lordosis of the lower lumbar vertebræ. As in spondylolisthesis, so here, the projection of the lumbar vertebræ into the inlet causes a great diminution of the antero-posterior diameter. To this class of deformities belongs the pelvis described by Olshausen, where the distance between the symphysis and the upper margin of the last lumbar vertebræ measured 8¾ centimètres, and that described by Gluge in Brussels, where the distance between the symphysis and the fourth lumbar vertebra also measured 8¾ centimètres.

Since attention has been called to these changes the diagnosis can no longer be difficult. The kyphosis is easily recognisable, and such a pelvis cannot be mistaken for an osteomalacic, to which it bears a certain resemblance as regards the outlet. The mistake was chiefly made because the movable joints of the kyphotic pelvis allowed an enlargement of the outlet during labour, just as occurs in the soft bones of the osteomalacic pelvis. Besides, the former is distinguished by the history of the case, by the entirely different shape of the sacrum and the iliac bones, as well as by the transverse diameter of the false pelvis, which in the osteomalacic pelvis is, as a rule, shorter whilst it is usually enlarged in the kyphotic. Usually, it is impossible to reach the promontory of a kyphotic pelvis. The outlet is to be measured in the way before described.

The prognosis depends chiefly upon the contraction of the outlet. There is danger to the mother when the outlet is very considerably contracted, and the child also has little chance unless the Cæsarean section be performed, though the deformity be not very great. Nevertheless, labour may terminate favorably for both mother and child, even with a very narrow outlet, when the pelvic joints are movable. An excessive degree of pendulous belly, which has been observed in several cases, may greatly disturb the progress of labour.

With regard to the treatment nothing definite is as yet known. Premature labour, if the opportunity is afforded, probably is indicated in all cases; the Cæsarean section will very rarely be necessary unless in very unusual contraction of the outlet. As the head can enter the brim without difficulty, labour may be left to nature as long as the state of the mother permits it. If operative procedures are required the forceps may at times be applied with success, but often it will be necessary to perforate the child.

Fig. 19 represents the pelvis described by Hönig. The excessive deformity was produced by lumbo-sacral kyphosis.

c.—*The Funnel-shaped Contracted Pelvis*

We have already stated that in the majority of instances the brim is the sole or at least the principal seat of the pelvic deformity, in consequence of which the pelvis becomes wider towards the outlet. The reverse of it, that is, an almost normal inlet with a gradual contraction towards the outlet, is not very frequent. This we have found to be the essential character of a kyphotic pelvis.

But there is another kind of deformed pelvis, in which the brim

Fig. 20.

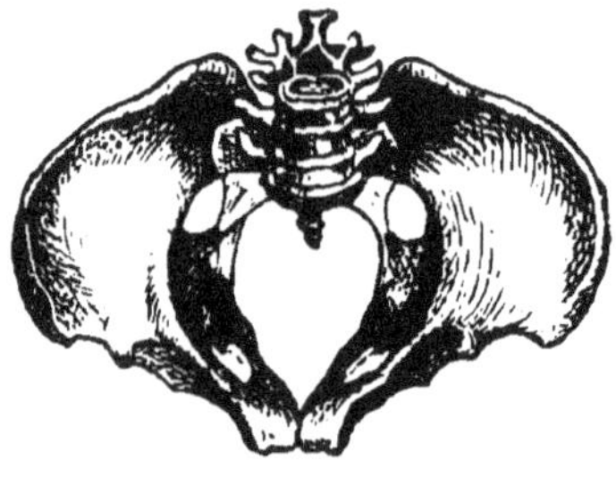

is almost normal, whilst the outlet is contracted either in the transverse or in the antero-posterior diameter or in both.

Contraction of the transverse diameters towards the outlet occurs, not only in the flat pelvis, as stated above, but also in those where the measurements of the conjugate are still within the normal limits. The sacrum has either quite the normal shape and position, so that the antero-posterior diameter of the outlet has almost the same size as the conjugate, or the sacrum has scarcely a longitudinal curve, and the antero-posterior diameter of the pelvic cavity is slightly, that of the outlet much, larger than the conjugate. The lateral walls of the true pelvis converge greatly towards the outlet, the result of which is that the ischial spines, and especially the ischial tuberosities, come very close together, and the pubic arch forms a very acute angle.

It is very seldom that the antero-posterior diameter of the outlet is alone or chiefly contracted, the sacrum being longitudinally extended and placed as in the kyphotic pelvis. This position of the sacrum is more frequently associated with a contraction of the transverse diameter, and such a pelvis bears a striking resemblance to a kyphotic one. These pelves are almost always more or less asymmetrical.

A diagnosis can be made without having recourse to instrumental mensuration when the contraction of the outlet is considerable. The descending pubic rami then pursue a parallel course, the pubic arch is narrow, and the ischial tuberosities are closely approximated. Contractions of lesser degree are not recognisable without instruments. They may be suspected when strong pains are unable to advance the head, which is already within the true pelvis. Careful exploration and mensuration will either serve to confirm or refute the suspicion.

In the slighter degree of the deformity labour may terminate without accidents, but if the contraction is more considerable both mother and child are exposed to certain danger. The head, by its continual pressure upon the pubic arch easily causes gangrene of the soft parts, which may be followed by cicatricial strictures of the vagina, or the pressure causes vesico-vaginal fistula and caries of the rami of the pubic arch. When the pelvis, the transverse diameter of which is contracted, has a wide antero-posterior diameter, the head may not be pushed so far into the narrow pubic arch.

Treatment.—If the outlet is greatly contracted the induction of premature labour is indicated. If the time for doing so is past, the question of turning during labour will rarely be raised, because the head stands, after the dilatation of the os, already low down within the pelvis. When the child is alive and the contraction not too great, the forceps may be applied. Forcible tractions are, however, to be avoided, else they may be followed by lacerations of the vagina, vesical fistulæ, caries of the bones and rupture of the joints. When steady but cautious tractions do not produce the desired effect, perforation must not be delayed too long. The dead child should be immediately perforated as soon as its expulsion is retarded at the outlet; the Cæsarean section will hardly ever be necessary.

The original of the pelvis represented in Fig. 20 is in the collection of the Erlangen Lying-in-Hospital. Its measurements are—

	Post. Diam.	Transv. Diam.
Brim .	11·4 .	12·6
Cavity .	13·6 .	11·2
Outlet .	10·7 .	8·5

D.—*The Anchylosed Obliquely Contracted Pelvis*

This deformity is not often met with, and up to the present time not more than fifty such pelves in the dried state have been described. The number of cases in which this deformity has been diagnosed on the living woman is still smaller. Nægele has described the pelvis of an Egyptian mummy which belongs to this class.

The deformity itself is produced by a far greater pressure acting upon one acetabulum than upon the other. The iliac bone of that side is thereby pushed upwards, inwards, and often backwards. The peculiarities of the pelvis are the following:

On the side of the anchylosis the lateral portion of the sacrum is smaller or has quite disappeared, the anterior foramina are narrower, and the auricular surface also is smaller than on the other side; the course of the sacro-iliac synchondrosis is shown by a ridge. The anterior surface of the lumbar vertebræ are somewhat directed to the side of the anchylosis. The iliac bone of that side is more elongated, higher up, and directed more backwards. The sciatic notch is thereby diminished, and the spines as well as the tuberosities of the ischium are approximated to the sacrum. The acetabulum

Fig. 21.

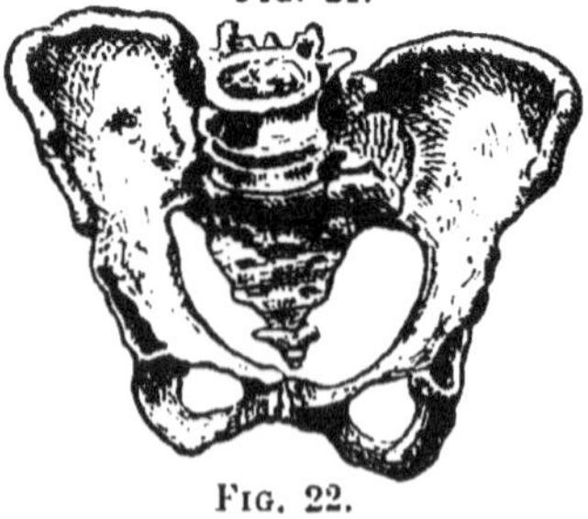

Fig. 22.

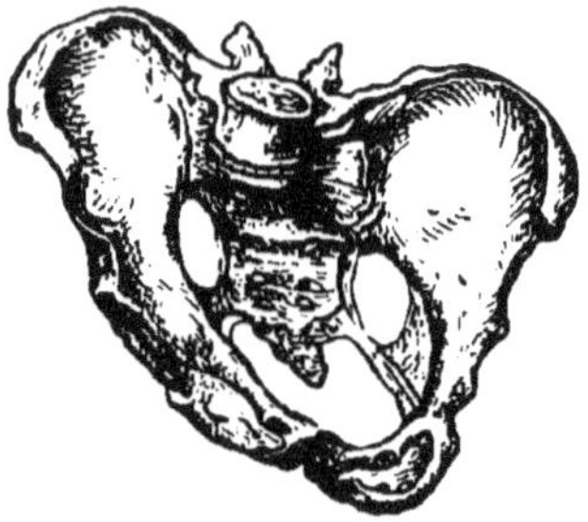

Fig. 23.

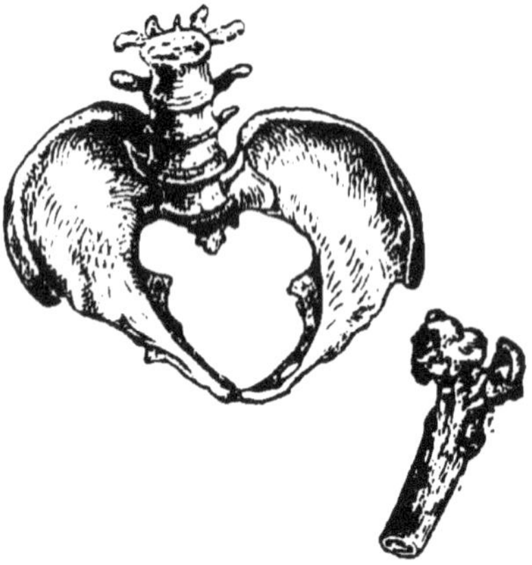

is also somewhat higher up, and looks more forwards. The opening of the pubic arch is directed to that side.

The symphysis is pushed over to the opposite side, the iliopectineal line on the anchylosed side is greatly elongated, that on the other side curved less behind and more in front.

The differences between the oblique diameters and the sacrocotyloid distances are usually very great. The true conjugate is enlarged somewhat on account of the oblique distortion, the transverse diameter of the brim is shortened, and this shortening increases towards the outlet.

Fig. 21 represents the pelvis described by Hecker-Paetsch.
" 22 represents the pelvis described by Litzmann.
" 23 represents the pelvis observed by Rosshirt and described by Litzmann.

Diagnosis. From the fact that the obliquely contracted pelvis has rarely been recognised in the living subject, the conclusion may be drawn that the diagnosis is beset with unusual difficulties. But this is so far only correct that with the usual methods of pelvic examination we may easily overlook such a deformity. We possess, however - at least, in the higher degrees of the deformity—diagnostic signs, which, when the suspicion has once been excited, allow us either to confirm or to refute it. Everything depends in a given case upon our thinking of the possibility of the presence of an anchylosed obliquely contracted pelvis.

This suspicion will rarely be entertained before labour has commenced, and only in those cases where the woman walks lamely, where one hip is higher than the other, where one side has been the seat of hip-joint disease, or where cicatrices of fistulæ exist on the buttock. The history also will occasionally show whether lameness or acute or chronic disease of the posterior region of the pelvis previously existed. However, in the majority of cases the deformity in question is met with in otherwise healthy primiparæ with conjugates of the normal size.

When labour has begun, suspicion is more easily excited. If in spite of powerful pains the head remains high up, and if by the measurement of the conjugate this delay cannot be explained, an obliquely contracted pelvis must be thought of when an abnormally large head can be excluded.

Attention to the following points will serve to point out or exclude the existence of this deformity.

1st. One iliac bone is situated higher up than the other. This can—with the necessary precautions—be determined when the woman is in the erect or the recumbent posture.

2nd. The posterior iliac spines of the anchylosed side are in most cases nearer to the middle line. Sometimes, however, the difference is so slight that it cannot be decided by measurements, or it may be entirely absent. If the posterior iliac spines of one side are much nearer to the spinous processes than those on the other the diagnosis is almost absolutely certain; a negative result, however, is by no means a proof of the absence of an anchylotic distortion.

3rd. By internal examination the pubic arch is found to be distinctly deviated to one side.

4th. The ischial spines may also be unequally distant from the lateral margins of the sacrum (*Ritgen*).

5th. On palpation of both lateral walls of the pelvis the straight course of the ileo-pectineal line on the diseased side will be remarked.

6th. If there is no objection to the introduction of two or three fingers, the promontory can always be felt. It will also be remarked that the promontory is on one side nearer to the ilio-pectineal line than on the other (inequality of the sacro-cotyloid distances).

Naegele has given various external measurements, which in the normal pelvis are equal on both sides, but differ from each other in the obliquely contracted pelvis. They are—

1. From the ischial tuberosity of one side to the posterior superior iliac spine of the other, on an average 17½ centimètres, or 6·9 inches.

2. From the anterior superior iliac spine of one side to the posterior superior iliac spine of the other, 21 centimètres, or 8·26 inches.

3. From the spinous process of the last lumbar vertebra to the anterior superior iliac spine of each side, 18 centimètres, or 7 inches.

4. From the trochanter of one side to the posterior superior spine of the other side, 22½ centimètres, or 8·7 inches.

5. From the middle of the lower edge of the symphysis to the posterior superior spine of each side, 17¼ centimètres, or 6·78 inches.

Measurements can pretty accurately be taken from the anterior and posterior spines (the skin is so firmly adherent to the posterior spines as to form a very visible depression). This is less certainty, however, with regard to the ischial tuberosities and the trochanters.

The differences between the two sides must, of course, be considerable (at least one centimètre) if they are to be of diagnostic value. It must also be taken into consideration that after disease of the hip-joint the bones and soft parts may somewhat atrophy.

By means of the signs just mentioned we may, in cases of considerable deformity, establish a diagnosis of oblique distortion of the pelvis; there are, however, far greater difficulties in deciding upon the degree of the contraction. This varies according to whether the pelvis was originally large or small, and also according to the extent of the dislocation.

An approximate estimate of the degree of deformity can, after all, be obtained when the various methods of examination are adopted which have been detailed above. Naegele's measurements are especially valuable in this respect, and if they show great differences between the two sides we may conclude that the deformity is also considerable. If all the measurements are either pretty large or remarkably small we have a clue as to the original size of the pelvis. The distances between the spines and the crests of the iliac bones are always smaller than the normal.

When there are traces of a previous inflammation we shall know for certain that the oblique distortion is attended by anchylosis. The presence of a synostosis is rendered very probable when the posterior iliac spine of the diseased side is very close to the spinous process.

It is possible that in a number of obliquely contracted pelves labour has terminated favorably, and that, therefore, the deformity has remained unnoticed; but where attention has been directed to them it has invariably been found that they gave an unfavorable prognosis for both mother and child. Litzmann states that of twenty-eight mothers twenty-two died in their first labour, and five of them undelivered; three died in their second and two in their sixth labour. Of forty-one labours six terminated naturally, and five of these occurred in the same woman. Of the forty-one children ten only were born alive (amongst these six by the same mother and two after the Cæsarean section). If we, therefore, do not take into account the woman with the six labours (whose pelvis was larger and only moderately deformed), we find the prognosis to be very sad.

The prognosis is more favorable when the pelvis was originally roomy. Then, even if the deformity is very considerable, a natural labour is possible; but if smallness of the pelvis is associated with great oblique contraction there is very little hope for the mother, and the child has only one chance, that is when the Cæsarean section is performed early. Breech positions are unfavorable to the child, because the extraction of the head is always very difficult; to the mother, however, they are less so than head positions, on account of the shorter duration of labour.

The mechanism of labour in an obliquely contracted pelvis is of great interest. The position of the promontory chiefly determines the way in which the head enters the brim. It most easily enters with the sagittal suture in the long oblique diameter, when the promontory somewhat recedes; but if the promontory projects more into the pelvis it approaches so much the hip-bone of the anchylosed side that the portion of the brim situated beyond the sacro-cotyloid diameter is completely lost for the mechanism of labour, the sacro-cotyloid distance of that side being so small that no portion of the head can enter it. In such a case the brim most resembles that of a generally uniformly contracted pelvis. The head then enters with the occiput very low down, sometimes most easily into the narrow half of the brim, and the sagittal suture then runs in the short oblique diameter. It is evident that, if the pelvis was small originally, and a considerable portion of the brim has been rendered useless for labour, it will not permit at all the entrance of the remaining portion of the head. But even when the head has entered the brim the difficulties of the labour are by no means overcome, for the pelvis becomes narrower in the transverse diameter. The head passes easiest of all through the outlet, with the sagittal suture in the short oblique diameter of the pelvis.

The after-coming head passes more easily through such a pelvis when the occiput is in the wider half.

The treatment depends, first, upon the original size of the pelvis; and, secondly, upon the degree of the deformity.

If we have the good fortune to recognise this deformity during a first gestation, or if a woman with such a pelvis again becomes pregnant after the successful termination of a previous labour, it will almost always be advisable to induce premature labour. This can then only be dispensed with when the obliquely distorted pelvis of a primipara is unusually large, or when previous labours have shown that even large children may pass through it without danger. The time for inducing premature labour is determined by the size of the pelvis.

If, however, the deformity is only recognised during labour, or in consequence of the impediment which it offers to it, if the head has already entered the brim, we must content ourselves with mere expectancy. As a rule, it will be necessary to shorten labour, and we shall then use the forceps when the child is alive, for it alone is able to save the mother and the child when the head is already in the pelvis.

As delivery is accomplished only by powerful traction, the use of the forceps is consequently attended with great danger. Another disadvantage of the forceps is that it may impede the necessary rotation of the head. When we have come to the conclusion that the head does not descend, even by very powerful traction, it will be the better plan to remove the forceps and to perforate immediately; for under such conditions a child is never born alive, and the prognosis for the mother becomes absolutely unfavorable when forcible traction is long continued; up to the present time in all the cases in which perforation has been performed the mothers have died; but there is little doubt that the operation had been delayed too long. If the child be dead and mechanical difficulties arise we ought to perforate immediately.

If, after the dilatation of the os, the head does not enter the pelvis, this is always so greatly contracted that the Cæsarean section appears to be the best means of delivery. If this can possibly be dispensed with, or, if the mother objects to the operation, we may endeavour to turn the child. Of course there is little prospect in such cases of delivering the child alive; the sole chance of preserving it depends upon the Cæsarean section. The mother, however, is certainly more benefited by version. If manual extraction of the head is impossible the child dies, and delivery is rapidly effected and with the best possible chance for the mother after the perforation of the after-coming head. It is preferable, however, to perforate the presenting head, whilst it is still moveable above the brim.

E.—*The Anchylosed Transversely Contracted Pelvis*

This deformity is the rarest of all. Its peculiarity consists in an anchylosis of both sacro-iliac synchondroses. The lateral portions of the sacrum are either entirely absent or only very rudimentary. The sacral vertebræ are small, the sacrum itself is transversely convex instead of concave, and its longitudinal curvature varies according to circumstances. The position of the sacrum also is changed.

FIG. 24.

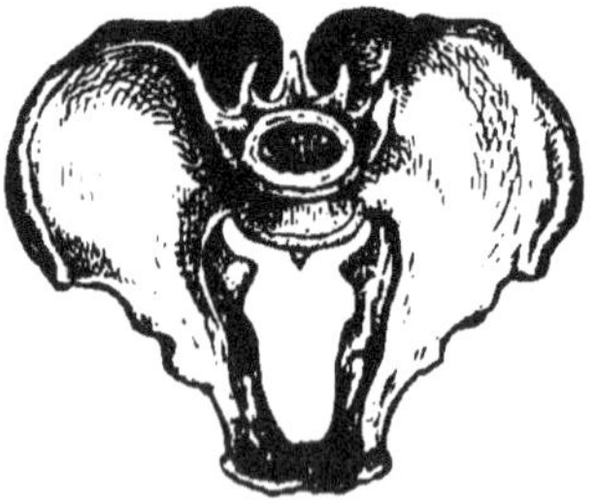

It has descended deeply into the pelvis, so as to produce an unusually great prominence of the posterior ends of the iliac bones. The posterior superior spines of the ilia greatly approach each other. The curve of the ilio-peritoneal line is very slight or almost completely absent. The iliac bones proceed from the anchylosed part directly forwards and unite in the symphysis at a very acute angle. The pelvis first described by Robert is symmetrical; in the others which have subsequently been made known asymmetries are found. The dimensions of the pelvis are, of course, greatly altered. The antero-posterior diameter of the brim would be greatly diminished by the descent of the sacrum, if part of that diminution were not counterbalanced by the greater distance between the promontory and symphysis from the want of the transverse extension of the iliac bones. The most important change consists in the great transverse contraction which increases towards the outlet, so as to form merely a long narrow fissure. The transverse diameter of the outlet has been in the cases above referred to from 6·1 cm. to 2½ cm. (2·4—1 inch).

The diagnosis will only rarely be difficult. The distances between the spines and the crests of the iliac bones, and especially between the trochanters, are abnormally small, whilst Baudelocque's diameter is of almost normal size. We must also examine the posterior surface of the sacrum. If the history shows that the anchylosis has not been preceded by a profuse suppuration we shall find the posterior superior spines of the iliac bones remarkably close to each other, whilst the spinous processes lie so deeply that they can scarcely be felt.

By an internal examination we shall immediately find the contraction of the transverse diameter and the almost parallel course of the descending rami of the pubic bones. The contraction is so great that such a pelvis can hardly be mistaken for any other. From the osteomalacic pelvis it is easily distinguished by the history of the case, by the difference of the posterior surface of the sacrum, by the position of the promontory and of the superior expansions of the iliac bones. The absence of kyphosis and the difference in the transverse diameters of the false pelvis and of the trochanters distinguish it from the kyphotic transversely contracted pelvis. When the transverse contraction is not very considerable there may, however, be some difficulty in distinguishing it from a very small pelvis with a funnel-shaped, transversely-contracted, outlet.

The prognosis is evident from the treatment adopted.

In the majority of the cases the Cæsarean section is the sole rational treatment, even when the child is dead. In seven well-authenticated cases it has been performed six times. In the gipsy pelvis of Prague, where the transverse contraction was least considerable, perforation and cephalotripsy were sufficient. The woman died of puerperal fever.

Fig. 24 is a copy of the pelvis described by Robert.

F.—*The Osteomalacic Pelvis*

Osteomalacic pelves greatly vary in size, and this variation depends upon the original conditions of the bones. Even very large bones, which have undergone the process of osteomalacic softening, may become much smaller by compression on all sides, and if we can suppose that during such a process the shape of the pelvis is maintained unaltered, a normal pelvis would be converted by osteomalacia into a uniformly contracted one.

Osteomalacic pelves are, moreover, distinguished by their remarkable lightness, although their bones may be thick. The individual bones are strongly bent, and even show infractions. Sections of the bones, as seen in the femurs, show a rarefied, diploë-like tissue, with a compact external osseous layer, which is often very thin. Not infrequently even this has entirely disappeared from parts of the pelvis, and the surface of the bones is rough and porous as if eroded.

The most important changes which take place in the bones are the following:

The sacrum is small, especially in its lateral portions. The vertebral bodies are (as in the highest degrees of rickets) pressed from between the auricular surfaces deeply into the pelvis. Sometimes the downward and forward traction upon the lateral portions of the sacrum produces distinct folds in them. The promontory has sunk deeply into the pelvis, and nearer to the symphysis and apex of the sacrum. In marked cases the promontory may approach so closely to the apex of the sacrum as almost to touch it, and this is due partly to the altered position of the promontory itself, partly to the considerable longitudinal concavity of the sacrum. This distinct incurvation,

FIG. 25.

most frequently situated at the upper portion of the third or even at the second sacral vertebra, is for the most part due to a distinct fracture. The sacral and also the lumbar vertebral bodies are compressed from above downwards and atrophied. By the descent of the promontory the lumbar vertebræ come nearer to the brim. The lumbar lordosis, compensated by a dorsal kyphosis, arches over the brim, similar to what is found in the spodylolisthetic pelvis. The obstetrical conjugate, therefore, lies between the symphysis and one of the last lumbar vertebræ. Scoliosis to any great extent is rare.

The iliac bones are at times small and in some places transparent, but in other cases they may be very thick. The distance between the anterior superior spines is usually somewhat smaller than the normal. The distance between the crests is, as a rule, much greater than that between the spines of the ilia. The iliac bones also show a distinct furrow extending from above downwards, which is sometimes bifurcated. The posterior superior spines of the ilia scarcely project beyond the sacrum; they are small and situated on a level with the spinous process of the last lumbar vertebra. The latter may at times be more prominent, and may also be bent to one side.

The anterior portion of the pelvic ring is compressed from the sides forwards, so that the ilio-pectineal tuberosities approach. The acetabula are directed upwards, forwards, and inwards. The brim is beak-like, pointed at the symphysis. The descending rami of the pubic and the ascending part of the ischial bones, as well as the tuberosities of the latter, have approached each other; usually, however, the latter are turned somewhat outwards. The weakest points become the seat of real incurvations. Sometimes the bones which form the pubic arch or the beak-like symphysis touch each other. More or less considerable irregularities are quite commonly observed in those places.

The dimensions of the pelvis are greatly altered. Usually, the outlet is narrower than the inlet; more rarely the reverse is the case.

Through the approximation of the promontory and the acetabula the brim assumes a three-cornered shape. In the highest degrees of deformity the brim is Y-shaped. The conjugate is frequently contracted and the transverse diameter almost always; this transverse contraction increases in the cavity and the outlet. The antero-posterior diameter of the cavity is not contracted, unless the apex of the sacrum is bent forwards, thus narrowing it. This circumstance may produce considerable contraction of the antero-posterior diameter of the outlet, and if in addition to that the rami of the pubic arch and the ischial tuberosities also approach each other, the outlet is almost completely obstructed.

At an early period of osteomalacia, when the characteristic changes of the pelvis are not yet marked, or only slightly so, a diagnosis may be almost impossible. In countries where osteomalacia is endemic it must always be thought of when pregnant or puerperal women complain of very severe rheumatic pains in the lower half of the body. If the height of the woman and the conjugate of her pelvis are repeatedly measured, diminution of those measurements may occasionally reveal osteomalacia at a time when all other diagnostic changes are as yet absent. This is the more important because slight degrees of contraction of the pubic arch, and of a beak-like projection of the symphysis, are difficult to recognise on the living woman with certainty.

When the disease has made further progress the characteristic changes produced by it in the pelvis can be recognised without difficulty, especially as in such cases the very peculiar history points to osteomalacia.

In examining the pelvis we shall frequently, but not always, find the external conjugate contracted, whilst we can almost always reach the promontory. The diameters of the false pelvis, and especially the distance between the trochanters, are small, and the iliac crests are greatly curved. Moreover, the beak-like symphysis, the parallel course of the rami of the pubic arch, and the ischial tuberosities close to each other, are characteristic of osteomalacia. The great incurvation of the sacrum can also be felt without difficulty.

We have already pointed out the differences in the kyphotic and in the anchylosed transversely contracted pelvis, for which the osteomalacic might be mistaken. The history of the case serves to distinguish it from a pseudo-osteomalacic pelvis.

It is extremely difficult to determine the degree of contraction, for neither the measurement of the diagonal conjugate nor that of any other diameter suffices for that purpose in this kind of pelvis. It is therefore advisable to introduce the hand into the true pelvis, and endeavour to obtain an approximate estimate of its capacity.

The prognosis is unfavorable. Complete cure of osteomalacia is extremely rare, and improvement takes place only when the woman remains a long time before again conceiving. But even then the changes in the pelvis persist. Winkel described the pelvis of a woman who died of rupture of the uterus, the bones of which were found to be firmer and harder than normal. If we consider, however, that in far advanced osteomalacia the pelvis may become so soft that it can again be dilated by the fœtal head forced into it by the action of the uterus, we must confess that, as far as it concerns our purpose, cure is followed by more unfavorable results than the further progress of the disease.

Since the highest degree of the disease rarely impedes conception, the majority of the patients die during, or from the consequences of, labour, whilst others perish in a more pitiable condition after prolonged suffering.

Litzmann has collected 72 cases to show the unfavorable influence of osteomalacia on labour. Of these 72 women who conceived during the course of the disease, 38 could not be delivered naturally, and 24 once only.

The result for the child varies. At the commencement of the disease, and when the pelvis is completely softened, the delivery of a live child *per vias naturales* is possible. On the other hand, since in osteomalacia it is very frequently absolutely necessary to perform the Cæsarean section, the child obtains the most favorable chances by that operation.

The treatment to be adopted depends upon a correct estimate of the pelvic capacity. We must endeavour to find out, by introducing the whole hand, whether an intact or, at least, a perforated head can pass through. In a great many cases the contraction is so considerable that delivery *per vias naturales* is absolutely impossible unless (which must always be thought of in the higher degrees of osteomalacia) the pelvis is so much softened that it can again be dilated. In a case of extreme contraction it may be very difficult to decide whether we shall immediately perform the Cæsarean section or whether we shall wait, trusting to the dilatability of the softened pelvis. If there is evidence that the pelvis can be dilated, it appears to be advisable to wait on account of the great dangers which arise from the Cæsarean section.

Fig. 25 represents an osteomalacic pelvis, which is interesting on account of the history connected with it. A woman named Charoubel, after two normal labours, was attacked by osteomalacia. The third labour terminated easily and successfully, in the fourth and fifth labours perforation was necessary, and in the sixth a dead child was delivered after version. In her seventh labour Kilian performed the Cæsarean section with success to mother and child; a portion of the omentum which had prolapsed the day after delivery was cut off. In her eighth labour the uterus ruptured spontaneously and the fœtus enveloped within the intact membranes escaped into the abdominal cavity. The dead fœtus was extracted by the Cæsarean section. In her ninth labour the uterus again ruptured and the Cæsarean section was, for the third time, performed with a successful result for the mother and child.

G.—*The Pelvis narrowed by Osseous Tumours*

Osseous tumours which impede labour are so very rare that a few cases only can be found in the literature of this subject. Fractures of the pelvic bones, followed by dislocation or deformed callus, may cause contraction of the pelvis.

As regards such cases no general rules can be given. The prognosis and the treatment depend upon the size and the position of the tumour.

Other tumours arising from the pelvic bones have already been mentioned above.

FIG. 26.

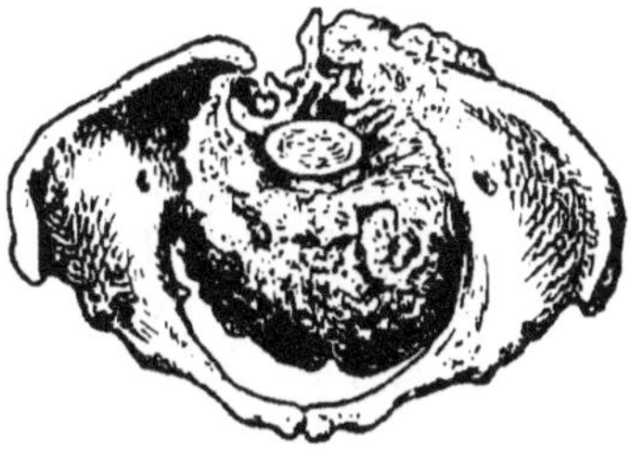

Fig. 26 represents a pelvis which has lately been described by Behm. The tumour filled up almost the whole pelvic cavity and the Cæsarean section had to be performed.

Literature.—Kussmaul, V. d. Mangel, &c., d. Gebärmutter. Würzb., 1859, pp. 167, 253.—W. J. Schmitt, Heidelb. klin. Annalen, I, p. 537.—Lachapelle, Prat. des Acc., T. 3, p. 298.—F. C. Naegele, Heid. kl. Ann., III, p. 492.—H. F. Naegele, Mogostokia e congl. orif. ut. ext. Comment. Heidelb., 1835, and Med. Annal., 1836, II, p. 185, and 1840, VI, p. 33.—Discussion üb. d. Congl. orif. auf d. Naturforscherversr. in Mainz, Neue Zeitschr. f. Geb., B. 14, p. 143.—Genth, e. l., B. 29, p. 118.—Arneth, Die geb. Praxis., p. 64.—Credé, Klin. Vorträge. I, p. 143.—E. v. Siebold, M. f. G., B. 14, p. 96.—Roth, M. f. G., B. 19, p. 144.—Martin, e. l., p. 254.—Wachs, e. l., B. 30, p. 46.—Chiari, Braun u. Spaeth, Kl. d. Geb., p. 226.—Schroeder, Schw., Geb. u. Woch., p. 80.—Winkel, Path. d. Geb., p. 155.—Cazeaux-Tarnier, Traité d'Accouch., 7 ed. Paris, 1867, p. 704.—Kuhn, Wiener Med. Jahrb., 1870, XX, B. IV, H., p. 24.—Kleinwächter, Prager Vierteljahrschrift, 1870, B. III, p. 109.—Salisbury, Boston m. and s. J., April 24, 1870.—Depaul, Gaz. Méd. de Paris, 1860, Nr. 22.—Arneth, Geb. Praxis, p. 66.—Hayn, Berl. klin. W., 1870, Nr. 10.—Kleinwächter, Prager Vierteljahrschrift, 1870, B. III, p. 110.—Latz, Berl. klin. W., 1870, Nr. 35.—V. Siebold, Neue Zeitschr. f. Geb. B. 11, p. 321.—Chiari, Braun u. Spaeth, Kl. d. Geb., p. 230.—Cazeaux-Tarnier, Traité des Acc., 7 éd., 1867, p. 690.—Roth, M. f. G., B. 19, p. 150.—Moritz, e. l., B. 13, p. 60.—Wachs, e l., B. 30, p. 54.—Schön, Allg. Wiener m. Z., 1868, Nr. 11.—Herzfeld, Wiener Med. Presse, 1868, Nr. 34.—C. Bell, Transactions of the Edinburgh Obstetrical Society, 1870, p. 116.—Dewees, Krankh. des Weibes, übers. von Moser, 1837, p. 25.—Credé, Verh. d. geb. Ges. in Berlin, IV, p. 57.—Kiwisch, die Geburtskunde, &c. Erlangen, 1854, I, p. 104.—Röbbelen, Deutsche Klinik, 1854, Nr. 10.—V. Scanzoni, Allg. Wiener Med. Z., 1864, Nr. 4.—V. Franque, Wiener Med. Halle, 1864, Nr. 50.—Edmond, Gaz. des Hôp., 1864, Nr. 52.—Fethersten, British Medical Journal, March 26, 1864.—Hubbauer, Zeitschr. f. Wundärzte u. Geb., 1863, XVI, 3.—Horton, Philadelphia Medical and Surgical Report, Nov., 1869, p. 314.—Holst, Scanzoni's Beiträge, B. V, p. 398.—Godefroy, Journ. d. Connaiss. Méd.-Chir., 1870, Nr. 3 und 4.—Leisenring, Philadelphia Med. Times, August, 1871, p. 395.—Cazeaux-Tarnier, Traité d. Acc., 7 ed., 1867, p. 689.—Neugebauer, Bresl. kl. Beitr. z. Gyn., III, p. 1.—Hanuschke, Chir. Oper. Erf., 1864, p. 182.—P. Müller, Würzb. Med. Z., VII, p. 61.—Puchelt, l. c., p. 225.—Meigs, London Medical Gazette, April, 1845.—Stoltz, Gaz. Méd. de Strasb., 1845, Nr. 1.—E. A. Meissner, M. f. G. B., 21, Suppl., p. 138.—Puchelt, l. c., p. 231.—Edinburgh Journal, IX, p. 281.—Ramsbotham, Medical Times, Jan. 1, 1859.—Hecker, Kl. d. Geb. II, p. 135.—Broadbent, Obstetrical Transactions, 1864, p. 44.—Charrier, Gaz. des Hôp., 1866, Nr. 6.—Puchelt, l. c., p. 157.—Litzmann, Deutsche Klinik, 1852, Nr. 38, 40, 42.—Playfair, Obstetrical Transactions, IX, p. 69.—Doumairon, Etudes sur les Kystes Ovar., &c., Thèse. Strasboug, 1869.—Barnes, Obstetrical Operations, 2 ed., p. 263.—W. Smellie, Collection of Cases. London, 1754, p. 367.—G. W. Stein d. ä, Kleine Werke zur prakt. Geb. Marburg, 1798, pp. 133 u. 157.—J. L. Baudelocque, L'art des Accouch., 8 éd.,

1844, T. I, p. 73.—Michaelis, Das enge Becken. Leipz., 1865, p. 81.—Crédé, Klin. Vortr., 1854, p. 620.—Schröder, M. f. G., B. 29, p. 30. Dohrn, M. f. G., B. 29, p. 291, B. 30, p. 241 und Volkmann's Sammlung Klin. Vorträge, Nr. 11.—Litzmann, Volkmann's Samml. Klin. Vorträge, Nr. 20.—Puchelt, l. c., p. 74.—Menzies, M. f. G., B. 5, p. 207.—Simpson, Sel. Obst. Works, I, 1871, p. 498.—Dieterich, der Gebärmutterkrebs als Compl. d. Geb., D. i. Breslau, 1868.—Puchelt, De tumoribus in pelvi part. imped. Comment. Heidelberg, 1840, pp. 107 und 116.—Pillore, Gaz. des Hôp., 1854, Nr. 137.—Lehmann, Nederl. Tydschr. for Genesk. Maart en April, 1854 (Schmidt's Jahrb., 1855, B. 85, p. 58).—Habit, Zeitschr. d. Ges. d. Wiener Aerzte., 1860, Nr. 41.—Hecker, Kl. d. Geb. II, p. 124 und M. f. G., B. 26, p. 446.—Breslau, M. f. G., B. 25, Suppl. p. 122.—Guéniot, Gaz. des Hôp., 1864, Nr. 43.—Toloczinow, Wiener Med. Presse, 1869, Nr. 30.—Magdelaine, Études. l. Tumeurs Fibreuses sous-périt., &c., Thèse. Strasbourg, 1869.—Lambert, Des grossesses compl. de myomes utérins. Thèse. Paris, 1870. (Sammlung von 165 Fällen.)—Oldham, Guy's Hospital Reports, April, 1844, vol. iii, p. 105, and vol. viii, p. 71 and Lambert, l. c.—Henr. v. Deventer, Oper. Chir. novum lumen exhib. obstetr., &c., Lugd. Bat., 1701, cap. 3, 27, deutsch; Neues Hebammenlicht, Jena, 1717, p. 196.—Guill. de la Motte, Traité compl. des acc. nat., &c. Paris, 1722, L. II, Ch. V, p. 202, and L. III, Ch. XIX, p. 418.—N. Puzos, Traité des acc. publié par Morisot Deslandes. Paris, 1759, Ch. I.—W. Smellie, Treatise on the Theory and Practice of Midwifery, vol. i, 3rd ed. London, 1756, p. 82, und a. a. O. and Tab. anatom., T. III, XXVII and XXVIII.—De Frémery, De mutat., fig. pelvis, &c., Lugd. Bat., 1793.—G. W. Stein d. j., Lehre der Geburtshülfe, 1, Th. Elberfeld, 1825. 1, Absch., 2 und 3, Kp. und in vielen anderen Schriften.—G. A. Michaelis, Das enge Becken, herausg. von Litzmann. Leipzig, 1851.—C. C. Th. Litzmann, Die Formen des Beckens, insb. des engen weiblichen Beckens. Berlin, 1861, and Volkmann's Samml. Klin. Vorträge, Nr. 23.—F. A. Kehrer, Beitr. z. vergl. n. experim. Geburtsk., 3, H. Pelikologische Studien. Giessen, 1869.—H. v. Deventer, l. c., cap. 3, 27.—G. W. Stein d. j., Annalen, 3, Stück, 1809, p. 23, and Lehre d. Geb., I, p. 78.—E. de Haber (Naegele), Diss. exh. cas. rar. partus, &c., Heidelb., 1830.—F. C. Naegele, Das schräg verengte Becken u. s. w. Mainz, 1839, p. 98.—Michaelis, l. c., p. 135.—Litzmann, l. c., p. 39.—Ries, Zur Kenntniss des allg. gleichm. verengten Beckens, D. i. Marburg., 1868.—Löhlein, Ueber die Kunsthülfe, &c., D. i. Berlin, 1870.—Betschler, Annalen der Klin. Anstalten. Breslau, 1832, B. I, p. 24, 60, B. II, p. 31.—Michaelis, l. c., p. 127.—Litzmann, l. c., p. 44.—Schroeder, l. c., p. 70.—M. Dionis, Traité Gen. des Acc. Paris, 1724, pp. 241 and 264.—M. Puzos, l. c., p. 4, sequ.—W. Smellie, l. c.—G. W. Stein d. j., l. c.—G. Michaelis, l. c., p. 122.—Litzmann, l. c., p. 47.—Halbey, Zur Kenntniss d. platten Beckens, D. i. Marburg, 1869.—Stanesco, Rech. Clin. s. l. rétréciss. du bassin basées sur 414 cas, &c. Paris, 1869.—Rigaud, Examen critique de 396 cas de rétréciss. du bassin. Paris, 1870.—Michaelis, l. c., p. 132.—Litzmann, l. c., p. 55.—Korten, De Pelvi Ubique Just. Min., D. i. Bonnae, 1853.—Halbey, Zur Kenntniss d. platten Beckens, D. i. Marb., 1869.—Smellie, Treatise, &c. London, 1752, p. 83, and Tab. Anat., T. III.—Stein d. j., Die Lehranstalt der Geb. in Bonn. Elberf., 1823, p. 184.—Clausius, Diss. s. cas. rariss, &c., 1834, and F. C. Naegele, Das schräg verengte Becken, p. 85, and T. XII and XIII.—Lange, Prager Vierteljahrsschrift, 1844, p. 5.—Hold, Zur Path. des Beckens, p. 78.—A. P. Kilian, De Rachitide, &c., D. i. Bonn, 1855.—Litzmann, l. c., p. 92.—Scanzoni, Lehrb. d. Geb., 4 Aufl., B. II, p. 437.—C. O. Weber, Erarr. cons. rach., &c. Progr. Bonn, 1862.—Rokitansky, Oest. Med. Jahrb., B. 19, 1839, p. 202.—Spaeth, Zeitschr. d. Ges. d. Wiener Aerzte., 10 Jahrg., 1 B, 1854, p. 1.—Kiwisch, Geburtskunde, II, p. 168.—Seyfert, Verh. d. Phys. Med. Ges. in Würzburg, III, p. 340, and Wiener Med. W., Januar, 1853.—Kilian, De Spondylolisthesi, grav. pelv. c., &c., Bonnae, 1853, and Schild. neuer Beckenf., &c., Mannh., 1854.—Breslau, Scanzoni's Beitr., 1855, B. II.—Breslau, M. f. G., B. 18, p. 411, and Billeter, Ein neuer Fall von hochg. Sp., D. i, Zurich, 1862.—Lenoir-Belloc, Archives Gén., 1859, p. 192.—Olshausen, M. f. G., B. 22, p. 301, and B. 23, p. 190.—Hartmann, M. f. G., B. 25, p. 465, and B. 31, p. 295.—Blasius, M. f. G., B. 31, p. 241.—Ender, M. f. G., B. 33, p. 247.—Hugenberger, Bericht u. s. w., Petersburg, 1863, p. 121.—Barnes, Lancet, 18th June, 1864, and Obstetrical Transactions, 1865.—Lehmann, Schmidt's Jahrb., 1855, Nr. 12, p. 328.—Blake, American Journal of Medical Science, July, 1867, p. 285.—Robert, M. f. G., B. 5, p. 81.—Lambl, Scanzoni's Beitr., B. III, p. 2.—

Herbiniaux, Traité sur divers acc., &c. Bruxelles, 1782, pp. 270 and 276.—Jörg, Ueber d. Verkr. d. menschl. Körpers, &c. Leipzig, 1810, p. 51.—Rokitansky, Med. Jahrb. des österreich. Staates. Wien, 1839, B. 19, p. 199.—Neugebauer, M. f. G., B. 22, p. 297.—A. Breisky, Zeitschr. d. Ges. d. Wiener Aerzte., 1865, I, p. 21.—J. Mohr, Das in Zürich bef. kyph. querverengte Becken. Zürich, 1865.—Hugenberger, Petersb. Med. Z., 1868, B. 15, H. 4.—Chantreuil, Déf. du bassin chez les cyph. Paris, 1869, and Gaz. Hebdom., 2 Sér., VII, 31, 1870 (s. Schmidt's Jahrb., B. 149, p. 178).—Höning, Beitr. z. Lehre v. kyph. v. Becken. Bonn, 1870.—Lange, Arch. f. Gyn., B. I, p. 224.—Baudelocque d. Neffe, Bello in Transact. Méd. Paris, Sept., 1833.—Enc. des Sc. Méd., T. XIII. Brux, 1833 (s. Wittlinger's Ann., I, 2, 1849).—Hayn, Beiträge zur Lehre vom schräg ov. Becken. Königsb., 1852.—Kiud, Pelv. inf. in ad., D. i. Marburg, 1854.—Lambl, Scanzoni's Beiträge, 1858, III, pp. 61 and 68.—Sinclair and Johnston, Practical Midwifery. London, 1858, p. 236, Nr. 28, p. 241, Nr. 69, and p. 502, Nr. 1.—Litzmann, Die Formen des Beckens, p. 64.—Birnbaum, M. f. G., B. 15, p. 102, B. 16, p. 67, and B. 21, p. 353.—Martin, Neigungen u. Beugungen, &c., p. 128.—Olshausen, M. f. G., B. 17, p. 255.—Chiari, Braun und Spaeth, Klinik d. Geb., p. 647.—Jenny, Würzb. Med. Z., VI, p. 335.—Schmeidler, M. f. G., B. 31, p. 31.—Elliot, Obstetric. Clinic., p. 251.—Stadfeldt, Med. Chir. Review, Nr. LXXXV, Jan., 1869, p. 24.—Br. Hicks, Obstetrical Transactions, X, p. 45.—F. C. Naegele, Heidelb. Klin. Annalen, B. X, p. 449, u. das schräg verengte Becken, &c. Mainz, 1839.—Betschler, Neue Zeitschr. f. Geb., 1840, B. 9, p. 121.—E. Martin, Progr. de pelvi obl. ov., &c., Jenae, 1841, u. Neue Zeitschr. f. Geb., B. 15, p. 49, and B. 19, p. 111, and Schmidt's Jahrb., B. 71, p. 360.—Unna, Oppenheim's Zeitschr. für die ges. Med. Hamburg, 1843, B. 23, p. 281.—Moleschott, c. l., 1846, B. 31, p. 441.—G. W. Stein, Neue Zeitschr. f. Geb., B. 13, p. 396, and B. 15, p. 1.—v. Ritgen, Neue Z. f. Geb., B. 28, p. 1.—E. Rosshirt, Lehrb. d. Geb. Erlangen, 1851, p. 305.—Kiwisch, Die Geburtskunde. Erlangen, 1851, II Abth., p. 173.—C. Hunnius (Walter), De pelvi obl. ov., D. i. Dorp., 1851.—Hayn, Beiträge zur Lehre vom schräg ov. Becken. Königsb., 1852.—Hohl. Zur Pathol. des Beckens. Leipzig, 1852.—Litzmann, Das schräg ov. Becken, &c. Kiel, 1853.—S. Thomas, D. schräg verengte Becken, &c. Leyden and Leipzig, 1861.—Olshausen, M. f. G., B. 19, p. 161.—S. Thomas, M. f. G., B. 20, p. 384.—Litzmann, M. f. G., B. 23, p. 219.—A. Otto, M. f. G., B. 28, p. 81.—M. Duncan, Obstetrical Researches Edinburgh, 1868, p. 113.—Kleinwächter, Prager Vierteljahrsschrift, B. 106, 1870, 2, p. 12.—Spiegelberg, Archiv f. Gyn., B. II, p. 145.—Siehe auch die Literatur zum querverengten Becken.—Die Casuistik bis zum Jahr, 1861, ist in den beiden Monographien von Naegele und Thomas enthalten. Seitdem s., Credé, M. f. G., B. 15, p. 258.—Hugenberger, Bericht, &c., Fall 72 (and 71, 73).—Kulp, De pelvi obl. ov., D. i. Bresl, 1866.—F. Robert, Beschreibung eines im höchsten Grade quer verengten Beckens, &c. Carlsruhe und Freiburg, 1842.—C. Kirchhoffer, Neue Zeitsch. f. Geb., 1846, B. 19, p. 305.—B. Seyfert, Verh. der Phys. Med. Ges. in Würzburg, 1852, B. III, H. 3, p. 324; and Lambl, Prager Vierteljahrschr., 1853, B. 2, p. 142, and 1854, B. 4, p. 1.—F. Robert, Ein durch mechanische Verl. und ihre Folgen quer verengtes Becken im Besitz von Herrn P. Dubois in Paris. Berlin, 1853.—Lloyd Roberts, Obstetrical Transactions, IX, p. 250.—Kehrer, M. f. G., B. 34, p. 1.—E. A. Martin, Ein während d. Geb. erk. querverengtes Becken, &c., D. i. Berlin, 1870.—Kleinwächter, Arch. f. Gynaek, B. I, p. 156.—P. Grenser, Ein Fall von quer verengtem Becken, D. i, Leipzig, 1866.—Litzmann, Die Formen des Beckens, p. 58.—G. W. Stein d. ä., Kleine Werke zur prakt. Geb. Marburg, 1798, VI, Abh. von d. Kaisergeburt. Dritter Fall, 1782, p. 283, Tab. 10, S. ebendaselbst, p. 325, auch die älteren Fällen von Cooper, 1776, and p. 327, von Vaughan-Atkinson.—G. W. Stein d. j., Annalen d. Geb St. I, p. 119, and St. II, and III, and Lehre d. Geb.. Elberfeld, 1825, Th. 1, p. 103.—F. C. Naegele, Erf. und Abh., &c. Mannheim, 1812, p. 409, and Clausius, Commentatio, &c. Frankof, 1834, p. 19.—H. F. Kilian, Beiträge zu einer genauen Kenntniss der allg. Knochenerweichung der Frauen, &c. Bonn, 1829, und das halisteretische Becken, &c. Bonn, 1857.—Litzmann, Die Formen des Beckens, p. 85, and p. 113.

CHAPTER III

ANOMALIES OF THE OVUM

I.—Anomalies of the Fœtus

A.—*Too great Development of the Fœtus*

In the introduction to the subject of pelvic deformity we have already stated that mechanical impediments to parturition are not only produced by the deformity of the pelvis, but also by an unusual bulk of the child, and especially of the head.

Although it never occurs that, independently of disease, a fœtus acquires such a bulk as to form an obstacle to labour of the same serious nature as often takes place in pelvic deformity, it may nevertheless happen that an unusually large child impedes labour, even in a quite normal pelvis, and that this impediment still further increases the difficulty of labour when the pelvis at the same time is slightly contracted. It is therefore important to know all the circumstances which influence a greater development of the fœtus. As the length of the fœtus alone is of no importance in the mechanism of labour, we shall not consider it, and shall only remark that in general it keeps equal pace with the weight of the fœtus. As regards the latter, we know from Gassner that heavy women also bear heavy children, and from Frankenhäuser that the weight of the children increases with the size of the mother (the influence of the father, which certainly has also to be considered, has not yet been determined in man by observation). The weight of the child also increases with the age, and especially with the number of the previous labours of the woman. Male children are generally heavier than female.

The bulk and consistency of the skull are of the greatest importance in the mechanism of labour. Both vary greatly, and usually correspond to the development of the child. On the whole it may be said that light children have small and, what is more important, flexible and compressible skulls, whilst heavy and greatly developed children have generally large and firm skulls. The heads of male children are larger than those of female, and the most important diameter of the head—the biparietal—increases quite out of proportion to the number of the labours and the age of the mother, so that the broadest skull may be expected in a male fœtus of a pluripara somewhat advanced in age.

The great bulk of the head cannot with certainty be recognised during labour. The suspicion may be excited by palpation of the head within the lower uterine segment, or by the length of the sagittal suture and the great distance between the two fontanelles; yet these signs are reliable only when they are extreme. Nor is it often possible to ascertain the sex of a child before birth, and it will therefore be necessary to consider all the circumstances above mentioned, and consequently not to expect very large heads in young primiparæ, whilst in an older woman who has often borne children a head of a considerable size may be looked for.

The prognosis and the treatment are just the same as in disproportion produced by moderate pelvic contraction. Only, since the large head has almost always lost compressibility, the prognosis here becomes somewhat more unfavorable for the child. This circumstance also is of great significance in regard to the extraction of the aftercoming head, for such large and hard skulls predispose to the spoon-shaped depressions, which greatly lessen the chances of the child's safety.

A too great development of the trunk of the fœtus only slightly retards the progress of labour when the pelvis is normal; but a contracted pelvis will not allow the ready entrance of the abnormally large trunk into the brim. The head, therefore, already at the outlet, is prevented from advancing, and its extraction may give rise to great difficulties.

B.—*Double formation of the Fœtus*

The redundant development of some fœtal parts or the formation of parasites is very rarely the cause of a serious impediment to labour. These appendages are almost always very soft and flaccid, and therefore capable of great compression, or they are sometimes torn off during labour.

Of far greater obstetric importance are twin-formations, which, in the highest degree, are two fully-developed children joined together by homologous parts.

The diagnosis of such an abnormality is impossible at the commencement of labour, and may be very difficult, even in its further progress. The presence of twins can usually be made out by the combined external and internal examination, but not that they are joined to one another, and this can only be thought of when an existing impediment to labour cannot be explained in any other way. Certainty is only obtained by introducing the hand into the uterus so as to feel the united parts.

In such cases head presentations are the most frequent, then breech presentation, whilst transverse positions are very rare.

In head presentations the mechanism of labour varies. If both necks are of an equal length and the fœtuses are well developed, the head of the one enters the brim, whilst the neck of the other is pressed against the symphysis. When the first head is born, it is directed upwards towards the mons veneris and the neck is pressed into the pubic arch. The trunk and breech of the first child then follow and the second enters the brim with the breech. The order in which they are delivered is therefore the head of the first, the breech of the first, the breech of the second, and, lastly, the head of the second. In other cases, especially when the neck of the second fœtus is shorter than that of the first, the second head is closely applied to the neck of the first child. The second head, therefore, follows immediately after the first, and the two trunks are born at the same time.

In breech presentations the mechanism of labour is less difficult. Both trunks are born at the same time and the heads, one shortly after the other, the one lying in the excavation of the neck of the other. In this way the passage of the double fœtus through the pelvis is much more easy.

To the mother the prognosis is not unfavorable, whilst on account of the malformation the fœtuses have, as a rule, little chance of life. Operative interference is not so often necessary as might *à priori* be expected. According to Hohl and Playfair, of 150 cases 85 terminated without assistance.

The treatment is to be conducted according to the usual principles. If the first head is born and the second impacted in the brim, the feet of the first, and then those of the second, child, must be brought down. If one head is not yet completely born and the hand cannot pass by it, it is to be extracted by means of the forceps or perforated if extraction is impossible. The second head may also have to be extracted by the forceps if it has entered the brim by the side of the neck of the first.

When in breech positions the malformation is early recognised, it is advisable to bring down at once all the four feet, in order to prevent an obstruction of the brim by the breech of the second fœtus. All the four feet must be drawn upon with an equal force. But it is better first to draw upon the fœtus situated behind, so that its head may be applied to the neck of the anterior fœtus whilst this remains above the symphysis.

In transverse positions the feet must be brought down by turning, and the subsequent extraction is performed as in breech presentations.

If these rules are strictly adhered to, operations for reducing the bulk of the children, and especially decapitation, will very rarely be necessary. When the labour is greatly impeded by extensive union between the fœtuses, embryotomy (decapitation of one) will have to be resorted to without hesitation, as they are not under such conditions capable of living.

The Cæsarean section is not to be performed on account of its danger to the mother. Moreover, if the union between the fœtuses is slight, and they are capable of living, the birth must be accomplished in some other way; but if the union is considerable, and the fœtuses, therefore, incapable of living, operations for the reduction of their bulk are preferable.

Schönfeld has performed intra-uterine separation of the fœtuses. There will seldom be an opportunity or the necessity for adopting such a practice. Not infrequently double monsters die within the uterus; the macerated state in which they are allows them to pass through the pelvis in any possible position.

c.—*Diseases of the Fœtus*

Acardiac monsters are known to develop from the anastomosis of the vascular systems of two fœtuses contained within the same chorion, and also of the same sex. If the blood-pressure of the one preponderates over that of the other the circulation of the latter is reversed, heart and lungs together with a variously large portion of the trunk atrophy, and the misformed fœtus is nourished by the well-developed one. The congestion of the umbilical vein may give rise to great hypertrophy and to œdematous swelling of the subcutaneous connective tissue of the monster. It is usually born in the footling position, within half an hour, or three, or twelve hours after the birth of a fully-developed child. The hypertrophy of its trunk may render extraction necessary, and sometimes this even is very difficult. In such cases it may be necessary to reduce its bulk by perforation.

Great distension of the cerebral ventricles by serum may, according to the quantity of the latter, and the time which has elapsed since its accumulation, produce either hemicephaly or hydrocephalus.

Hemicephalic or Anencephalous monsters are frequently very well developed and have particularly broad shoulders. The small head is immediately attached to the trunk. The face is turned upwards, the ears rest on the shoulders, the eyes are widely opened, and the tongue sometimes projects from the mouth. The quantity of liquor amnii is frequently very considerable. Anencephalous monsters sometimes present with the base of the skull or with the face, but transverse and footling positions are just as frequent. When the skull presents it is recognised by the sella turcica being felt, whilst in face presentations the smallness of its component parts, together with the signs just mentioned, call for an examination of the skull. If the head comes first, as its smallness does not prepare the way for the breech and shoulders, their breadth may prove a considerable impediment to the labour. But since they can easily be drawn through the pelvis when the breech comes first, podalic version is indicated. Or, if this is impossible, the fœtus is extracted by the head, or by a finger, or a blunt hook placed in its mouth, or by one or both of its arms, which are brought down by the side of the head. In one or the other way extraction is accomplished.

More important than hemicephaly are the cases where dropsy of the ventricles has not produced atrophy, but distension of the cranial vault, constituting hydrocephalus. Hydrocephalic children may be repeatedly met with in the same woman.

It is rare, however, for a hydrocephalic head to cause any real obstruction to labour. Such a head bears a great disproportion to the face. The forehead is unusually prominent, the cranial bones thin, their edges radiated, and separated from each other by wide interspaces; the sutures are broad and the fontanelles of an enormous size.

If the head is not very tense and the bones are soft and pliable, it may gradually accommodate itself to the brim by assuming a pointed shape and thus pass through the pelvis. Or if it ruptures, the watery contents flow off and the head, now diminished in bulk, is expelled. In the great majority of instances, however (sixty-three times in seventy-seven cases, Hohl), an operation was necessary. The most frequent position is with the head presenting. Often, however, the head is not directed quite straight towards the brim, but somewhat laterally, the forehead or occiput resting upon one margin of the brim, so that by this its entrance into the pelvis is still more impeded. Pelvic positions are, indeed, more favorable, and also occur more frequently in hydrocephalus than under normal conditions.

The diagnosis of hydrocephalus is not always easy. The enlarged head can often be felt through the abdominal walls, and in pelvic positions this is indeed the sole sign. In head presentations a tense bag of waters is felt, especially during a pain, and in an interval the broad sutures and fontanelles and the radiating bones can be distinguished; sometimes the sutures are made to project. The diagnosis is easier if the cranium is less distended, and it can then only be mistaken for the head of a macerated fœtus. If there is still any doubt one or two fingers or the whole hand may be introduced, and then the bulk of the head and its proportion to the face will be recognised.

The prognosis is very unfavorable to the child. Even if the labour is successfully accomplished, such children can only live a short time. The chances for the mother greatly depend upon the kind of assistance she receives. Rupture of the uterus easily takes place, especially if unsuitable operative procedures are employed.

The treatment must be directed to the safety of the mother. The sole and most rational treatment consists in the reduction of the abnormally great size of the child's head. And as, in extra-uterine life, tapping by means of a fine trocar is adopted for the treatment of hydrocephalus, this must be done in all cases as soon as the head can be reached. After tapping the feet are to be brought down, if possible, and the child extracted. After evacuation of a quantity of serum the aftercoming head can much more easily be drawn through the pelvis, whilst the pains are only with difficulty able to propel the flaccid presenting head. If the head has already so far entered the pelvis that version is impossible after tapping, its expulsion must be left to the forces of nature. The forceps ought never to be applied, for the blades are very liable to slip when strongly drawn upon, and this is the most frequent cause of rupture of the uterus. In pelvic positions, unless the head comes spontaneously, or can easily be extracted, it must be tapped through one of the lateral fontanelles.

Enlargements of the abdomen will only impede labour when they are very great. This may be the case in ascites, distension of the bladder by urine, and enlargements of the kidneys and liver. But whilst slight enlargement would only retard the progress of labour, this will by the higher degrees be brought to a standstill, or only proceed after the rupture of the abdominal walls. As a rule the diagnosis is only made when, after the birth of the head, or (what is more frequent) after that of the breech, the obstruction is looked for. If the child does not follow after traction with the hand, the fluid is to be evacuated by means of a trocar. In solid tumours embryotomy is to be performed.

Tumours of the fœtal body may also impede labour. They are—Lipomata, Carcinomata, and the so-called cystic Hygromata, which may acquire the size of a child's head. They are situated at the anterior surface of the neck, in the nape, in the axilla, upon the pectoral muscles, or in the perineal or sacral regions. Joulin records a case where a serous cyst of the abdominal walls as large as a man's head greatly obstructed labour. Spina bifida also occurs in consequence of a partial division of the spinal canal. Projection of the abdominal viscera through a fissure in the walls of the abdomen only embarrasses the diagnosis, as in such cases the liver may present. Yet Költsch records a case where the liver contained in an umbilical hernia obstructed the course of labour. The above-mentioned tumours may be so great an impediment that it becomes necessary to reduce their bulk.

D.—*Malposition of the Fœtus*

One of the most frequent and, practically, also, one of the most important impediments to labour arises from the fœtus being in such a position that it cannot pass through the pelvis. This is the case when the axis of the fœtus does not coincide with that of the uterus, so that neither the head nor the breech can enter the pelvis. The fœtus lies then more or less transversely within the uterus. During pregnancy and at the commencement of labour, especially as long as the membranes are still intact, the position of the fœtus may be quite transverse. Head and breech are on about the same level. No fœtal portion, or only some portion of the back, or, more frequently, a smaller fœtal part, can be felt within the os. After the rupture of the membranes, unless the transverse position is converted into a longitudinal, one shoulder always enters the brim. The position then ceases to be quite transverse, and the breech is somewhat higher up than the head. The attitude of the fœtus also is not quite normal. The non-presenting side appears to be compressed, and the presenting one thereby becomes more convex. Besides, with the rupture of the membranes the arm corresponding to the presenting shoulder is very easily separated from the chest, and falls into the vagina.

The transverse position varies in relation to the uterus. We distinguish a first transverse position when the head of the fœtus is in the left side of the uterus, and the second transverse position when the head is in the right side of the uterus.

We also make subdivisions according to whether the back of the child is directed to the abdominal or dorsal surface of the mother. The former we call the first, the latter the second subdivision. The first subdivision of the first transverse position, that is, head to the left, back in front, is the most frequent. The position with the head to the left is 2·6 times as frequent as that with the head to the right, and the back is 2·9 times more often in front than behind.

To understand the etiology of transverse positions we must refer to the causes of the normal position of the fœtus during pregnancy. We have already stated that in primiparæ at term the head is, under normal conditions, already within the true pelvis. In exceptional cases, most frequently produced by pelvic contraction, the head is at the commencement of labour, still movable above the brim, or it has deviated somewhat to one side. It is similar to what, as a rule, is met with in pluriparæ.

In these the commencing pains force the head, which stands above the brim, into the pelvis. If the head has deviated to one side it is, as a rule, placed longitudinally. Transverse positions may, therefore, be expected in cases where the head has deviated to one side, and certain causes oppose its being placed longitudinally. Lateral deviation is very frequent in great flaccidity of the uterus, as often occurs in pluriparæ; but it is most pronounced in pelvic contraction after previous very tedious labours. A great quantity of liquor amnii increases the mobility of the child, and allows the head to deviate from the brim. As has already been said, in the majority of instances, the commencing contractions of the uterus, which diminish its transverse diameter, place the head upon the brim. This natural correction of position may, however, be prevented by the following circumstances:

Firstly: The sudden escape of the liquor amnii before the contractions have properly begun may fix the head in the position which it just then occupied, because, after the complete escape of the waters, the uterus closely surrounds the fœtus. If at that time the head had deviated to one side, it would remain there, and the presenting shoulder is forced into the brim. Secondly: The posture which the woman occupies at the commencement of labour has a decided influence upon the position of the head. If, for example, the head is somewhat to the left, and the woman either in the erect or in the recumbent posture, the commencing pains will, doubtless, be able to push the head into the brim. If she lies on the left side the fundus of the uterus containing the breech of the child would also fall over to the left, and thus the head very easily comes to the brim; but if she were to lie on her right side the breech also would fall over to the right, and the head deviate still more to the left. The consequence of this posture would be a complete transverse position. Thirdly: Purely mechanical conditions, such as the seat of the placenta in placenta prævia and tumours of the brim may prevent the head from being placed upon the pelvic inlet. Fourthly: The second twin, which, by the impediment offered by the first, is very frequently not in a longitudinal position, may, after the expulsion of the first, be easily fixed by the contracting uterus in its faulty position.

Although most of the transverse positions have originated during pregnancy and continued during labour, it yet happens that the head which presented at the commencement of labour is made to deviate subsequently. Hence the transverse position is secondary, and this is chiefly the case where the brim is too small for the head, or the head too large for the brim. The head then slides to one side and the shoulder is pressed into the brim. A secondary deviation of the head is greatly facilitated by an unsuitable posture of the woman, as has already been explained above.

We will also call attention to the fact that dead, and especially macerated fœtuses, are in faulty positions on account of the loss of the normal attitude, and of the change in the position of the centre of gravity.

The diagnosis of transverse positions is easy; they can be recognised with certainty by an external examination alone. On inspection the uterus is found to be increased in width, and palpation shows the fœtus to be in a transverse position.

Delivery in transverse positions is only possible without operative interference when the presenting part is still movable above the brim. Very often the commencing pains replace the head, which during pregnancy had somewhat deviated to one side upon the brim. This process, which has been called spontaneous version, is seldom effectual when the transverse position is more pronounced, and then only when the pains are at first slow and feeble; for by a considerable force at the commencement the presenting part would be pushed into the brim and fixed there. After rupture of the membranes the pains rarely accomplish the replacement of the child; they succeed most easily in those cases in which the woman assumes a suitable posture, for example, lying on the left side if the head has deviated to the left. In all these cases the prognosis is not worse than in primary longitudinal positions, with the sole exception that after rupture of the membranes the funis is liable to prolapse, thereby lessening the chances of safety to the child. Exceptionally the breech is nearest to the brim; by the same agency a breech position may be established.

Spontaneous version, however, is extremely rare, when the shoulder has once been pressed into the brim by the pains. Yet under favorable circumstances the head may rise a little higher up, the shoulder leaves the brim, and the breech enters it. The forces of nature are then, as a rule, not adequate to terminate the labour. It is possible only, when the pains are very strong, and the pelvis is sufficiently roomy (especially in its transverse diameter, notwithstanding a slight contraction of the conjugate), or the fœtus very small (immature, or the second of twins). This process is greatly facilitated by the softness and compressibility of the child, also in dead, and especially macerated fœtuses.

The mechanism of spontaneous evolution is the following:—The powerful pains press the shoulder lower down into the pelvis, and being the advancing part it turns forwards for the reasons above stated. Whilst the head is still within the false pelvis, the shoulder passes under the symphysis. The longitudinal axis of the child is thereby so much curved, that head and breech lie close together. The whole trunk is then pushed down past the shoulder, so that first the side of the chest, then the pelvis, then the legs, and at last the head is born.

Another process of spontaneous evolution is more rare. It is facilitated by the great compression which a fœtus that has been dead a long time can undergo. Here the shoulder also advances, but the head is forced into the true pelvis by the side of the chest, so that first the shoulder, then the head deeply pressed into the thorax, and last of all the breech is born.

Spontaneous evolution is almost always fatal to the fœtus, for only small immature ones, which die immediately after birth from weakness, can, without injury, pass through the pelvis in this way.

A case recorded by Kuhn, where a child, weighing 4½ lbs. and 17¼ in., was delivered by spontaneous evolution and remained alive, is certainly an extremely rare exception. The mother also may suffer, unless the fœtus be very small or macerated, from the mechanical impediments which self-evolution encounters.

If a transverse position is left to nature in cases where the pelvis is contracted or the uterine activity inefficient, the woman will invariably die undelivered either from inflammation of the abdominal organs, rupture of the uterus, loss of blood, or from subsequent exhaustion.

By a suitable and, above all, by a timely treatment the prognosis of transverse positions becomes much more favorable to the mother.

The treatment consists exclusively in operations, the procedures of which have already been described in the previous chapters.

If the examination of a pregnant woman reveals a transverse position, this may at once be converted into a longitudinal by means of external manipulations. As a rule, it easily succeeds, but the longitudinal position will rarely last up to the setting in of labour.

If a transverse position is met with at the commencement of labour, it is by no means advisable to wait until the os is sufficiently dilated in order to turn by internal manipulations. The earlier the longitudinal position is established the better, and the operation is therefore to be proceeded with immediately. If the membranes are still intact, and the head not too far away from the brim, it is often accomplished by a suitable posture, the woman lying on that side to which the head has deviated. The fundus uteri and with it the breech, which is higher up than the head, falling over to that side, the head goes to the opposite side, that is to the brim. A pillow so placed as to meet the object in view will greatly assist in bringing about the change in the position of the head. If this fails to bring the head upon the brim, or if the child lies quite transversely, external manipulations must be adopted. Continued perseveringly they often prove successful even after the escape of the waters. But if in this way the position of the child cannot be changed, version is to be effected, when the os is as yet little dilated, by the combined external and internal manipulations of Braxton Hicks. Either head or breech is made to present. When all these attempts have proved unsuccessful it will be needful to wait until the os is sufficiently dilated to admit one hand. The child is then turned by internal manipulations, and, as a rule, by the feet.

If when the woman is first seen the os is already so far dilated as to admit one hand, it is nevertheless advisable to try version by the combined external and internal manipulations, and only in case of failure to introduce the whole hand into the uterus.

If, however, the shoulder is so deeply pressed into the true pelvis that version seems to be impossible, it may yet, in apparently hopeless cases, succeed under the influence of chloroform. Or, if the pelvis is roomy, at least in its transverse diameters, and the fœtus small, and if, moreover, the state of the mother does not urgently require immediate delivery, spontaneous evolution may be waited for, or its occurrence hastened by traction upon the presenting part. Otherwise, embryotomy is to be performed by decapitation when the neck is accessible, or by evisceration. The fœtus is then extracted, imitating the mode of spontaneous version or of spontaneous evolution.

E.—*Faulty Attitude of the Fœtus*

The faulty attitude of the child in face presentations, the prolapse of an arm in transverse, and that of one or both feet in pelvic positions, have already been mentioned above. We shall only speak here of the prolapse of an extremity when the head presents and the pelvis is normal.

By no means rarely a hand can be felt through the membranes lying at the side of the head, when this is still movable. As a rule, however, it is withdrawn either before, or at least at the time of the rupture of the membranes. Treatment therefore is unnecessary. If the hand remains there after the membranes have ruptured, the head is forced down past it. The retraction of the hand may be facilitated by placing the woman on the side opposite to the one on which the extremity has prolapsed.

It is different, however, when a greater part of the extremity has prolapsed in front of the head. The treatment varies according to whether the head is still movable above the brim, or whether it has already entered the pelvis. In the former case, as there is no knowing whether head and arm can together enter the brim, reposition is always to be performed. The arm is pushed into the uterine cavity past the face, and the operation always succeeds under these circumstances. If, however, the head is still so movable that after reposition the arm will again prolapse, the head is to be pushed into the pelvis, adopting Kristeller's method of pressure. In the normal pelvis this also succeeds without difficulty.

But the labour is left to nature when head and arm are already within the pelvis. The fact that both could simultaneously enter the pelvis is the surest proof that it is large enough for the passage of the head and arm together. Extraction of the head by means of the forceps is never required. Only complications, and most frequently feeble uterine action, call for the application of that instrument.

Prolapse of the lower extremity in front of the head is extremely rare in the normal pelvis and with a mature fœtus. It is more frequently seen when the fœtus is immature. The foot must then be pushed back as far as possible, and the head placed upon the brim. If it fails the fœtus is turned by the feet, and this may sometimes be done by means of the combined manipulation.

F.—*Pathological Conditions in Twin Labour*

When the first twin presents with the breech and the second with the head, it may happen that the presenting head of the second enters the brim before the aftercoming head of the first. The first child is born without difficulty up to the navel, or even to the shoulders, but the passage of its head is now greatly obstructed by that of the second, which is so applied to the neck of the first, that either both lower jaws or both occiputs are hooked together, or that the face of the one lies in the nape of the other.

Delivery may be effected by the natural forces. After the birth of the trunk of the first twin the second is born, and then only the head of the first. As a rule, however, operative interference will be necessary. It is useless to attempt to push back the head of the second and to draw upon the trunk of the first. The second head must be extracted by means of the forceps, and the first then follows spontaneously. In case of failure the second head must be perforated, as the first cannot be reached.

It is far more rare for both heads to enter the pelvis together—the head of the second applied to the neck of the first. The impediment to labour is not so great in such a case, and it can scarcely be recognised before the birth of the first head. As both heads can be successively extracted by means of the forceps, this procedure appears to be most rational and quite sufficient.

II.—Abnormalities of the Fœtal Appendages

The birth of the fœtus is rarely, and in a slight degree only, impeded by abnormalities of the membranes.

Great thinness of the membranes may favour their premature rupture. The consequence is that the greater portion of the liquor amnii escapes, whilst the presenting part is still movable above the brim. Usually, however, so much water remains within the uterus that the mechanical process of labour is not interfered with. But where labour is greatly protracted by complications—and especially by a pelvic contraction—air may enter the uterus during the repeated manual examinations and displace the waters contained in it. The maternal soft parts may thereby become dry, the expulsion of the fœtus is delayed, and thus irritation and inflammation of the uterus, vagina, and vulva, are produced. In the narrow pelvis unnecessary examinations must therefore be avoided when the membranes have prematurely ruptured. A moderately filled caoutchouc bag should be placed in the vagina, and the woman should lie on her side or back. The rest of the liquor amnii is thereby pent up and only a small quantity escapes during a pain, and the uterine action is increased.

Great thickness of the membranes and their firm adhesion to the lower uterine segment may greatly impede the course of labour. The causes and treatment of this anomaly have been already mentioned.

Too small a quantity of liquor amnii will scarcely be able to delay labour under otherwise normal conditions, whilst too great a quantity may impede it in various ways. It especially favours malpositions of the fœtus, and the contractions of the uterus are usually deficient.

Labour may be unfavorably influenced by a too short umbilical cord. This is either absolute or produced by coils around the fœtal body. In head positions before the birth of the head, the cord can scarcely exert such an influence, because in each pain the contracting uterus follows the progress of the head. At least the symptoms, which are said to point to a short funis, as recession of the head in an interval, fixed pain, and discharge of blood, are not at all reliable. The head always recedes after a pain, when the soft parts of the floor of the pelvis offer a great resistance, which is especially the case in primiparæ. A fixed pain at one spot of the uterus is not rare. It is probably due to an irritation of a circumscribed spot of the peritoneum, and it is very doubtful, indeed, whether it can be produced by dragging upon the umbilical cord. The escape of blood may proceed from lacerations of the os, but, independently of that cause, it is merely a sign of partial or complete detachment of the placenta, which is very rarely effected by a too short umbilical cord. If, however, the funis were too short prior to the birth of the head, it would rather tear or cause detachment of the placenta, than prove an impediment to labour. We admit that this anomaly of the cord may be instrumental in producing an abnormal rotation of the head as well as a short delay in labour. Usually, however, it is after the birth of the head, and especially of the shoulders, that the complete expulsion of the child, which is accomplished by the vagina, is delayed by a short funis, unless the placenta is detached. The tension of a coil will show that the shortness of the funis is due to it, and an examination will show if the cord is absolutely too short. In such cases the funis is divided before the child is completely expelled. The fœtal end is either ligatured or compressed by the fingers.

A too short cord is more frequently met with in breech positions, especially when it passes between the thighs (the child rides on it). After the birth of the breech it is often easy to push the cord aside over one buttock; but if this be neglected the funis may be so stretched along the perinæum of the child, that it must be divided before the trunk and head can be born. When the cord is in its natural position, it is rarely so short as to be stretched in cases where the breech presents. The treatment is here also the same; it is evident that after the division of the cord the child is to be immediately extracted. We have spoken elsewhere of the delay in the labour caused by the retention of the placenta.

Literature.—Hhol, Die Geburten missgestalteter, kranker und todter Kinder. Halle, 1850.—Playfair, Obstetrical Transactions, VIII, p. 300.—Hohl, l. c.—Joulin, Des cas de dystocie app. au foetus. Paris, 1863.—Denman, Londoner Medical Journal, Vol. V, 1785, Art. V, p. 371—Douglas, Explanation of the Real Process of the Spontaneous Evolution, &c., 2 ed. Dublin, 1819.—Gooch, Medical Transactions, VII. London, 1820, X, p. 230.—W. J. Schmitt, Rheinische Jahrb., B. III, St. 1. Bonn, 1821, p. 114.—Hayn, Ueber die Selbstwendung Würzburg, 1821.—D. W. H. Busch, Geburtsh. Abhandl Marburg, 1826.—Betschler, Ueber d. Hülfe d. Natur z. Beendig. d. Geb. bei Schiefl. d. Kindes, Klinische Annalen, II, p. 197.—Birnbaum, M. f. G., B. 1, p. 321.—Haussmann. M. f. G., B. 23, p. 202 and p. 361.—O. Simon, Die Selbstentwicklung, D. i. Berlin, 1867.—Barnes, Obstetrical Operation, 2 ed., p. 107.—Kleinwächter, Arch. f. Gyn., B. II, p. 111.—Credé, Verh. d. Ges. für Geb. in Berlin, IV, p. 153.—Pernice, Die Geb. mit Vorf. d. Extrem. neben d. Kopf. Leipzig, 1858.—Kuhn, Wiener Med. Woch., 1869, Nr. 7—15.—Hohl, Neue Zeitschr. f. Geb., B. 32, p. 1—Joulin, Des cas de dystocie app. au Fœtus. Paris, 1863, p. 83.—Kleinwachter, Lehre von den Zwillingen. Prag, 1871, p. 167.

CHAPTER IV

ANOMALIES OF PARTURITION WHICH DO NOT IMPEDE THE MECHANISM OF LABOUR

1. Compression of the Umbilical Cord

When the umbilical cord lies in the usual way, and we must consider its coiling round any part of the fœtus as usual on account of its very great frequency, it is very seldom so much compressed that the circulation in it is impeded, and the fœtus in consequence becomes asphyxiated and dies. As long as the membranes remain intact, as Lahs correctly observes, no such one-sided pressure takes place within the membranes. The general internal pressure of the uterus remains the same, whether it acts directly through the walls of the uterus or mediately through the liquor amnii. The form restitution power of the uterus exerts, indeed, a one-sided pressure, and although it may be able to compress the funis lying between the breech and the fundus, or between the head and the lower uterine segment, yet it is hardly strong enough to stop the circulation in the umbilical vessels; therefore, as long as the membranes remain intact, the cord is only exposed to compression when by the powerful action of the abdominal pressure the head, surrounded by the lower uterine segment, is forced into the small pelvis. The funis is then tightly squeezed between the head of the fœtus and the pelvis covered by the wall of the uterus. If parturition has so far advanced that the head has passed out of the uterus into the vagina, the cord lying close to the head may also be compressed between the head and the pelvis by the force of the uterine contractions. This most frequently occurs in the stage of expulsion, when the funis is coiled around the neck, and this is pressed for some time against the posterior surface of the symphysis. Veit has shown that disturbances of the fœtal respiration are two to three times more frequent when the cord is coiled than in any other position, but that death in such cases is only relatively rare, because the danger occurs a short time before birth.

The prolapsed funis far more frequently suffers such compression as to endanger the life of the fœtus.

The funis is said to be prolapsed when, after rupture of the membranes, one or more coils are lying in the os, the vagina, or outside the external genitals. It is said to present when, before the rupture of the membranes, it can be felt lying in the os behind the membranes.

The prolapse of the cord is of varying significance, according to the presentation of the fœtus. It will, accordingly, be necessary to consider each separately.

In head presentation in primiparæ the lower uterine segment normally surrounds the head, which at the commencement of labour has already entered the small pelvis, so closely that a prolapse of the cord is actually impossible. The entrance of the head into the pelvis, and the accurate adaptation to it of the lower uterine segment, are, in some cases, impeded by an abundant quantity of liquor amnii or by twins, most frequently, however, by a contracted pelvis. Consequently prolapse of the cord in head presentations in primiparæ is almost always dependent upon a contracted pelvis. In pluriparæ the head is often at the commencement of labour still high up and easily movable, even over the normal inlet, and it consequently happens that in pluriparæ the cord is often projecting behind the membranes. When there is not too much liquor amnii and the membranes rupture at the proper time the presenting cord is, as a rule, pushed aside by the head entering the pelvis; but the presenting cord prolapses if there is hydramnios, which retards the entrance of the head, or if the membranes rupture too early, before the head is closely surrounded by the lower uterine segment. The same occurs when, for some other reason, as an upper extremity lying close to the head, the lower uterine segment is prevented from closely surrounding the head, and thus allows the necessary space for the prolapse of the cord by the side of the head. Shortness of the cord and loops in it hinder its prolapse, and this the more easily the higher up in the fundus the placenta is inserted. On the contrary, low insertion of the placenta plays an important part in the causes predisposing to prolapse of the funis. As a rule the cord prolapses when several of the predisposing causes occur at the same time; for instance, when the membranes are ruptured the head is situated on one side of a contracted pelvic inlet, the cord is long and without loops, and inserted low down in the uterus.

Diagnosis.—When the membranes are tightly stretched it may not be possible to feel the presenting cord, but in the interval between the pains the soft pulsating cord can almost always be detected behind the membranes. Prolapse of the funis is, of course, much more easily detected, yet it may also be overlooked in an insufficient examination, when, for instance, only one coil projects close to the periphery of the head.

Except in those rare cases where the projecting cord is so tight that the advancing head exerts considerable traction on the placental insertion, prolapse of the funis is without the slightest consequence to the mother. To the child it is one of the most dangerous complications, and in head presentations, without suitable assistance, it almost always causes its death. The danger to the child does not follow from the abnormal position of the cord, but because the prolapsed cord is much more frequently and more easily than one in a normal situation exposed to pressure, which stops the circulation.

In head positions the head is so closely applied to the pelvis, that the cord lying between them is always compressed; moreover, the passage of the head through the pelvis takes, as a rule, so long that the child dies during the time from the impediment to the placental circulation.

The treatment varies greatly, according to the progress of the labour.

If the cord presents behind the membranes, everything must be done to delay the rupture of the membranes as long as possible. For, as long as the membranes are intact, the cord can still be pushed aside, and when prolapse has actually taken place it is much less dangerous when the os is fully dilated than when it is only slightly so. The examination must be made gently, and not at all during a pain. A moderately filled caoutchouc bag is to be placed in the vagina in order to offer counter-pressure to the membranes. Much is gained if the rupture of the membranes is delayed until the os is sufficiently dilated. For usually the presenting cord spontaneously retires when the head enters the pelvis. Exceptionally it does not retire, and it is then exposed to dangerous pressure between the head and the pelvis. If the membranes are still intact, attempts must be made to push the cord back, or the head must be pushed to one side in order to remove the pressure from the cord. If all these attempts fail, nothing remains but to rupture the membranes, and to act upon the principles about to be given.

Matters are worse when the cord is prolapsed after the premature rupture of the membranes, and the os is only slightly dilated. Immediate delivery is impracticable without the greatest danger to the mother. The preservation of the child depends alone in protecting the cord from dangerous pressure until the woman is delivered by the forces of nature or by art.

Reposition of the cord is the most natural way of preventing the pressure. Under these circumstances this may be very difficult. If the os will admit only one or two fingers the cord may be pushed back a little, but on withdrawal of the fingers it always comes down again. It must therefore be carried higher up into the uterus by means of instruments, in the expectation that the cord will adhere to something, or that the head may in the meantime be firmly impacted in the inlet, or that the cord by a suitable posture of the woman may be prevented from again coming down. To effect reposition a number of instruments have been devised, of which Braun's is the most simple and the most practical. It consists of a small caoutchouc rod, with a hole at the thinner end, through which a double cord passes. The coil of the funis is placed in the latter, and the loop of the cord is put over the tip of the rod. The rod with the umbilical cord is now introduced high up into the uterus. In withdrawing the rod the loop of the cord is displaced from the tip of the rod, and the coil of the funis falls out. In the dorsal position of the woman, however, the funis only remains in the uterus when it happens to fall over one of the extremities, or when in the meantime the head has firmly entered the pelvis. Roberton's instrument appears to be more effective, because by it, as in similar older methods, the funis is permanently retained at the fundus. It consists of a thin caoutchouc tube about forty centimètres long (if necessary also a simple elastic catheter), mounted on a stilet. From its upper lateral eyelet-hole a loop of cord passes out. The coil of the prolapsed funis is loosely placed in that loop, and both are carried past the presenting part into the uterus. The stilet is withdrawn, but the elastic catheter remains, and is expelled only after the child, together with the placenta. The advantage is that the funis, which remains within the loop, is permanently retained in the uterus by the elastic catheter, which remains at the fundus.

In all other methods the umbilical cord, following the laws of gravity, again prolapses after reposition. To avoid this the position of the woman requires full consideration. She must be so placed that the lower uterine segment is no longer the deepest part of the organ. This is best effected by the knee-elbow posture. The lateral posture is not so suitable. Therefore if practicable the woman is to be placed in the knee-elbow posture for the reposition of the funis, and when this has been performed in that posture, attempts may be made to facilitate the entrance of the head by Kristeller's method of pressure. If that posture cannot be assumed, the lateral posture must be retained. The woman is placed on that side towards which the head has deviated. Thus, if the head lies to the left, the left lateral posture is assumed, though the funis may have come down on the right side. This posture is only altered when it becomes unsuitable.

When the os is insufficiently dilated and reposition impossible the life of the child can only be saved if the dangerous pressure can be taken off the umbilical cord. The danger arises when the head is pressed against the inlet. The danger is prevented by making the head deviate to one side, when the head position is converted into a transverse, or, still better, into a breech position.

Version with an undilated os is best made according to the method advocated by Braxton Hicks. The artificially produced transverse or breech position is then treated in the way to be mentioned below.

When the os is so far dilated as to admit one hand it is, doubtless, the most rational treatment to replace the funis, and to re-establish the normal relations. Reposition is always to be performed by the hand corresponding to the side towards which the funis has prolapsed; that is, the left hand, when the funis lies towards the right side of the mother; in the knee-elbow posture, however, the right hand for the right side. The funis is then carried by the fingers, formed into a cone, past the head of the fœtus, and, if possible, appended to one of the lower extremities, or placed in its natural position in front of the thorax and abdomen. If the funis remains there the fingers are immediately withdrawn; if not, they must be kept there for some time. All this time attention must be paid to the pulsations of the cord. During a pain the hand may be slowly withdrawn, and the head so placed by means of external manipulations as to render prolapse impossible.

Schmeisser recommends fastening the coil of the funis to the base of a properly constructed caoutchouc bag, and then to replace it. The bag is then to be inflated, and in this way it is retained within the uterine cavity.

However rational it may appear to replace the prolapsed cord, yet this must depend upon certain considerations. If the cord has not been at all compressed, and the fœtal circulation is quite undisturbed, there is nothing to say against reposition; for if the worst happens, that it cannot be replaced, nothing is lost by it. If, however, the cord has been already compressed, even if reposition is rapidly and successfully effected, as happens in the most favorable cases, yet the life of the child may not be preserved; for the cord, without again entirely prolapsing, may come to lie so as to be exposed to fresh pressure. Examination of the sounds of the fœtal heart by the stethoscope will show, in such a case, that their frequency at first increases, but that after some time it gradually diminishes, thus furnishing a sure sign that the child is in danger. In still other cases, even if the funis lies so that it can by no means be compressed, the heart's sounds remain irregular. At one time they are extremely frequent, at another very slow, until they suddenly cease. It seems in such cases as if the circulation, once disturbed, would not again return to its previous course, although the cause of the interruption has ceased to exist.

Reposition, however, by no means always succeeds so rapidly and successfully. In many cases it is, as Boër says, "really a work of the Danaides." As often as we attempt to replace it, it comes down; and if, after much trouble, one coil has been luckily replaced at one spot, another coil has already prolapsed at a second place. The more time we lose in fruitless attempt to replace the cord, the greater becomes the danger to the child. We have, therefore, considered whether it would not be preferable, in all such cases in which the child is already in danger, and the circumstances give no reasonable prospect of an easy operation, to neglect for the moment the funis entirely, and immediately to turn and extract. Under not too unfavorable circumstances, in almost less than a minute, the child can be brought into a position to inspire atmospheric air, and is thus freed from the dangers which threaten it from a prolonged stay in the uterus, and if asphyxia already exists the necessary treatment can immediately be instituted. If under such conditions version is immediately performed, without previous trials at reposition, the child will not have more unfavorable chances than when reposition succeeds, but far more favorable ones than when, after unsuccessful trials at reposition, version must be performed.

If in prolapse of the funis the head is already in the true pelvis, unless delivery is very rapidly terminated, the child dies from compression. In primiparæ, therefore, when the head is in the pelvis, a pulsating loop of the funis will very rarely be felt. In pluriparæ it more frequently happens that when the head is high up the funis prolapses on the rupture of the membranes, and the head, at the same time, drops down. Unless the head is immediately born the forceps must be applied in order to extract at once. Under such circumstances the head is sometimes born so rapidly that there is no need of applying the forceps.

The treatment of prolapse of the funis in face presentation is essentially the same. It is better, however, at once to turn rather than to attempt reposition.

The case is different, however, in breech presentations when the funis presents or has already prolapsed. For here everything favors prolapse, because the breech never fills the inlet so completely as the head. For the same reason the danger also is much less. The soft breech rarely exerts so much pressure upon the funis as to stop its circulation, and, if it has prolapsed at the side of the promontory, it is scarcely exposed to any pressure at all. Usually reposition is unnecessary, although it can often be successfully done, there being always room enough for it to come down again. But, because during the course of labour there may be possibly a greater pressure upon the funis, care must be taken so as to be able to terminate labour should danger arise. If the breech has not yet completely entered the true pelvis, it is urgently necessary to bring down one or both feet. Not only is the circumference of the breech diminished by it, but it also affords a good hold for extraction. During the progress of the labour it is, of course, necessary to watch carefully the pulse of the child and to extract it as soon as the frequency diminishes. When the breech has completely entered the pelvis with the legs high up, extraction must be tried as soon as asphyxia of the child commences. All that can be done here is to extract with all possible rapidity in order to save the child.

In pure footling presentations no further treatment is required, since the cord is not compressed.

In transverse positions it occurs much more frequently that the funis prolapses on the rupture of the membranes, but only in greatly neglected cases is it exposed to any dangerous pressure. If the os is still closed its dilatation is patiently awaited, and then podalic version and extraction made. Cephalic version with simultaneous reposition may also succeed under favorable circumstances. But it is difficult, and a favorable issue so little probable, that, indeed, it will very seldom be performed.

Another question remains to be decided, viz. is an operation indicated, when there is very slight, if any, pulsation in the funis. In the first case decidedly so; in the second there is no need of an operation if we are sure the pulsation has ceased for any length of time. For by rapid extraction we may save a child in a profound state of asphyxia, but in whom life is not yet quite extinct. The more immature a child is, the longer it can live in the uterus after the interruption of the placental circulation, but there is less probability of its living after its birth.

2. Lacerations of the Soft Parturient Passages

A.—*Lacerations of the Uterus*

Solutions of continuity of the uterus are either incomplete, that is not penetrating through the whole thickness of the uterine parenchyma, or complete, that is, perforating into the abdominal cavity. The latter differ clinically, according to whether they are due to a sudden rupture, or to a gradual tearing through of the parenchyma of the uterus and of the peritoneum.

The non-penetrating or incomplete lacerations of the uterus are almost always situated in the cervix, and those of slight extent arise in the physiological process of labour. In each labour the os uteri is torn at the sides and this can be felt after delivery. These rents do not give rise to special symptoms, and can afterwards be detected with great difficulty if at all. Only in succeeding pregnancies when the portio vaginalis is greatly swollen and œdematous, the old cicatrices which are not extensible are distinctly perceptible, and they form one of the best diagnostic signs between primiparæ and pluriparæ. It is very rare, indeed, that these small longitudinal rents of the cervix acquire a pathological significance. This only occurs in great rigidity of the cervix, most frequently in carcinomatous degeneration, and in very violent pains. Transverse rents of the cervix are very rare when the tissue is normal. In very exceptional cases the anterior lip of the os uteri is so violently bruised between the head and the symphysis that it is more or less completely torn from the uterus.

The ordinary rents in the cervix are not attended by any symptoms after delivery, but the larger longitudinal or transverse lacerations cause severe disturbances. There is inflammatory fever; the parts near the wound inflame and become infiltrated with the products of inflammation or with blood, the neighbouring connective tissue also participates in the inflammation, and the healing only slowly proceeds. If the lacerations extend almost to the peritoneum, perimetritis or even general peritonitis may be the consequence. During labour they cause little disturbance. The hæmorrhage, also, is usually moderate, for the lacerated tissue is exposed to so great a pressure from the head pressing against it that the vessels are more or less completely closed. Sometimes, however, much blood is effused into the tissue itself, forming a tumour which is distinctly felt in the vaginal vault. On the whole, delivery is effected in the usual way, and only after birth the lacerations cause profuse hæmorrhage, which is very difficult to stop. The bleeding spot is, in such cases, not easily detected, since uterus and vagina form an irregular cavity, with flabby and sinuous walls, over which the blood continually pours. The treatment is, therefore, usually limited to injections of cold water and vinegar, and they mostly suffice, because the hæmorrhage is parenchymatous.

Sudden perforating ruptures of the uterus perhaps never occur when the uterine tissue is healthy. They require a predisposition of the tissue and an exciting cause. In an otherwise normal uterus there may be some places which are feebly developed. Most frequently anomalous development, interruption of the normal tissue by interstitial fibroids or cicatrices, separation of the muscular fibres by submucous fibroids or by projecting small parts of the fœtus, or inflammatory softening of some portions of the parenchyma during pregnancy, predisposes to ruptures. The exciting cause proceeds from considerable impediments to the progress of labour. They are chiefly narrowness of the pelvis, hydrocephalus of the fœtus, transverse positions, and a sudden and great increase of the contents of the uterus, as by introducing the hand in turning; it thus becomes intelligible why, in breech positions, the uterus seldom ruptures. A perforating rupture is easily produced by such an exciting cause when the predisposition already exists in the tissues. It is not improbable that a rupture may also proceed from a small spot in the uterine wall, which had been injured by the pressure between the head and the pelvis, and that from this circumscribed lesion the lacerations start under the influence of a strong pain.

Usually the rent is so large that the child passes through it into the abdominal cavity. In the dead body it appears very small on account of the contraction of the muscular tissue. The rent is, as a rule, in the neighbourhood of the cervix, and then frequently has either a transverse or, at least, an oblique direction.

The symptoms of perforating rupture of the uterus are generally very characteristic. At the height of the uterine contraction the woman experiences a severe piercing pain, and at the same time feels a sudden turning of the child. She feels an oppression at the chest, a darkness comes over her eyes, a cold perspiration breaks out, her limbs tremble, the pulse becomes very small and frequent, and all the symptoms of an internal hæmorrhage set in. Uterine action suddenly ceases, the presenting part completely recedes or is, at least, more movable, and blood flows from the genitals. Almost always the child is partially or entirely expelled from the uterus into the abdominal cavity, the uterus then firmly contracts. For that reason the hæmorrhage is sometimes not very considerable, and the patient may sometimes recover from the first shock (especially if she has been delivered in the mean time).

In other cases, most of these symptoms may be absent. The rupture has taken place with scarcely any symptoms, and the cessation of the uterine action is almost the sole apparent sign. It is remarkable indeed to see how little sudden disturbance such a grave lesion sometimes produces. The pulse, however, in these cases always becomes frequent and small, and the subjective symptoms are, at least, never entirely absent.

From the clinical aspect above given a diagnosis can easily be made, but if the symptoms are as just described it will be somewhat difficult to interpret them correctly. A sudden cessation of the uterine action, when the pains have been very strong should always greatly excite the suspicion of the accoucheur.

If the presenting part until then immovable at the brim has receded from it and blood escapes from the vagina, the diagnosis is certain. The two latter signs, however, may be absent when the head is impacted in the brim. In such a case the diagnosis is made certain by palpation. The fœtal parts are now much more distinctly felt than before beneath the abdominal walls, and near them a large hard ball—the contracted uterus. The general condition also gives very important indications, for there is rarely an entire absence of anxiety, nausea and attacks of syncope. Whilst the temperature is of no diagnostic value in this case, the state of the pulse is of great importance. The pulse which shortly before had been full and not very frequent has now become small, threadlike, and very frequent, and indicates a grave lesion.

Almost all cases of perforating rupture of the uterus are fatal to the child, for, unless it be very rapidly extracted, the placenta is detached by the contractions of the uterus. The mother may rapidly bleed to death, but usually dies from consecutive peritonitis in childbed. When the child remains within the peritoneal sac, a favorable issue is extremely rare, yet it is possible that the case may terminate in all the various ways of extra-uterine pregnancy. If the child be extracted, the rent may heal up.

It is unfortunately extremely rare that precautionary measures can be taken. Usually the predisposing conditions are unknown, and the exciting causes cannot or dare not be removed. Thus in contracted pelvis very strong pains are required to overcome the obstacle, or the sole possibility of delivery may depend upon the introduction of the hand into the greatly contracted uterus. Rupture is never preceded by any definite symptoms, and therefore its occurrence can only very exceptionally be prevented. Occasionally some places in the uterine wall feel remarkably thin, and during a pain are distended into a bladder by the liquor amnii. In such a case it is decidedly most urgently necessary to abbreviate labour, although after its happy termination no one can say with certainty that a rupture of the uterus has been prevented.

When rupture has taken place immediate delivery is most urgently indicated. Delivery *per vias naturales* almost always succeeds when the child is partly or wholly within the uterus. The forceps need only be applied in those rare cases where the head is quite impacted. Otherwise turning is performed by one or both feet and the child extracted. When the child has passed into the abdominal cavity, delivery *per vias naturales* may then become impossible, since the rent has been too much narrowed by the contractions of the uterus. As a rule a rent through the thinned tissue of the cervix (and still more so in perforating ruptures of the vagina) remains wide enough, though the rest of the uterus be firmly contracted, to allow of delivery *per vias naturales*, and this, if practicable, is the most favorable way, only there is the danger that coils of intestine may be drawn into the uterus. If the child is alive, and cannot be delivered in the natural way, the Cæsarean section is then the sole operation by which it can be extracted. When the child is dead, seeing that mere waiting gives decidedly unfavorable results, it is also advisable to extract the child by the Cæsarean section, and to close again the abdominal cavity after having carefully removed all clots, &c. Frequently the placenta is detached; its removal becomes necessary, for the hæmorrhage only ceases with the contraction of the uterus. Coils of intestine which may have passed through the rent into the uterus are to be carefully replaced.

The treatment becomes exceedingly difficult when, after removal of the placenta, the uterus does not contract, but remains flaccid, and the hæmorrhage continues. Attempts may then be made to produce contractions by introducing one hand into the uterus, and after removal of the coagula to rub it externally with the other hand. If this also fails in arresting the hæmorrhage, small pieces of ice are placed in the uterus, or cold water is carefully injected. The patient is placed on the side opposite to the rent, lest blood or injected water enter the abdominal cavity. Plugging would only impede the outflow of the blood, and is therefore not to be employed. The consecutive peritonitis is to be treated, as in traumatic cases, with cold and opium, and not with laxatives, as otherwise is done in the lying-in state.

Perforation of the uterus into the abdominal cavity also occurs from gradual friction of its tissue and of the peritoneum at a circumscribed spot. This occurs almost exclusively in a contracted pelvis, when the same spot of the uterus has for a long time been exposed to great pressure between the head and the pelvis. When the bones are smooth their pressure will not easily produce the perforation. It is almost always pointed or sharp projecting portions which exert that pernicious pressure. The promontory very frequently presents such a sharp ridge; in other cases rough places exist on the anterior wall of the pelvis. But because even the sharpest ridges are covered by soft parts it requires the persistent counter-pressure of an equally very hard body to cause "usure" of the uterus. The fœtal head pressed by the contractions of the uterus against such a spot can only then produce it by the compression it exerts when the pressure lasts a very long time and the pelvis is very narrow. A far greater pressure is exerted upon the dangerous spot by the application of the forceps or of the cephalotribe. Those instruments exert their noxious influence in two ways. Either they are so situated that one of their hard unyielding edges crushes the uterine substance by direct pressure against the bony pelvis, or they abolish the compressibility of the head, for whilst they compress the head in about the direction of the transverse diameter of the pelvis they prevent its enlargement in that direction in which alone the head can yield when the conjugate is contracted. They thus convert the head into a hard unyielding object, which is capable of exerting most dangerous pressure. If long-continued and powerful traction is necessary the soft parts situated between both prominent hard spots are crushed. First and most of all the uterine parenchyma suffers so that sometimes it alone is found to be bruised and softened; but by long-continued pressure the peritoneum also is bruised, and then a funnel-shaped opening is seen, the wider end of which looks into the uterine cavity, and the tip is formed by a small hole in the peritoneum. Complete perforation, that is, a free communication between the uterus and the abdominal cavity, frequently takes place during the puerperal state when the scab covering the bruise is detached.

The symptoms of this kind of perforation are completely different from those above described. The characteristic signs of rupture or perforation are absent. There are also the symptoms peculiar to protracted and exhausting labours in general, and especially disturbances of the general state, and even they are not regularly present. The state of the pulse is again of the greatest importance. A thready, very frequent pulse indicates a grave injury to the peritoneum. Hecker and Jolly have also observed in such cases an elastic blood tumour in the connective tissue between the bladder and the uterus and vagina. It is said to project through the anterior wall of the vagina, and may easily be distinguished from a cystocele by the application of the catheter. Such a tumour, however, may arise also independently of a perforation of the peritoneum, and although its consequence is very grave, it is not absolutely fatal. The small perforation of the peritoneum, which is regularly followed by death, cannot with certainty be recognised during life.

There may also be a usure at the anterior part of the uterus so that perforation does not take place into the peritoneal sac but into the bladder, forming a vesico-uterine fistula.

Death from peritonitis is always the termination in the first mentioned case, because the edges of the wound bruised to such an extent do not heal. Sometimes the patients apparently recover after delivery if the hæmorrhage is only inconsiderable, but the consecutive purulent peritonitis causes death.

Whilst speaking of the contracted pelvis we have already indicated the prophylactic measures to be taken. As perforation never occurs suddenly but always gradually through protracted pressure, the chief indication is to shorten that pressure. Great care must be taken not to convert the favorable chances which the compressible and relatively soft head offers into the contrary, through the blades of the forceps or of the cephalotribe. Version and perforation are here innocent operations, whilst the use of the forceps and the cephalotribe may prove fatal.

B.—*Laceration of the Vagina*

Lacerations of the vagina are of varying significance, according to their depth, position, and extent.

The upper portion of the vagina may be the seat of sudden perforations into the abdominal cavity, and they are closely allied to the perforating ruptures of the uterus. They chiefly occur when, after complete dilatation of the os, the vagina is greatly overstretched by powerful pains, the head not being able to enter the pelvis. In such cases the vagina almost always tears transversely; sometimes it is completely torn from the uterus. The symptoms, prognosis, and treatment of these ruptures are just the same as those of perforating ruptures of the uterus, especially of those seated in the cervix.

The same causes which gradually produce the boring through the uterus may also give rise to loss of substance of the vagina. Usually the gangrene commences during labour, and the mortified tissue is thrown off only in the puerperal state. If, as is usually the case, this place is situated in the anterior wall of the vagina, a vesico-vaginal fistula easily forms.

In the upper portion of the vagina longitudinal rents occur together with those in the cervix uteri, and pass more or less deeply through the mucous membrane. The symptoms are perfectly like those of lacerations through the middle portion of the vagina.

These rents are always longitudinal, situated on the right or left side, and produced by the great distension which the vagina suffers from the head. They most frequently occur in primiparæ, and very easily when the narrow vagina is distended not by the head alone, but by the head and the forceps. Now, as a rule, the blades of the forceps do not lie quite flat upon the head, but project with one edge. If the vagina is predisposed to rupture this will always take place where the edge of the blade presses against the mucous membrane.

These rents are almost always several centimètres long, and usually only extend through the mucous membrane. Sometimes they extend deeper into the submucous connective tissue.

The smaller rents give rise to no symptoms except a slight febrile disturbance, and they are discovered only on very close examination. They heal by suppuration, the products of which are lost in the lochia. But if the lacerations extend deeply into the connective tissue they may give rise to very high fever and be followed by extensive infiltrations. These lacerations are not painful in themselves, but if the finger during an examination happens to touch them an insupportable pain is produced. Usually they heal by granulation without injurious consequences, and the product of exudation into the subjacent connective tissue is reabsorbed. Narrowing of the vagina may at times be the consequence. Treatment is unnecessary; only much attention should be paid to cleanliness, and the stagnation of pus should be prevented.

It is very rare that these lacerations of the vagina give rise to profuse hæmorrhage. The bleeding spot can scarcely be detected immediately after delivery. The speculum only shows a wide cavity continually overflowed by blood. Treatment is, as a rule, limited to injections of cold water or to the placing of pieces of ice in the vagina. If these means do not control the bleeding, and the blood is supposed to come from the middle portion of the vagina, the bleeding spot may be compressed by a suitable tampon.

c.—*Lacerations of the Vulva*

The vulva may be the seat of numerous lacerations, which, of themselves, are of no consequence.

The ostium vaginæ is so narrow that it can rarely be stretched by the head without causing some laceration. Small rents, penetrating only the thickness of the mucous membrane, are, as a rule, found in primiparæ and very frequently also in pluriparæ. Independently of rupture of the perinæum they are very frequently met with in primiparæ behind the intact frenulum, consisting of rents through the mucous membrane of the fossa navicularis. In individual cases large shreds of mucous membrane are quite torn off and are loosely hanging. Smaller rents are almost constantly found in the labia minora and at the sides of the urethra. In pluriparæ one frequently sees nothing else but radiating rents in the mucous membrane.

Lacerations of the mucous membrane between the clitoris and the urethra are of greater importance. Even a very superficial tear in the vascular cavernous tissue may be followed by great loss of blood. Sometimes several arteries spout, at others the blood comes or wells up as from a compressed sponge. These hæmorrhages are important because they cause great danger, and are doubtless often mistaken for flooding from a contracted uterus. They are only recognised, and then very easily, by an accurate local examination. The arteries cannot easily be ligatured in the brittle tissue, since the ligatures will not hold. A not too profuse bleeding is sometimes arrested by a stream of cold water. If this fails, it is best to dip a piece of lint into a solution of perchloride of iron and to press it against the bleeding spot and then bring the thighs closely together. This styptic leaves a healthy looking wound, with good granulations. For some cases acupressure or sutures may be necessary.

The frenulum and the perinæum are far more liable to be lacerated.

Slight lacerations of the frenulum are very frequent, especially in primiparæ. As the result of accurate examinations of the genital fissure, the frenulum is preserved in 39 per cent. of primiparæ and torn in 61 per cent.; in pluriparæ the frenulum is not torn in 70 per cent. In primiparæ also the lacerations of the perinæum are more extensive, Whilst in pluriparæ we found only 9 per cent. of actual lacerations, in primiparæ there were 34½ per cent.

The lacerations are produced by the great distension to which the ostium vaginæ is subject from the passage of the head, and occasionally also from that of the shoulders. The narrower the genital fissure, and the larger the head and the more rapid its passage, the more easily, *cæteris paribus*, the perinæum ruptures. Very much, of course, depends upon the condition of the tissues composing the perinæum, and Hecker correctly observes that sometimes a perinæum is met with that tears like tinder. The posture of the woman has a great deal to do with it. In the dorsal posture lacerations easily occur, because the expelling forces and the whole weight of the child press upon the perinæum, and the head is forced in the stage of expulsion to ascend, contrary to the effect of its own weight. These unfavorable circumstances do not exist in the lateral posture, but, for the reasons already stated, a kneeling or sitting posture is the most favorable for the passage of the child through the genital fissure. The frenulum is rarely found torn in women who have been surprised by labour or who have secretly given birth to a child in a crouching posture (we have examined eight such cases, all of them primiparæ, and have not found a single rent). For the same reasons a very slight pelvic inclination, especially if this is still more lessened by a pillow placed under the nates, favours rupture of the perinæum, while this rarely occurs in very considerable pelvic inclination. A too narrow pubic arch also may greatly press the head against the perinæum and rupture it.

In many cases the frenulum only, which is stretched during labour, is torn; in others the rent extends to the sphincter ani, and in very rare cases also through it into the rectum. Sometimes the skin only is superficially torn, more frequently the rent is deeper. With a very high perinæum and very slight pelvic inclination the head is so pressed against the perinæum that the skin in the middle of it begins to burst, and the frenulum is only torn later on in the course of the labour. It even happens that this remains intact, whilst the middle of the perinæum ruptures. In very rare cases the whole child even may pass through an enormous perinæo-vaginal fistula, rectum and frenulum remaining intact (central rupture).

Lacerations of the perinæum very rarely cause profuse hæmorrhage, and a spouting artery is very seldom seen. In childbed even very large lacerations do not, as a rule, give rise to any other symptoms than a burning pain in the wound and fever. When the sphincter, too, is lacerated, there may be involuntary passage of intestinal gases, and even fæces.

The diagnosis of these lacerations requires a very careful examination, even by inspection if necessary. This must on no account be omitted, for neglected laceration of the perinæum may be followed by tedious, and even very sad consequences, such as prolapse of the posterior wall of the vagina and the uterus, recto-vaginal fistula, and lasting incontinence of fæces, &c.

Prophylactic treatment is of the greatest importance with regard to laceration of the perinæum. If the perinæum is lacerated, the wound cannot be left alone. As a rule, it does not heal by first intention, but gradually cicatrizes by suppuration, and at least causes an enlargement of the vulva. The wound is, therefore, to be closed by sutures.

According to the length and depth of the wound, usually one to six sutures are required, more or less superficially. The legs of the woman are then tied together with a towel, and the sutures removed in from four to seven days. If removed earlier the still tender cicatrix breaks down on active movement. No inconvenience arises from allowing the sutures to remain so long, and even if the little holes suppurate at first they easily close after the removal of the sutures. In the great majority of instances there is union by the first intention if the woman is a little cautious and the sutures are carefully placed. The after-treatment does not require emptying of the bladder by means of the catheter or injections into the vagina. When the wounds are larger, reaching to the rectum, the sutures must be very skilfully placed, sometimes through the vagina, the perinæum, and the rectum.

3. Hæmorrhage during Labour

A.—*Hæmorrhage before the Expulsion of the Child*

There is an essential difference between hæmorrhages occurring before the expulsion of the child and those met with in the period of the afterbirth.

As regards the former, it has already been said that they may be produced by lacerations of the uterus, the vagina, or the vulva. With the exception of these and other slight hæmorrhages due to the separation of the decidua when the lower uterine segment is drawn back over the ovum, almost all hæmorrhage before the expulsion of the child is due to partial or entire premature separation of the placenta. This is most frequently the case when the placenta has an anomalous seat, constituting what is called placenta prævia. Here partial premature separation is the rule; this only exceptionally occurs when the placenta is normally situated.

1. *In Normal seat of the placenta.*—Hæmorrhage in consequence of the separation of the placenta whilst in its normal seat at times occurs towards the end of pregnancy; this secondarily arouses the uterine action, but at other times it only begins with the commencement of labour.

The causes which have already been stated to produce on the part of the mother hæmorrhage during pregnancy are here also active in producing hæmorrhage before the membranes are ruptured.

The blood may escape externally or it may remain within the uterus. The latter occurs when the placenta remains attached throughout its whole circumference, and the blood is then effused into the detached central part; or the presenting part may be surrounded so closely by the lower uterine segment that the escape of the blood from the uterus is thereby prevented; also when the bleeding is not very considerable, and takes place near the fundus. As a rule, a bloody serous fluid escapes by drops when, after rupture of the placenta or of the membranes, the blood has entered the cavity of the ovum.

These internal hæmorrhages, if at all considerable, give rise to very alarming symptoms. The signs of acute anæmia suddenly appear, the patient is collapsed, and severe pains are felt in the abdomen. The uterus rapidly increases in size, so much so that women at an earlier period of gestation may appear to be at term, or to be pregnant with twins; the distension may even cause intense dyspnœa. There are either no labour-pains whatever or only very weak ones. We know from experience that a diagnosis of this condition is not easily made, yet the symptoms are so evident that one can only think of an internal hæmorrhage; in doubtful cases the diagnosis, at least, may be arrived at by the method of exclusion. The diagnosis is facilitated if, as occasionally happens, an external hæmorrhage is connected with it, or if a bloody serum escapes. The sudden alteration in the general state and the signs of internal hæmorrhage may lead one to think of rupture of the uterus. But in rupture the uterus is felt either diminished in size, or, at least, not enlarged, and the presenting part recedes unless it be very tightly fixed in the pelvis.

The prognosis for the child is very bad, because, as a rule, it dies unless delivery be very rapidly accomplished by nature or by art. For the mother also the prognosis is unfavorable, and much more so than in placenta prævia.

As the hæmorrhage itself cannot be stopped, treatment must be directed to emptying the uterus as quickly as possible of its contents, in order to close the bleeding spot by the contraction of the uterus. When the os uteri is sufficiently dilated the woman is to be immediately delivered by means of the forceps if the head has descended, and if not, by turning and extraction. If the os is not as yet sufficiently dilated free incisions are to be made into it. If labour has so little advanced that an immediate delivery appears inadvisable ergot is to be given internally, and attempts are to be made by strong friction to make the uterus contract. If that is of no avail and the bleeding persists, and the patient is still collapsed, the membranes must be ruptured. By this the volume of the uterus is diminished, and if after the rupture of the membranes the uterus is energetically rubbed externally it will then duly contract and cause the bleeding to stop. Of course, there is the risk that the emptied uterus, if it remains flaccid, will only afford room for a larger quantity of blood.

There are other cases, where the blood escapes externally, whilst the ovum remains intact. The diagnosis then offers no difficulties, the prognosis is more favorable, and the treatment more simple.

Since the uterus is only very exceptionally capable of any great distension before a partial emptying of its contents, it is in general an almost absolute certainty that before the rupture of the membranes the hæmorrhage can be stopped by means of a tampon placed in the vagina. And this may be done with the greater unconcern when by friction of the abdomen a flaccidity of the uterus has been prevented. If, unexpectedly, the external hæmorrhage has been converted into an internal one, the same treatment is to be followed as stated above. If the os is sufficiently dilated it will be in the interest of the child to deliver immediately. Hæmorrhage occurring when the uterus is partially emptied of its contents has an essentially different clinical aspect. The placenta is then partially or entirely separated by the contractions of the uterus far more frequently than by the other causes above mentioned. We know that, physiologically, after the birth of the child, the placenta is separated by the contractions of the uterus. The same takes place very easily when the volume of the uterus is greatly diminished before the child is entirely delivered, and especially when very much liquor amnii was present, or the whole quantity has flowed off suddenly, or in twin cases after the birth of the first child. For the same reasons there is danger to the child in pelvic positions when the birth of the aftercoming head is retarded.

Under such conditions it is very rare that the hæmorrhage is internal, unless the lower uterine segment very tightly grasps the head. But even then the blood is during a pain forced out in jerks past the head, or at least a thick stream flows out when attempts are made to push the head somewhat backwards. Most frequently the bleeding is external, and then, certainly, there can be no difficulty in diagnosis.

By these hæmorrhages the life of the child is always much exposed, because usually a large part of the placenta is already separated, or the separation gradually increases. The prognosis for the mother essentially depends upon the efficacy of the treatment. Since after the partial emptying of its contents the uterus can easily again be distended to its previous volume, great caution is required in the use of the tampon after the rupture of the membranes, lest the external hæmorrhage be converted into an internal one. The tampon may, however, be employed when immediate delivery is impracticable. The re-distension of the uterus must be prevented by friction of the fundus and by the internal use of ergot of rye. If there is a possibility of an immediate delivery, the os being either sufficiently dilated or able to be artificially dilated in a way not too dangerous to the mother, the child is to be extracted. It is urgently necessary, even if the hæmorrhage is not very considerable, in the interest of the child; and if the hæmorrhage is profuse in the interest of both mother and child.

2. *Placenta prævia.*—Hæmorrhages far more frequently occur during labour when the placenta has an anomalous position on or above the internal os—placenta prævia—than when it has its usual seat.

This important and dangerous anomaly arises when the ovum is imbedded close to the internal os, so that the decidua serotina is developed there, and later also the placenta. When the ovum is inserted so close to the os that in the course of its growth the villi of the fœtal chorion entirely extend over the os, and the finger, easily passing through the cervix, can only feel within the internal os placental tissue, and nowhere membranes, it is called *placenta prævia centralis;* whilst if the placenta is felt on one side of the os it is called *placenta prævia lateralis.* The causes of the anomalous insertion of the ovum are, at any rate, too great width of the uterine cavity and abnormal smoothness of its mucous membrane. It is, *à priori*, evident that they must facilitate the descent of the ovum towards the internal os, and that it really is the case is shown by the circumstance that placenta prævia most frequently occurs in multiparæ, in whom the uterine cavity is wide, and after a previous leucorrhœa, during which the mucous membrane has become smoother.

If the placenta is situated centrally the vascular connexion, even over the closed os, is so loose that not seldom abortion sets in within the first few months. As soon as the internal os is somewhat dilated—that is, in the last months of pregnancy—hæmorrhages occur almost regularly. Hæmorrhages are far more rare at an early period of pregnancy when only a portion of the placenta extends to the internal os.

In the former case hæmorrhage usually occurs in one of the latter months, suddenly and unexpectedly, and without any known acute cause. It may be very profuse and cause syncope and even pulselessness; in other cases it soon stops spontaneously, and the patient again attends to her duties, until a fresh hæmorrhage brings fresh and usually increased danger. Sometimes the bleeding does not entirely stop, and a slight yet continuous flow of blood leads to a high degree of anæmia. Most commonly after hæmorrhage, if it is at all profuse, pregnancy is prematurely interrupted. As a rule, in the cases where the placenta has a lateral insertion, the hæmorrhage first commences when the pains begin; this only rarely occurs when the placenta is placed centrally.

The effused blood is derived from the vascular system of the mother, because at each separation of the placenta the large lacunæ must be opened up, in which the maternal blood circulates between the villi of the chorion. The fœtus itself only very rarely loses any considerable quantity of blood, for although the villi of the chorion may be torn, especially by an incautious examination with the finger, yet the fœtal vessels contained in the villi are so small that they cannot be the source of profuse hæmorrhage.

When the cervix is pervious for the finger the diagnosis is easily made, because the uterine surface of the placenta has a characteristic feel, which cannot easily be mistaken. There is sometimes hæmorrhage from a laterally situated placenta when it does not extend into the os. Then the thickened membranes are at first only felt, but when the finger is passed higher up it comes to the margin of the placenta, which is situated laterally. In profuse hæmorrhage, with the cervix firmly closed, placenta prævia must be thought of, and in the mean time, until by the dilatation of the cervix the diagnosis is rendered certain, such a treatment is to be followed as if it actually were present. Great swelling and thickening of the lower uterine segment increases the suspicion, but does not suffice for a sure diagnosis. The prognosis essentially depends upon the way in which the treatment is conducted. By suitable treatment, in the great majority of instances, the life of the mother is, at any rate, preserved, whilst, certainly, many children perish.

The danger to the mother lies in the loss of blood; the danger to the child in suffocation, which follows when the portion of the placenta not yet separated, but still continuing its function, does not suffice for the respiration of the child. This insufficiency occurs more easily when the separation takes place suddenly. It is certainly very rare that the child bleeds to death from the torn vessels of the chorion. The dangers are, of course, far greater to mother and child when the placenta is central than when one cotyledon only projects into the os.

When the hæmorrhage occurs during pregnancy, and the quantity lost is inconsiderable, the woman must remain quiet in the dorsal posture, and must not be exposed to injurious influences. If the quantity of blood lost is more considerable the tampon is applied to the vagina, either by means of the colpeurynter or, what is more sure, by pieces of lint, as already described. This treatment is continued or repeated until the bleeding stops. Very frequently uterine action is aroused by it.

During labour also, when the cervix is not yet dilated and the os not at all or only slightly open, the correct treatment is the application of the tampon to the vagina. But if the pains increase, and the os is so much dilated that two fingers can pass through it, attempts may be made in lateral insertion of the placenta to turn by one foot, as Braxton Hicks recommends. If this succeeds, all the indications are satisfied. The lower extremity within the cervix acts as a tampon from above upon the bleeding spot, and the more the os dilates the thicker the tampon becomes when the foot is drawn upon. The dilatation may even be hastened by drawing upon the foot, and delivery terminated in the shortest possible time without any more loss of blood.

In central placenta prævia one cotyledon is to be separated from the side of the os, and after the central has artificially been converted into a lateral the treatment is the same as just stated. Of course the placenta is separated at that side where the smallest cotyledon is situated, and if that cannot be made out it is separated at that side where the external examination shows the feet are placed, in order to facilitate version.

The treatment, therefore, of placenta prævia, when podalic version succeeds, is very simple, viz.—1. A tampon to the vagina until the cervix is so dilated that the foot can pass through it. 2. One foot is brought down by the combined manipulation, and gradually the foot is more strongly drawn upon.

If version does not succeed at this stage, and the fœtus is in the longitudinal position and the pelvis normal, the membranes are to be ruptured. This is done that the presenting part may be pressed by the pains against the os, thus acting as a tampon. If the pains are at all strong the hæmorrhage stops and delivery is terminated by the natural forces. We would not, however, recommend the rupture of the membranes at so early a period, because it is not absolutely certain that the bleeding is stopped by those means; and if it does not succeed, a tampon must be applied, which after rupture of the membranes may give rise to an internal hæmorrhage. We doubt whether this fear is as little justified as Kuhn supposes from his and C. Braun's experience.

It is therefore advisable to continue with the tampon without rupturing the membranes until the presenting head be firmly pressed against the lower uterine segment, thus stopping the hæmorrhage, or until the os is so far dilated as to admit the hand for turning and extraction. The lower uterine segment being generally soft and yielding in placenta prævia, this often succeeds very early. The *accouchement forcé*, that is, the artificial dilatation of the unprepared os by incisions (which in placenta prævia may easily give rise to dangerous hæmorrhage), or by other forcible manipulations, is always to be avoided. Instead of the usual tampon, Barnes's fiddle-shaped india-rubber bags, the smaller middle part of which is to lie within the cervix, may be used for its dilatation. Barnes and Elliot have obtained good results from their use.

It is always best for the mother, and especially for the child, to turn as soon as it is practicable, and only to omit doing so when the head is so forcibly pressed against the lower uterine segment as to stop the bleeding. When the os is sufficiently dilated, so that the head can enter it, it may be extracted, if necessary, by means of the forceps.

When the child is not in danger, or is clearly dead, it is in the interest of the mother that extraction, in whatever way it is carried out, be performed as slowly as possible. It is known that a rapid delivery, in the case of very great anæmia, may cause the sudden death of the mother (doubtless from acute cerebral anæmia). But if the life of the child is in danger it will then be necessary to extract it as rapidly as possible.

If the child is demonstrably dead Radford's and Simpson's plan deserves to be recommended. The placenta is to be entirely separated and to be delivered before the child, for it is known that after entire separation of the placenta the hæmorrhage almost always stops.

Immediately after the birth of the child attempts are to be made, as usual, to remove the afterbirth by gentle pressure externally.

If the woman is very much exhausted immediately after delivery wine must be liberally given. Syncope is best avoided by placing the head low. In extreme cases transfusion may become necessary.

3. *Hæmorrhage from the vessels of the umbilical cord.*—The velamentous insertion of the umbilical cord may also give rise to a peculiar kind of hæmorrhage during parturition; for if the cord is so inserted that its vessels run in that portion of the membranes which are situated within the os, the vessels may then be torn on the rupture of the membranes. The child often bleeds to death in consequence of this; but even when, after the rupture of the membranes, the presenting head firmly compresses the vessels, the child dies, nevertheless, of asphyxia, on account of the impediment to the exchange of gases. On examining the bodies of children who have thus died we find the signs both of anæmia and of suffocation.

This anomaly is recognised by a pulsating vessel of the size of a raven's quill, which is felt in the somewhat thickened membranes. The rupture of the membranes is then to be delayed as long as possible by the introduction of a moderately filled india-rubber bag; for if the membranes rupture whilst delivery is still impossible the child is lost if a vessel is torn; but if the os is sufficiently dilated when the membranes are ruptured the life of the child can be saved by immediate extraction.

B.—*Hæmorrhage in the Afterbirth Period*

Insufficient contraction of the uterus is by far the most frequent cause of hæmorrhage after delivery. Such atony of the uterus is especially observed after a rapid emptying of its cavity, whether artificially or naturally produced, also after a previous and very considerable distension. It therefore occurs after very rapid deliveries, too early turning and extraction, in hydramnios, and twins. The hæmorrhage is sometimes also due to general debility and too feeble development of the uterine muscles (either congenital or depending upon previous very difficult labours). In still other cases this atony is always met with in the labours of the same woman, without there being any plausible reason for it. Partial adhesions of the placenta to the uterine wall, which, however, are rarely caused by real connective-tissue bands, may also give rise to profuse hæmorrhage, because the separated places in the vicinity of the adhesions can only imperfectly contract.

The natural consequence of the defective or insufficient contraction of the uterus is that the placental vessels remain open. The blood flows from them, and the quantity is so great that the woman dies in a very short time. These hæmorrhages can very easily be recognised, even when the internal os is occluded by great anteflexion or by the placenta or by larger coagula, so that only a little blood escapes externally; for the uterus is so flabby that by palpation it is either not felt at all or only as a soft, flaccid body, whereas, normally it ought to be felt as a hard ball.

Very simple treatment almost always suffices. The flabby uterus is energetically rubbed with the hand. After slight friction it is distinctly felt to be harder; by continued friction, and, at the same time, by grasping it with the whole hand, it is brought to such a degree of contraction that the hæmorrhage is arrested. If from the reasons above mentioned the hæmorrhage was chiefly internal, large coagula, very frequently together with the placenta, are at first expelled by compression of the uterus, followed by a great rush of liquid blood, and then the bleeding ceases. There is, therefore, no need to be uneasy if, in consequence of the compression, the flow of blood at first appears to be greater, since the escaping blood formed the contents of the flabby uterine cavity, and was already lost to the system.

Although we almost always succeed in making the uterus contract in the way just described, so that the bleeding ceases, nevertheless as long as the placenta is not yet expelled the uterus is continually liable to again become flaccid, and to bleed during this flaccidity. It will, therefore, be necessary to practise Credé's manipulation until the afterbirth has been squeezed out. This always succeeds, except in the very rare cases when there are firm adhesions between the placenta and the internal walls. It is equally rare also, before the expulsion of the placenta, for the uterus not to contract by friction, and the hæmorrhage, therefore, to persist, or that the hæmorrhage continues after the uterus has contracted. In such cases the artificial manual separation of the placenta becomes necessary.

For that purpose, the woman lying on the back or the side, the hand, well oiled, with the fingers closely kept together, is introduced and passed along the umbilical cord to the insertion of the placenta. Whilst with the other hand counter-pressure is made externally, the adhering portions of the placenta are separated by the ulnar side of the little finger. If this fails, and if thick, cord-like adhesions are felt, they must be broken through with the finger-nails. The placenta, now being entirely separated, is taken into the hand, and thus removed. Sometimes, especially when previously ineffectual attempts have been made to remove the placenta, the uterus is irregularly contracted. The internal os then forms a stricture, through which the hand cannot pass. By a little patience, the use of chloroform in case of need, and cautious dilatation of the os, this impediment is always overcome.

When the expulsion of the placenta is delayed for many hours we must never wait for it to be expelled naturally. The afterbirth should be artificially removed, because the os closes more and more firmly. In some cases the retained placenta remains a remarkably long time well preserved, and is often later on expelled naturally; but, as a rule, it putrefies, and may consequently give rise to the most dangerous diseases during childbed.

The treatment of hæmorrhage persisting after the removal of the placenta will be given as we proceed.

4. Inversion of the Uterus

It very rarely happens in the period of the afterbirth that the uterus undergoes that peculiar change of position which consists in either a partial or a complete turning inside out, so that the inverted fundus passes through the os. The inversion is said to be incomplete as long as the fundus is still above the os, and complete when it has passed through it into the vagina. In the highest degree prolapsus of the inverted uterus takes place, so that the completely inverted organ lies outside the external genitals.

The period of the afterbirth is most favorable for the occurrence of this rare complication. It requires a dilated and flaccid uterus and a dilated os. In the presence of these conditions the fundus may be inverted by traction downwards or by pressure from above. It was formerly most frequently produced by midwives drawing on the umbilical cord, the placenta being inserted into the fundus; it may exceptionally occur in spontaneous labours, especially whilst standing or kneeling, when the weight of the child drags on the still firmly attached placenta. The uterus is very rarely inverted when Credé's manipulation is performed in an unsuitable way in the flaccid organ, or by the pressure of the abdominal muscles.

The sudden inversion of the uterus always leads immediately to great general disturbance; the heart's action is disturbed; syncope, convulsions, vomiting, &c., may sometimes be caused by the shock to the nervous system produced by the sudden change of the position of the uterus. More frequently those symptoms depend upon acute cerebral anæmia, to which the sudden emptying of the uterus of its contents already predisposes, but which must be still greater when not only the contents of the uterus, but the whole organ, itself passes out of the abdominal cavity; the blood then rushes into the vessels of the abdominal cavity, which are suddenly under a greatly diminished pressure, and the cerebral anæmia is due to the scanty supply of blood which the upper half of the trunk now receives.

If the placenta is still completely attached to the uterus there may be no hæmorrhage at all; but almost always it is partially separated, and in this insufficient contraction of the uterus the hæmorrhage is considerable, and may lead to immediate death. In other cases the hæmorrhage somewhat remits, or does not threaten life, and only the succeeding inflammation brings danger. If the os firmly contracts there may be strangulation of the fundus and gangrene.

A recent inversion of the uterus is recognised without any difficulty; the tumour felt in the vagina or seen in front of it is characteristic, and can only be mistaken for a polypus. By the combined examination, however, the uterus is found in its usual place in the case of a polypus, whilst in inversion it is absent from the abdomen.

Immediate suitable assistance renders the prognosis not unfavorable, though never entirely free from danger. In the absence of assistance the prognosis is very bad.

The treatment consists in the reposition of the uterus, which, as a rule, can always be performed without difficulty. If the placenta is not at all separated, or only to a slight extent, it is best to replace it with the uterus. If this does not succeed there is always time enough to separate it and replace the uterus without it. Reposition is easy as long as the os is not yet firmly contracted; otherwise great difficulties may be met with, and repeated trials under chloroform may become necessary. When reposition has been accomplished the hand is to remain within the uterus until it has sufficiently contracted. If the uterus remains very flaccid ergot of rye is given internally and the cold douche also is usefully employed. To prevent a return of the inversion the woman must remain for a long time in the dorsal posture and avoid every effort which would bring into play the abdominal muscles.

5. Convulsions of Parturient Women; Eclampsia

All kinds of convulsions may occasionally occur during parturition. Thus, epileptic women may have an attack during labour, or delivery may be complicated by hysterical convulsions. There may also be epileptiform seizures in consequence of meningitis or cerebral tumour, apoplexy, or from a profuse and sudden loss of blood. But there is a special kind of convulsions which most frequently occurs during parturition. Each attack is perfectly like an epileptic seizure; the convulsions, however, recur after a longer or shorter interval, and whilst after the first attacks, consciousness returns, as the convulsions continue stupor comes on in the pauses. This form of convulsions is called eclampsia.

Nothing definite is as yet known as to its etiology. After Lever, Devilliers, and Reynauld had observed that the convulsions were associated with albumen in the urine, Frerichs brought forward the opinion that eclampsia is the most prominent symptom of uræmia, that is to say, that in eclampsia the blood is charged with excrementitious matter which is usually excreted by the urine. As he was not able to demonstrate the presence of urea in the blood, Frerichs supposed that the urea was converted into carbonate of ammonia, and as such exerted the deleterious influence.

The principal support of this opinion was derived from the fact that in almost all cases albumen was found in the urine of the patients. At the post-mortem examinations the kidneys were in various stages of Bright's disease. In most cases the slight amount of the renal disease was out of proportion apparently to the grave symptoms which were said to have been produced by the anatomical lesion, and then it was said that the acute occurrence of the disease accounted for it. But lately, since much more attention has been paid to the examination of the urine of eclamptic patients, cases have accumulated where the kidneys were not diseased at all or only slightly so at the commencement of the seizures, and where the albumen appeared in the urine during the course of the eclampsia, so that there is reason to suppose that in some of the previous cases also the albumen found in the urine had appeared there only in consequence of the pains and of the attacks of the eclampsia. It is therefore evident that Frerichs' theory is insufficient to account for the whole number of cases. Halbertsma has lately modified that theory. He is of opinion that most cases of eclampsia are caused by the retention of the constituents of the urine, and this retention is not due to primary disease of the kidneys, but to the pressure of the uterus upon the ureters. Hyperæmia of the kidneys and diffuse nephritis favour the occurrence of eclampsia.

Traube has given a more satisfactory explanation of those cases of eclampsia in which the kidneys have been found to be healthy. He does not ascribe the so-called uræmic symptoms in renal disease to the retention of excrementitious matter in the blood, but believes that the loss of the albumen and the consequent hydræmia cause, by the simultaneous hypertrophy of the left ventricle, a greater pressure in the arterial system, and this leads to œdema of the brain, which shows itself as coma when the cerebrum is œdematous, or as convulsions when the middle portions are affected. The convulsions, therefore, are not said to depend upon the disease of the kidneys, but upon two other conditions—a deficiency of albumen in the blood and an increased pressure in the arterial system, which, though commonly produced by diseases of the kidneys, may under circumstances arise independently of them. Rosenstein has made use of Traube's theory of the so-called uræmic symptoms for the explanation of eclampsia, and it must be admitted that in the present state of science it gives by far the most satisfactory explanation of the whole series of cases.

The two conditions which, according to Traube and Rosenstein, are essential for the occurrence of convulsions are regularly present, at least to a very slight degree in parturient women. For a watery condition of the blood is peculiar to pregnant women, and during the pains also, since the muscles of the body are in a state of activity, the pressure is increased in the arterial system. If one or both of these conditions reach a very high degree there may be an attack of eclamptic convulsions. The hydræmia frequently becomes very intense in consequence of the loss of albumen through the kidneys, and it is therefore easily intelligible that eclampsia especially frequently (but not exclusively) occurs in parturient women who suffer from Bright's disease. The pressure in the arterial system may also be increased, according to Rosenstein, by the action of carbonic acid upon the heart, the quantity of which is increased by imperfect respiration.

According to this theory the pathological process of eclampsia is briefly the following:—An increased pressure suddenly produced in the aortic system, in cases where the blood is in a high degree of hydræmia, leads to hyperæmia of the brain. But on account of the watery condition of the blood œdema of the brain is the necessary consequence of this hyperæmia. The passage of water into the tissues, again, exerts a mechanical pressure upon the blood-vessels, and leads consecutively to anæmia of the brain. The effect of this acute cerebral anæmia is observed as an epileptic seizure.

We, finally, must not omit to state that the etiology of those convulsions which occur in the period of reproduction in the woman, and are here described under the symptomatic name of eclampsia, is by no means sufficiently known, and that it is highly probable that the clinical picture here called eclampsia may in future be broken up into several morbid processes, widely differing in an anatomical and a pathological point of view.

Eclampsia is not very frequent, and occurs once in about 500 deliveries. Primiparæ are chiefly attacked, and, as a rule, in such cases the uterus is very considerably enlarged and the uterine action greatly augmented. Œdema and loss of albumen during pregnancy especially predispose to it. Most frequently it occurs during parturition, but it may occur also during pregnancy or only after delivery.

Of 316 cases the eclampsia occurred 62 times during pregnancy, 190 times during labour, and 64 times in the lying-in state, and here it always occurred within the first two days.

The attack itself sometimes sets in quite suddenly and unexpectedly. In other cases premonitory symptoms—uneasiness, headache, nausea, twitchings, sudden vertigo (abortive attacks, so to say)—have preceded. The pulse is hard and full. The attack itself is just like an epileptic fit.

There is complete loss of consciousness, with widely dilated pupils and tonic and clonic contractions. There is foaming at the mouth, the tongue is bitten, and the respiration becomes stertorous. The attack sometimes lasts only one or two minutes, frequently much longer; the convulsions cease, perspiration breaks out, and after a comatose condition of varying duration the patient awakes, greatly fatigued and with pain in all the limbs, in order to go through a new attack after a varying interval. Not infrequently the pauses between the attacks, which may be very numerous (as many as thirty, or even seventy or eighty), become gradually shorter, the patients remain comatose also in the pauses, and at the height of an attack death may take place from œdema of the lungs or apoplexy. In favorable cases the pauses become longer, the attacks themselves more imperfect and shorter, until they finally cease.

The pains are usually very strong during eclampsia, and frequently delivery is terminated in an unexpectedly short time.

On post-mortem examination little that is characteristic is found. Most constantly there is anæmia and œdema of the brain and flattening of the convolutions; there are, besides, very frequently changes in the kidneys, from a congested condition up to the highest degree of parenchymatous nephritis.

The diagnosis of eclampsia is, as a rule, easily made. A single attack does not differ from an epileptic seizure; but in epilepsy there have previously been attacks, whilst in eclampsia a second and a third soon follow. Hysterical convulsions may be recognised from the history of the case, and the attack itself is different. There is never complete loss of consciousness, nor is it followed by coma. Apoplexy of the brain is followed by symptoms of paralysis, and convulsions caused by acute anæmia are also easily recognised.

The prognosis is, in the natural course of the disease, grave for the mother and bad for the child. The treatment about to be mentioned renders the disease less dangerous to the mother, but the life of the child is always greatly exposed to danger. In consequence of eclampsia the mother may suffer from kidney disease and mania.

There are two methods of treatment, viz. by the abstraction of blood or by the use of narcotics.

If the theory of Traube and Rosenstein be correct a sudden depletion of the vascular system, by which the pressure is diminished, must stop the attacks. From experience it is known that after venesection the quantity of the blood soon becomes the same through the serum taken from all the tissues, whilst the quality is greatly deteriorated by the abstraction of blood. A short time after venesection we shall expect to find the former blood-pressure in the arterial system, but the blood far more watery than previously. From this theoretical consideration it follows that abstraction of blood, if the above-mentioned conditions really cause convulsions, must be attended by an immediate favorable result, and under certain circumstances the whole disease may surely be cut short by it. But if all other conditions remain the same the blood-pressure will, after some time, again reach its previous height; the quality of the blood has, in the mean time, been greatly deteriorated, and consequently the danger of the disease will be increased. Experience speaks in favour of this opinion. Venesection has often given favorable and exceedingly rapid results; but frequently the attacks have soon recurred, and then taken a far more unfavorable course.

If it be assumed that an attack of eclampsia is due to an increased pressure in the arterial system, produced by great muscular activity during the pains, it follows that each attack, itself consisting of tonic and clonic contractions of all the muscles, still more increases the arterial pressure, the venous hyperæmia of the brain, and the consecutive œdema, so that an attack becomes in its turn again the cause of another attack. A rational treatment must, therefore, be directed to paralysing the activity of the voluntary muscles, and this can be obtained by the exhibition of narcotics. It is especially to be insisted upon that narcotics are given in such doses as not only to stop the movements of the woman during a pain, but to paralyse the muscles so that their contraction is impossible during an attack. In this way the convulsions are stopped with certainty for many hours if the narcosis is continued so long; and there is this advantage besides, that the blood itself is not deteriorated. Scheinesson has also shown by way of experiment that chloroform diminishes the blood pressure in the arterial system.

Experience is decidedly in favour of this treatment, only one must not be contented with having ordered a certain quantity of narcotics. Treatment is then only effectual when the narcosis is absolute, that is, when the patient is quite unconscious, so that the voluntary muscles no longer contract. As long as an eyelid quivers or the body is in any way thrown about, another dose is requisite. Such narcotism can be obtained by means of chloroform, but since it has to be maintained for some time (six to twelve hours) it is better to employ morphia. If we wish to procure narcosis as quickly as possible, chloroform may be first used and afterwards replaced by subcutaneous injections of morphia. The quantity of the narcotic need not be considered, for the narcosis must be absolute, and to obtain this result very different doses are required in different individuals. Recently, hydrate of chloral has also been used with advantage in eclampsia. It may be given internally, but the subcutaneous application recommended by Rabl-Reickhard appears to be preferable. Martin has given with good result two enemata of half a cup of mucilage of starch with thirty-two grains of hydrate of chloral in one ounce of decoction of Althæa.

It is self-evident that when narcotics are used all excitants, such as cold applications, irritating clysters, sinapisms, &c., are not only unnecessary but even injurious, for they retard the setting in of absolute anæsthesia. Venesection is certainly very seldom necessary, and may safely be omitted. If everything depends upon cutting short an existing attack or in counteracting a threatening one, venesection may successfully be employed. But it is always to be borne in mind that venesection deteriorates the state of the blood, and that it strengthens the determining cause of the attack.

No other treatment, and especially no kind of obstetric manipulation, is required for the safety of the mother. The condition of the woman contra-indicates obstetrical manipulations, and every irritation of the uterus, such as introduction of the hand for the purpose of turning, application of the forceps, or even mere friction of the fundus is apt to cause a fresh attack by reflex action.

In the interest of the child, however, active procedure may become necessary. Theoretically it is easily intelligible, and experience has confirmed it, that eclamptic seizures of the mother bring danger to the life of the child, and the more so the more frequently the attacks follow each other and the longer they last. If this danger can be avoided by the induction of narcosis, the latter itself may not be without danger to the child. Since even slight anæsthesia does the child harm, there is little doubt that complete and absolute narcosis, prolonged for hours, may prove fatal to both mother and child. If the child can be delivered easily by turning or by the forceps, it is certainly in the interest of the child to do so.

Appendix

I.—*Death of the Mother during Labour*

The mother may die during parturition from various causes. Exhaustion from too prolonged labour or too great exertion, perforating ruptures of the uterus, eclampsia, apoplexy, or a sudden loss of blood, may cause death more or less rapidly. Sudden death of parturient women from embolism of the pulmonary arteries or from entrance of air into the uterine veins will be considered in the pathology of the lying-in state. When labour has once begun it is desirable, even if the woman dies, to terminate it, whether the child be alive or dead. In the first case the termination of the labour is legally demanded. If it is practicable to deliver by turning and extraction by the feet, or by extraction by means of the forceps, it is to be done in that way; if that is impossible, the Cæsarean section is immediately to be performed. Even if the child be dead it is certainly best not to make the maternal body the coffin of the child, and medical politics require delivery to be accomplished, especially in those cases where very great difficulties have existed, and which have caused the death of the mother.

II.—*Premature Respiration and Death of the Child during Parturition*

In the previous chapters mention has frequently been made of dangers to the child without giving particulars as to the way in which those dangers arise. An accurate knowledge of the mechanism by which the fœtus dies in the uterus is of the very greatest importance to the obstetrician.

The tissue changes in the fœtus are the same as those in an adult. The organs and life of the fœtus can, then, only be preserved in their integrity when the organism is supplied with oxygen and assimilable nutritive material. For the various expressions of animal life depend upon the free access of oxygen to the highest organized combinations and upon their destruction into lower ones, by which latent chemical forces are set free. Now, man in extra-uterine life receives oxygen through the lungs and nutritive material through the intestinal canal; the fœtus is supplied with both by the maternal blood through the placenta and umbilical cord. The products of the retrogressive metamorphosis, at any rate carbonic acid, and probably also simple nitrogenous combinations, pass in the placenta from the blood of the fœtus into that of the mother, and in their stead highly organized combinations are received from the plasma, and oxygen from the red blood-corpuscles of the mother. If this connection with the maternal blood-vessels is interrupted at any one spot the fœtus is asphyxiated and starved; but because the want of nutritive material can be borne much longer than that of oxygen, the fœtus dies only with the symptoms of asphyxia. Normally, the fœtus exists in the liquor amnii in a state of apnœa; that is to say, it has no need to respire, for it receives the necessary amount of oxygen by some other channel.

Normally this state is altered only after the birth of the child. Though it must be admitted that the irritation produced by the cold air upon the moist skin may at times cause the first inspiratory movement, yet the necessity for respiration is not dependent upon that. The healthy child involuntarily respires very soon after birth, because the source of the supply of oxygen has ceased to exist when by the contraction of the uterus the placenta has been detached. Pfluger is correct when he states that not the accumulation of carbonic acid within the fœtal blood is the peculiar exciting agent of respiration, but the want of oxygen, in consequence of which the insufficiently burnt products of the tissue metamorphosis exert a very rapid and intense irritation upon the medulla oblongata.

In exceptional cases, such as the death of the mother, or a great loss of blood, or compression of the umbilical cord, or premature separation of the afterbirth, the placenta may have ceased to supply the fœtus with oxygen before it is born. Then just the same takes place as normally occurs after birth. The fœtus makes the first inspiratory movements within the uterus, and this is attended by the same immediate consequences as the normal respiration. First of all, the placental circulation is either entirely stopped or at least greatly weakened, because the blood pressure within the right ventricle, which is maintained by the passage of the blood through the ductus Botalli, is greatly diminished by the sudden and considerable enlargement of the area of the pulmonary artery; and, secondly, the medium existing in front of the respiratory openings of the child enters its air-passages. The important difference consists in this, that the medium, under normal conditions, is the atmospheric air, the oxygen of which oxidizes the fœtal blood within the pulmonary vesicles, whilst so long as the fœtus is contained in the genital canal this medium consists of liquor amnii, mucus, and blood. In the latter case, therefore, the oxidation of the fœtal blood in the lungs is wanting; fresh inspiratory movements follow, drawing the blood of the right heart into the thorax. The products of the retrogressive metamorphosis, neither oxidized in the placenta nor in the lungs, accumulate more and more within the blood, the irritability of the medulla oblongata decreases so that the usual irritations no longer suffice to produce respiratory movements, and on account of the paralysis of the cardiac nerves the heart's action stops—the fœtus dies.

Experience has shown that pressure upon the brain during labour may be attended by the most serious consequences to the child. It remains to be seen in what way these unfavorable results of cerebral pressure can be explained. It may well be doubted whether pressure upon the medulla oblongata so irritates it as to produce the first inspiratory movement; at any rate, prolonged cerebral pressure, through irritation of the vagus, slackens the pulse and diminishes the irritability of the medulla oblongata because the exchange between the maternal and the fœtal blood is impeded, and, consequently, the blood circulating in the fœtus is poorer in oxygen. By cerebral pressure, therefore, the child becomes comatose, and this may assume such a degree that the usual irritations are no longer able to produce inspiratory movements. The child is exposed to such a danger by compression of the head within a contracted pelvis or by the firmly compressed forceps. Intra-cranial hæmorrhage also may produce the same effect. If the blood is effused upon the hemispheres, and not in too great quantity, it is known by experience that it is unexpectedly well borne by the new-born child, the pressure not acting immediately upon the medulla oblongata. But an effusion of blood at the base of the brain is followed by very fatal results, and, since the pressure continues after birth, every attempt to save the apparently stillborn child by artificial respiration usually fails.

Inspiratory movements made within the uterus can be recognised in the dead body by the signs of death by asphyxia and drowning. By intra-uterine inspiratory movements liquor amnii, together with vernix caseosa, or mucus, blood, and meconium, which, as a rule, is voided at the commencement of the coma, enters the air-passages and is found at the post-mortem examination. It is very exceptional indeed not to find these foreign bodies, and this is the case only when the respiratory openings are closed by a piece of the membranes or by the walls of the genital canal.

Another very important condition is also observed in the dead bodies of these children. There is such intense hyperæmia of the lungs that it is always followed by an effusion of blood. In this active dilatation of the thorax the lungs cannot follow the traction of the walls of the thorax, since the want of air prevents the alveoli from unfolding, and thus blood is drawn into the thoracic space as if by a pump. All the branches of the pulmonary artery are overfilled and blood is extravasated from them—the so-called ecchymosis of Bayard. These extravasations are of various sizes (up to that of a lentil), and are situated beneath the pulmonary and costal pleura as well as beneath the pericardium, and are quite regularly found in children who die suffocated within the uterus.

The term "asphyxia" or apparent death, applied to the results of impeded diffusion of gases or to those of intra-uterine respiration, signifies only the highest degree of that state where the child shows no other signs of life but the impulse of the heart; and yet even the preliminary stages of the condition, for which there is no generally accepted term, are of the greatest moment. Even very slight cases, where the child has once only respired within the uterus and is born in full health, require careful attention on the part of the medical attendant. For with the first inspiration various foreign bodies, capable of producing fatal inflammation, may have entered the lungs. It is, therefore, advisable to designate all those transitions from full health where râles only indicate the presence of foreign bodies within the lungs up to the highest degree of apparent death as "premature respiration."

A timely diagnosis of premature respiration is of the greatest importance for the prognosis and the treatment. In spite of very valuable symptoms, the diagnosis is often not without difficulties. The impeded diffusion of gases causes, as a rule, by the irritation of the vagus, retardation of the frequency of the pulse. Since the sounds of the fœtal heart can be ascertained by means of the stethoscope, this proves a very valuable aid. But irritation of the vagus may follow from causes which do not as yet produce the first inspiratory movement. It has therefore to be settled what changes in the frequency of the pulse are sufficient to indicate that respiratory movements have taken place. It depends upon intra-uterine respiration, which greatly diminishes the energy of the placental circulation, and still more impedes the interchange of gases. Asphyxia without previous inspiratory movement is, indeed, possible, but at any rate developes very slowly and very seldom causes death.

It is very difficult to explain how the frequency of the pulse may be greatly diminished without previous inspiratory movements. Sometimes an ordinary pain greatly diminishes the frequency of the pulse; this is still more the case with strong pains occurring in rapid succession. It is impossible to state the definite number of fœtal heart's sounds which accompany premature respiration. What is useful to note is, whether the heart's sounds regain their previous frequency, or whether the retardation progressively increases. If at the commencement, before a pain, the heart's sounds were 150 per minute, during a pain 120, and if this proportion constantly decreases in about the following way—

140 to 110,
130 „ 100,
120 „ 90,
110 „ 80,

the child is certainly in danger. One single diminution in the frequency of the pulse, however, even below 100, does not prove anything. We will not omit to state that from our own experience we can confirm the observation of Hohl, Stoltz, Cazeaux, Kiwisch, and Hüter, that after disturbance in the interchange of gases within the placenta the frequency of the fœtal pulse is considerably increased. According to Schultze, this increase is due to a paralysis of the vagus, and is always preceded by a stage of irritation of the vagus, that is, a decrease in the frequency of the fœtal heart's sounds.

Another very important sign of the commencing coma is the escape of meconium. It is less due to the pressure of the descending diaphragm than to the relaxation of the sphincter produced by the coma. This sign, however, is of little value when in breech presentation the breech is in the small pelvis, for under those circumstances the meconium is mechanically pressed out. In head and transverse positions the escape of liquor amnii with meconium in it is so far important as it indicates the comatose condition of the fœtus, which most probably has already made respiratory movements.

The prognosis for the child in premature respiration is always grave. Even in the slightest cases, where children are born in apparently good health, the inspired foreign bodies in the lung tissue may produce a fatal lobular pneumonia. In other cases they may, by obstructing the air-passages, mechanically impede the access of air to the alveoli. If there be actually asphyxia the irritability of the medulla oblongata may be so small that only weak and slight inspiratory movements are made by which an insufficient quantity of air enters the lungs, or there are no respiration movements whatever, the heart's sounds become less and less frequent, and cease immediately after birth.

The prognosis is more favorable if a rational treatment be adopted. There are three indications for treatment, viz. (1) the child, must be brought as rapidly as possible into a position to inspire atmospheric air; (2) the inspired foreign bodies must be removed from the air-passages; and, (3) if the irritability of the medulla oblongata has been so weakened that no spontaneous inspirations or only very weak ones are made, the normal condition of the central organ must be restored by artificial respiration.

To satisfy the first indication, that treatment must be instituted which has repeatedly been mentioned in the general therapeutics; the two other indications can be more or less perfectly met in various ways.

The catheterization of the air-passages in the treatment of premature respiration is very useful and almost indispensable for the removal of inspired masses. The little finger is passed into the pharynx and the epiglottis pressed forwards, then the catheter is passed into the trachea, using as little force as possible.

The foreign bodies within the air-passages of the child are drawn into the catheter by applying the mouth to the end and sucking, so that they can be easily removed when the catheter is withdrawn, and this procedure must be repeated as long as the catheter still fills on sucking. Usually the irritation caused by the catheter produces spontaneous inspirations, and thus frequently artificial respiration is rendered unnecessary. If the child is apparently dead air is blown into the lungs with moderate force through the catheter, and again expelled by gentle pressure upon the thorax. This is to be continued until the child respires spontaneously.

The introduction of the catheter, as a rule, easily succeeds, and the inspired masses are very perfectly removed in that way. In those cases in which the liquor amnii contained large quantities of meconium it is advisable to use the catheter also on children in whom, though breathing spontaneously, yet the presence of râles shows foreign bodies within the air-passages.

It is certainly beyond all doubt that liquor amnii, blood, and mucus may be absorbed by the mucous membrane of the air-passages without noxious consequences. We also frequently see the child by a powerful expectoration eject those liquids with pieces of meconium; but it also happens that by premature respiration the pieces of meconium enter so far into the smallest bronchi, and so entirely fill them up that no air can pass by them, even by the most powerful inspiration. They remain there, and, as we have frequently seen in post-mortem examinations, cause lobular pneumonia in their vicinity. Such foreign bodies can only be removed by aspiration.

To satisfy both these indications Schultze recommends another method of introducing air into the lungs, and it is, doubtless, very efficacious. It consists of the following manipulations:—The child is so held between the legs of the accoucheur that the thumbs are placed upon the anterior surface of the thorax, the index finger in the axilla, and the other fingers along the back; the face of the child is turned away from the accoucheur. The child, thus grasped, is then swung upwards, so that the lower end of the trunk turns over towards the accoucheur, and, by bending the trunk in the region of the lumbar vertebræ, the thorax is greatly compressed. By such passive expiratory movements the inspired liquids pass abundantly out of the respiratory openings. A very powerful inspiration is then produced by extending the body of the child by swinging it backwards, so as to return it to its previous position. In this way expiration and inspiration are repeated until they become spontaneous.

Schultze's method is so far preferable that it can easily be followed out in all cases, and that inspiration and expiration are made quite in accordance with the natural process; but we are not so perfectly convinced that the inspired foreign bodies are so completely removed by it as by catheterization.

If no foreign bodies have been inspired, or if they have already been removed, artificial respiration can be performed also in another way.

Thus, whilst the child is bathed the thorax can be dilated by supporting the back, the head, pelvis, and arms, being allowed to fall backwards; a powerful expiration is then obtained by bending the child over the abdominal surface, thereby compressing the thorax. Spiegelberg's experience leads him to recommend Marshall Hall's method. The child is first placed upon the abdomen, and then brought into the lateral posture. The thorax is then dilated, and air enters the lungs. It is then again placed upon the abdomen, and thus the air, together with the foreign bodies, as Spiegelberg states, are forced out of the air-passages. These manipulations are repeated until normal respiration is established.

Pernice has recommended faradization of the phrenic nerves. By this, however, not only do the foreign bodies remain within the air-passages, but not even respiration is always established. There is, besides, an apparatus required, which is not always at hand. This method is inferior to that recommended above.

After respiration has been once established the child must be watched until it has gained its natural red colour, moves the limbs actively, and cries with a loud voice.

Literature.—Michaelis, Abhandlungen, &c. Kiel, 1833, p. 263.—Schuré, Procid. du cord. ombil. Strasb., 1835.—Kohlschütter, Quædam de fun. umb., &c. Lipsiæ, 1833 (s. Wittlinger's Analekten, I, 1, p. 142).—Hecker, Kl. d. Geb., I, p. 165, and II, p. 183.—Hildebrandt, M. f. G., B. 23, p. 115.—Massmann, Petersb. Med. Z., 1868, II, 3 and 4, p. 140.—Deneux, Essai sur la Rupture de la Matrice, &c. Paris, 1804.—Mme. Lachapelle, Pratique des acc., T. III, Mém. VIII.—Bluff, Siebold's Journ., 1835, B. XV, p. 249.—Duparcque, Hist. compl. des Rupt., &c. Paris, 1836, deutsch von Nevermann. Quedlinburg and Leipzig, 1838.—Lehmann, M. f. G., B. 12, p. 408.—Chiari, Braun and Spaeth, Kl. d. Geb., p. 184.—Radford, Obstetrical Transactions, VIII, p. 150.—M'Clintock, Dublin Quarterly Journal, May, 1866.—Schroeder, Schw., Geb. u. W., p. 160.—Winkel, Path. u. Ther. d. Wochenbettes, 2 Aufl., p. 50.—Klaproth, M. f. G,, B. 11, p. 81, and B. 13, p. 1.—Winkel, Path. u. Ther. d. Wochenbettes, 2 Aufl., p. 108.—Schroeder, Schw., Geb. u. W., p. 165 and 166.—Müller, Scanzoni's Beiträge, B. VI, p. 148 and 156, and B. VII, p. 201.—Kleinwächter, Prager Vierteljahrsschrift, 1871, B. 3, p. 14.—Schultze, M. f. G., B. 12, p. 241.—Hecker, Kl. d. Geb., I, p. 141.—Preiter, Ueber Dammrisse, D. i. München, 1867.—Winkel, Path. u. Th. d. Wochenbettes, 2 Aufl., p. 37.—Schroeder, Schw., Geb. u. W., p. 163.—Habit, Wiener Med. Wochenschrift, 1866, No. 39, 40.—Goodell, American Journal of Obstetrics. Vol. II, 2, p. 281.—Holst, M. f. G., B. 2, p. 81, u. s. w.—Simpson, Select Obstetrical Works. London, 1871, p. 177.—Seyfert, Prager Vierteljahrsschr., IX, 1852, B. II, p. 140.—Chiari, Braun and Spaeth, Kl. d. Geb., p. 151.—Greenhalgh, Obstetrical Transactions, VI, p. 140.—Kuhn, Wiener Med. Presse, 1867, No. 15, &c.—Fränkel, Berl. klin. Wochenschrift, 1870, No. 22 and

23.—Benckiser (Naegele), De Hæmorrh. int. part., &c., D. i. Heidelb., 1831.—Ricker, Siebold's Journ., B. 12, p. 506.—Hüter, Neue Zeitschr. f. G., B. 12, p. 48.—Hecker, Klin. d. Geb., I, p. 162.—Chiari, Braunand Spaeth, Kl. d. Geb., p. 183.—Hüter, M. f. G., B. 28, p. 330.—Cazeaux-Tarnier, Traité de l'art des acc., 7 éd. Paris, 1867, p. 771.—Hyrtl, Die Blutgefässe der menschl. Nachgeburt. Wien, 1870, p. 63.—Meissner, Frauenzimmerkrankheiten, 1843, B. 1, p. 732.—Crosse, An Essay, &c. Transactions of the Provincial Medical and Surgical Association. London, 1845.—Lee, American Journal of Medical Sciences, October, 1860.—Betschler, Breslauer kl. Beitr., I, p. 1.—Veit, Krankh. der weibl. Geschlechtsorgane, 2 Aufl. Erlangen, 1867, p. 342.—v. Scanzoni, Beiträge zur Geb. und Gyn., B. V, p. 83.—Litzmann, Deutsche Klinik, 1852, No. 19—31, 1855, No. 29 and 30, and M. f. G., B. 11, p. 414.—Wieger, Gaz. de Strasb., 1854, No. 6—12 (Schmidt's Jahrb., B. 87, p. 57).—Braun, Chiari, Braun and Spaeth, Kl. d. Geb., p. 249.—Hecker, Kl. d. Geb., B. II, p. 155, and M. f. G., B. 24, p. 298.—Rosenstein, Path. u. Ther. der Nierenkrankh. Berlin, 1863, p. 57, and M. f. G., B. 23, p. 413.—Brummerstädt, Bericht a. d. Rostocker Hebammenanstalt, &c. Rostock, 1866, p. 89.—Dohrn, Z. Kenntn. d. heut. Standes d. Lehre v. d. Puerperal-Eclampsie. Programm. Marburg, 1867.—Elliott, Obstetrical Clinic. New York, 1868, p. 1, &c.—Simon Thomas, Ned. Tijdschr. v. Geneesk., II Afd., 4 Afl., p. 321, 1869, s. Schmidt's Jahrb., 1871, B. 149, p. 290.—Hall Davis, London Obstetrical Transactions, Vol. XI, p. 268.—v. Mieczkowski, Fünfzig Fälle von Eclampsie, D. i. Berlin, 1869.—Spiegelberg, Arch. f. Gyn., B. I, p. 383.—Schroeder, Geburtshülfe, 3 Aufl.—Krahmer, Deutsche Klinik, 1852, No. 26.—Hecker, Verh. der Berliner geb. Ges., 1853, VII, p. 145.—Schwartz, Die vorzeitigen Athembewegungen. Leipzig, 1858, u. Arch. f. Gyn., B. 1, p. 361.—Boehr, Henke's Zeitschrift f. d. Staatsarzneikunde, 1863, 1 Heft, and M. f. G., B. 22, p. 408.—Pernice, Greifswalder Med. Beiträge, 1863, B. II, H. 1, p. 1.—Schultze, Jen. Zeitschr. f. M. u. N., 1864, B. I, p. 240, Virchow's Archiv, 1866, B. 37, p. 145, and Der Scheintod Neugeborener. Jena, 1871.—Poppel, M. f. G., B. 25, Suppl., p. 1.—For the treatment see Hüter, M. f. G., B. 21, p. 123.—Olshausen, Deutsche Klinik, 1864, No. 36, &c.—Stempelmann, M. f. G., B. 28, p. 184.—Schroeder Schw., Geb. u. W., p. 128.—Löwenhardt, M. f. G., B. 30, p. 265.—Spiegelberg, Würzb. Med. Z., B. V, p. 150.—Seydel, M. f. G., B. 26, p. 284.—Schultze, Jenaische Zeitschr. f. M. u. N., B. II, p. 451.

PART VIII

THE PATHOLOGY OF CHILDBED

In the present section all those diseases of childbed will be considered which have a causal connection with the processes occurring during parturition or regularly after it, viz. the changes in the genital organs and the development of the mammæ for the maintenance of the child.

To speak of the affections of each organ separately would appear the best course to follow in the description of these diseases. But they make up together a group of diseases which is distinguished by its acutely fatal course or at least by a tendency to such, and they are all dependent upon one common cause. They are due to infection by septic material, and are commonly comprehended under the one name of "puerperal fever."

These affections are of such great practical importance, that it is advisable to give a uniform and comprehensive description of them from their common etiological point of view. Of course, this would be a deviation from the plan which it was proposed to follow in arranging the matter for the present work, viz. to consider separately the affections of each organ. Thus, whilst in the former parts the various diseases of each organ have been spoken of, now the pathological condition of various organs are taken together as being due to one common cause. The importance of the subject will serve as our excuse for thus attempting to describe the uniform aspect of the diseases of childbed depending on septic infection.

CHAPTER L

PUERPERAL FEVER

Undoubtedly cases have occurred at all times where lying-in women have been attacked by septic infection and died, and mention of such is made by even the most ancient writers. Hippocrates gives the history of cases which, in all probability, were due to it, and so do Galen, Celsus, Avicenna, and other authors down to the seventeenth century.

But epidemics proper are only mentioned since special lying-in hospitals or special departments in general hospitals have been established. The first institution of that kind, in which men like Mauriceau and De la Motte received their obstetric education, was established in the Hôtel Dieu of Paris. Peu tells us that the mortality amongst lying-in women in that institution was sometimes immense, and especially in the year 1664. At the post-mortem examination the bodies were found to be full of abscesses. De la Motte also mentions an epidemic in the Hôtel Dieu in 1678, and another at the commencement of the eighteenth century in Normandy, especially in Caen and Rouen. In other towns also, where special obstetric departments had been established, the epidemic occurrence of puerperal fever was soon observed. Thus, in 1750 in Lyons; 1760 and 1761 in the British Hospital, London, and in a small private lying-in institution; 1765 and 1766 in the Copenhagen Lying-in Hospital; and in 1767 in that of Dublin, ten years after its establishment. In Germany the first epidemic was observed in 1770 at the Hospital St. Marx of Vienna. In 1772 it occurred in Edinburgh, 1777 in Berlin, 1781 in Cassel, &c. &c.

The greatest ravages were caused by puerperal fever in the Maternité (the separate obstetric department of the Hôtel Dieu) where it raged year after year. In 1829, of 2788 lying-in women, 252 died; in 1831, of 2907, 254 died. In February, 1831, six to seven women were delivered in one day, all of whom died. In Vienna also puerperal fever raged extensively. In March, April, May, 1823, out of 698 lying-in women, not less than 133 died (19 per cent.), and every two days three deaths occurred; in 1842, in the Vienna hospital out of 3287, 518 died (almost 16 per cent.); in 1846, of 4010, 459 died; in 1854, of 4393, 400 died. From these few data, which can easily be multiplied by a great number of similar ones, it will be seen with what severity puerperal fever has sometimes raged in lying-in hospitals.

It would lead us too far to consider all those theories which have been brought forwards to explain the occurrence of puerperal fever, were it only from an historical point of view. Eisenmann and Silberschmidt discuss the subject very completely. More recently two opinions especially have contended for supremacy. According to the one, puerperal fever is due to a miasma formed by the crowding together of puerperal women, and according to the other it is due to the absorption of septic material. The purely miasmatic theory that puerperal fever is due to infection with a specific material formed under atmospheric, cosmic, and telluric influences, which, acting exclusively upon puerperal women, causes puerperal fever, so that puerperal fever rarely becomes a malarial fever—is quite untenable, and now almost universally abandoned. The opinion, however, is still somewhat prevalent, that puerperal fever is, like typhus of miasmatic origin, and that under suitable conditions the diseased organism may reproduce the virus and propagate it to other predisposed individuals without the original miasma being still active; so that in the course of the miasmatic disease a contagium is generated and propagated to others as the determining cause of the disease.

Latterly, the view that the origin of puerperal fever is due to the absorption of septic material from the surface of the wound has gained more and more adherents, and this view will be accepted in the present work. It is based upon a large number of observations, which, as Hirsch remarks, "have partly, at least, the conclusive evidence of experimental demonstration; and if, in the etiological research, with a correct appreciation of the known facts, the mathematical precision of a proof is for the present renounced, there are few questions of etiology which are able to be solved in a more unbiassed and certain way." In fact, whoever, after he has carefully perused the treatises of Veit and Hirsch on the subject, still doubts the possibility of the origin of puerperal fever by the absorption of decomposed organic materials, is not be convinced. It was, therefore, thought unnecessary to give even a digest of the numerous reasons and observations, and only a few striking cases will be related, which differ from experiments performed on animals only in so far that they were performed unconsciously, and not with the intention to infect.

The opinion which we follow originated with the English, and has been still furthered worked out by German physicians.

Denman was the first who contended that physicians and midwives who attended to cases of puerperal fever propagated that disease to other puerperal women. A great number of proofs of manual propagation soon accumulated in England, and many observations were made known where puerperal women had been infected by accoucheurs who attended, not only puerperal fever cases, but also cases of phlegmonous erysipelas and of ichorous wounds. It had, therefore, become the custom with some English physicians to give up practice for some time, if one of their puerperal patients suffered from puerperal fever. Semmelweiss has the merit of having worked out the subject carefully and pointedly, with special reference to lying-in hospitals, and he deserves to be mentioned amongst the first of the benefactors of man. In 1847 he first stated that puerperal fever depended upon an infection with cadaveric poison, an assertion which can easily be seen to be onesided and insufficient; but he himself corrected his opinion afterwards, and what is now known of the etiology of puerperal fever is essentially due to Semmelweiss.

A.—*Definition and Origin of Puerperal Fever*

Under the term puerperal fever we place all those diseases of puerperal women which are caused by the absorption of septic matter, that is, organic substances in the process of decomposition.

That absorption may take place, a fresh wound is required by which the septic poison can enter. Through the intact skin or mucous membrane, through the lungs or intestinal canal, septic materials, as a rule, never enter the blood as such, although a long sojourn in an atmosphere loaded with gases the product of organic decomposition can give rise to chronic illness. But fresh wounds exist in every puerperal woman. In every one some of the maternal blood-vessels are opened up by the detachment of the placenta, and in almost all there are slight lacerations of the cervix and vulva.

The sources from which the infecting matter is derived are chiefly twofold; either it belongs to the infected organism itself—auto-infection, or it is introduced from without—external infection.

Auto-infection is possible in all cases in which parts of the maternal organism decompose during or immediately after delivery. This decomposition occurs very readily and rapidly in disintegrating new growth, as carcinoma of the cervix; it may also be produced by mortification of the maternal soft parts which have been exposed to great pressure, or through pieces of the membranes or placenta which have remained within the genital canal. The latter, however, is a comparatively rare cause of auto-infection, because, to produce the infection a recent solution of continuity is required, which, as a rule, does not occur a few days after delivery, when the ichorous decomposition begins. The wound at the placental insertion is then closed, and the lacerations of the mucous membrane have united by first intention, or have become granulating ulcers; but the proliferating granulations prevent absorption. Under exceptional circumstances a wound may remain capable of absorption, or a fresh excoriation of the mucous membrane, or the destruction of the granulations, may lead to inoculation.

The most favorable conditions for auto-infection exist at the time when the wounds are fresh, that is, when at the birth of the child there are already decomposed materials. This is especially the case (1) when the dead fœtus, still contained in the uterus, has, after the rupture of the membranes, been exposed for a long time to the influence of the atmosphere. (According to our experience, a macerated fœtus, if the access of air has been prevented, does not infect.) (2) When the pressure upon the maternal soft parts has lasted so long that gangrene sets in before delivery is terminated. (3) In carcinoma of the cervix, where the new growth readily undergoes putrefaction. Auto-infection rarely takes place when the bruised maternal soft parts decompose somewhat later, since the recent wounds have then ceased to be capable of absorption. This is so because the mortified parts are separated from the healthy tissues by the line of demarcation which does not absorb. The partial gangrene of the hymen sufficiently proves that, as a rule, the line of demarcation prevents the absorption of putrefying substances by the neighbouring healthy tissue. In every primipara it can be seen that by the pressure of the child's head some portions of the hymen are changed into a bluish-black mass. These portions mortify in a few days; and later, instead of the prominent lips, small ulcers are found. But since the gangrenous parts are thrown off a few days later infection does not follow. Billroth has shown by experiments that granulating sores do not absorb as long as the infecting agent does not destroy the granulations.

Infection from without takes place when septic materials are brought to the recent wounds of the genital organs by means of a sponge or of linen used for cleaning the parts, or by instruments, and very frequently also by the examining finger. It is possible also that septic substances floating in the air of the lying-in room may come in contact with the recent wounds; yet those substances never exist as gaseous miasms, but rather as organic compounds suspended in the air. On the whole, there are no cogent reasons for such an assumption.

The infecting matter is derived from a great variety of sources. It is formed everywhere where organic compounds decompose; therefore it is derived from dead bodies, from suppurating wounds, disintegrating neoplasms, and especially from the secretions of diseased and sometimes also of healthy puerperal women. Puerperal fever, therefore, is nothing else but poisoning with septic matter from the genital organs.

This definition first of all shows that there is nothing specific in puerperal fever. Where the products of putrefying organic matter enter living tissues the consequences are essentially the same. Puerperal fever is quite the same state which is frequently observed in surgical wards, and designated as erysipelas, pyæmia, ichorrhæmia, and septicæmia. A specific difference does not exist, although there are modifications in the symptoms; but they are due in a greater measure to the peculiar place where the septic material enters, and in a lesser degree to the changes of the genital organs in the puerperal state.

Infection with septic materials arising from the female genital organs occurs as a rule only in the puerperal state. Exceptionally it is seen also after gynæcological operations. It is then followed by the same changes as in the puerperal state, as is shown in the very interesting cases communicated by Buhl. Changes were found in every respect like those following puerperal fever in the post-mortem examinations of two girls upon whom episioraphy was performed, and in two others in whom the vaginal portion of the uterus was amputated on account of carcinoma.

That there is nothing specific in puerperal fever is shown also by the fact that puerperal women can be infected and puerperal fever produced in them by patients with phlegmonous erysipelas or with suppurating and ichorous wounds. It has also been observed in lying-in hospitals that, at a time when puerperal fever was prevalent, phlegmonous erysipelas developed in pregnant women and nurses, who, having slight excoriations, came in contact with the diseased puerperal women. The septic infection can also be transferred to the newly born child, and its action is most frequently favoured by the umbilical wound. The consequences are erysipelas of the abdominal walls, disintegration of the thrombi of the umbilical vessels, even peritonitis and inflammation of the subperitoneal connective tissue and ichorrhæmic metastasis into other organs. It is rather common to see during an epidemic of puerperal fever phlegmonous inflammations arising from small excoriated places on the hands and feet of new-born children. From all these facts we may conclude that puerperal fever is really not contagious, for by a contagious disease is meant one in which a specific poison is produced within a diseased organism, and which, transferred to other individuals, always produces the same specific disease, such as measles, scarlatina, smallpox, syphilis, &c. Although the secretions of puerperal fever patients transferred to other puerperal women may in them produce puerperal fever, there is nevertheless nothing specific in the secretion. Those secretions have only all the characters of the products of decomposing organic compounds, and the phlegmonous inflammation and the septicæmia which they produce in any wound exactly corresponds to the puerperal fever of puerperal women. If, therefore, puerperal fever cannot be considered contagious in the usual sense of the word, yet it must be admitted that it is manually transferable.

B.—*Pathological Anatomy of Puerperal Fever*

We shall first consider the changes at the place of infection, and the way in which the disease extends by continuity of the tissues.

It has already been stated that the infection may start from various places. No doubt the part where the placenta was inserted is able to absorb septic matter, yet this rarely takes place, because the infecting agent does not commonly reach it.

Infection takes place much more frequently through the lacerations of the cervix. They are, though slight, scarcely ever absent in any case of labour, and the examining finger, as the bearer of the infection easily passes up to them. In the great majority of instances the infection is favoured by the slight rents in the mucous membrane, to which all primiparæ and most pluriparæ are subject when the head appears in the vaginal outlet. In every case of labour the examining finger of the accoucheur or midwife frequently passes through this portion of the vagina. The septic matter brought by the finger is easily deposited there, and, if slight solutions of continuity exist in the mucous membrane the poison is absorbed. It only exceptionally occurs that infection is produced by the absorption of the mortifying portions of the maternal soft parts by the surrounding living tissues; it is met with more frequently when the wounds are recent over which the putrid secretion passes.

Very often the edges of the wound begin to ulcerate, forming the so-called puerperal ulcer. This is, of course, situated at the place of the wound, usually, therefore, in the vulva, on both sides of the labia, or behind the frenulum. Lacerations of the perinæum also are easily changed into ulcers, and heal up later on by second intention. We often find, however, ulcers in the vulva, whilst the lacerations of the perinæum have healed by first intention. Then the inoculation had not taken place at the perineal wound, and it was able to unite before the putrid secretion passing over it could convert it into an ulcer. The same can be seen in the lacerations of the cervix, and in considerable endometritis also at the placental insertion. The ulcer has tumid edges, the base is covered with a dirty yellow coat; it has a tendency to extend, and heals only very slowly when the coating is thrown off and fine granulations appear at the base. Still more advanced changes, in the cervix and at the placental insertion discovered at the post-mortem examination, show the ulcers to be covered by a diphtheritic, brownish-green scar.

If the inflammation extends along the surface the mucous membrane of the vagina becomes swollen, and the submucous tissue is the seat of a more or less extensive œdema. The mucous membrane in colpitis feels soft and infiltrated; the greatest swelling is around the ulcers, and usually considerable œdema of the labia pudendi is met with. The process is the same as takes place in the skin, and is known as erysipelas, and Virchow, therefore, has called it *erysipelas malignum puerperale internum.* It is not improbable that the same process sometimes extends also to the labia pudendi, the nates, and the neighbouring parts of the thighs; yet in women recently delivered the diagnosis is very difficult, as those parts are already, as a rule, of more than a rosy colour. The great œdema of the labia and the sloughing which sometimes arise after a few days, attended by high fever, favor that view. Then between this inflammation, extending along the surface of the mucous membrane and the phlegmonous process, attacking the subjacent connective-tissue layers, there exists no radical difference. Both consist in the infiltration of the tissue by small cells (derived, according to former views, from the proliferation of the connective-tissue cells, and, according to Cohnheim's observations, due to the migration of the white blood-corpuscles from the vessels), attended by great hyperæmia and serous transudation. Volkmann's and Steudener's researches have shown that in simple erysipelas this process does not remain limited to the rete Malpighii, but that it extends into the subcutaneous connective tissue, as in the phlegmonous variety of erysipelas.

The inflammation produces only palpable changes in the uterine mucous membrane. When it is very intense in the normal puerperal state the inner surface of the uterus has the appearance of intense catarrhal inflammation. When portions of the uterine mucous membrane rapidly mortify the endometritis assumes a diphtheritic character. The aspect of the mucous membrane varies according to the extent and the depth of that necrosis. If large shreds of the decidua have remained behind they become œdematous and form rounded projections.

The superficial layers mortify in patches or in larger masses, and between the normal mucous membrane yellowish-brown places can be seen, from which masses of detritus can be scraped off with the knife. The place of the placental insertion mostly projects into the uterine cavity, but is, on the whole, little changed. If the entire inner surface of the uterus becomes necrosed, there will be found everywhere, according to the state of the serous transudation into the organ, either brownish particles or a smeary mass of chocolate colour, after the removal of which the deeper layers of the mucous membrane or the muscular fibres themselves are laid bare. The place of the placental insertion also participates in these changes. It is then covered either by a thick scar or after the disintegration of the thrombi it shows the smooth muscular layer and the gaping openings of the veins.

Under such conditions the uterus itself is never quite unaffected. It is either only slightly contracted or its whole substance is œdematous. In the severe forms of the disease we almost always find, either in the uterine parenchyma or close to its margins, distended lymphatics with purulent contents, the origin of which cannot infrequently be traced to the unhealthy ulcers of the cervix. The lymphatics are sometimes found to have partial dilatations of the size of a nut, which are filled with pus. They look like abscesses in the uterine parenchyma, and can only be distinguished from such by their smooth walls, since the afferent and efferent vessels cannot always be found.

As a rule, the inflammation does not extend to the mucous membrane of the Fallopian tubes, and even in very intense affection of the lining membrane the tubal mucous membrane remains normal, or only in a state of slight catarrhal inflammation. Occasionally a purulent salpingitis occurs, and this may lead to peritonitis, either by the propagation of the inflammation or by rupture of the tube through purulent effusion. However, in a great many cases the peritonitis arises from other causes.

If the whole of the lining membrane of the uterus is converted into a pultaceous mass the parenchyma also becomes implicated in the process. If the endometritis is more intense there are usually also changes in the uterine parenchyma which constitute metritis. They consist of an œdematous condition and a cloudy swelling of the whole organ. The uterus is then badly contracted, and is so soft that the pressure of the intestinal coils resting upon it leave their impression. The transudation is either purely œdematous, more frequently, however, it is cloudy, albuminous, of a finely granular aspect, sometimes also tinged with blood. If the ichorous endometritis extends deeper, a portion of the uterus also mortifies (putrescentia uteri), and this may lead to perforation into the abdominal cavity.

In other though very rare cases a copious proliferation of cells in circumscribed spots of the uterine parenchyma leads to the formation of abscesses, which terminate in caseous inspissation or in perforation. They are distinguished from the lymphatic dilatations filled with pus by the absence of a smooth lining membrane and by their more winding and less round shape, but have, no doubt, been frequently mistaken for them. Although under certain circumstances the inflammation may be propagated from the uterus to the surrounding connective tissue, yet parametritis usually arises in another way. The inflammatory process due to the infection extends less along the surface of the mucous membrane, but follows from the place of infection (which is, as a rule, the vaginal outlet) the course of the connective tissue situated around the vagina and uterus.

This affection of the connective tissue is perfectly identical with the phlegmonous process not infrequently observed in the connective tissue of the extremities. It consists in an acute inflammatory œdema. The connective tissue swells and becomes somewhat cloudy and opaque; in severer forms the swelling increases, the whole tissue is tumid, and their interstices filled with serum, frequently also with a gelatinous, semi-coagulated material. There is besides a copious infiltration of the tissue with small cells.

This diffuse acute œdema occupies all the pelvic connective tissue. It commences, as a rule, around the vagina, extends higher up, and also attacks the whole subperitoneal tissue, so that the peritoneum itself appears to be somewhat raised from its basal structure. It may also extend to the iliac bones, behind to the connective tissue around the kidneys, even to the diaphragm, and to a large portion of the anterior abdominal wall. More rarely the process extends to the connective tissue accompanying the larger vessels and nerves of the lower extremity, and produces the swelling of the thigh known as phlegmasia alba dolens. The most considerable swelling is usually under the portion of the peritoneum lining the pelvis, for here all the conditions appear to favour the swelling. After delivery the circumference of the uterus has greatly diminished, and therefore the pelvic portion of the peritoneum is still too large, so to speak, for the pelvic organs, although the elasticity of the membrane prevents the formation of folds. The cellular tissue between the two layers of the ligamentum latum is, on account of its looseness (during pregnancy the broad ligaments unfold and cover the sides of the uterus), especially liable to swell from the inflammatory œdema. As these places are easily accessible to the combined external and internal examination, the diagnosis of inflammation of the pelvic connective tissue can be made from the condition of the broad ligaments. The peritoneum, however, adheres closely to the uterus, and a considerable exudation is therefore impossible.

In mild uncomplicated cases the process always terminates in recovery, and the œdema rapidly disappears. Where the formation of the cell-element is slight hardly a trace of a change is left behind. If, however, cellular elements accumulate in larger masses they, as a rule undergo fatty degeneration; and whilst the liquid constituents are rapidly absorbed, a hard tumour remains, consisting of finely granular detritus, which under favorable circumstances may also be absorbed in the course of a few weeks. In comparatively rare cases the cellular infiltration becomes so thick at a limited spot that an abscess forms, which terminates in the way described below.

A still more intense infection may be followed by a kind of necrotic softening of the subserous connective tissue. The peritoneum is then found to be raised from the subjacent structures, the connective tissue itself infiltrated with an opaque serum, of a dark brownish-red or chocolate colour, due to an imbibition with hæmatin and to putrefaction, evolving a putrid odour. The necrotic connective tissue may also be thrown off, as has been observed in a case where the patient succumbed to a putrid pulmonary abscess, whilst an intense general peritonitis tended to recovery. The retroperitoneal connective tissue above the left ilio-psoas muscle was found as a white necrosed shred about one foot long, perfectly free, in a cavity beneath the peritoneum.

In many cases of parametritis thrombosis of the lymphatic vessels is found within the inflamed spot. The coagulated lymph either uniformly fills the vessel or gives the appearance of a string of beads. Sometimes also single larger dilatations of lymphatic vessels are seen, like those of the uterus described above. The thrombosis may be due to the direct influence of the infecting matter, but more frequently it is caused by the inflammation of the connective tissue around the vessel. The products also of the inflammation of the connective tissue have a tendency to coagulate, and the contents of the lymphatic vessels participate in the process.

The significance of the thrombosis of the lymphatic vessels has been variously interpreted. Whilst Hecker and Buhl ascribe to the inflammation of the lymphatics all the pernicious characters of puerperal fever, Virchow, on the contrary, was the first to show that the lymphatic thrombosis is to some extent a favorable incident, since the occluded vessels are thus prevented from transporting the infecting substances; and, in fact, the nearest group of lymphatic glands is alone found to be inflamed; the inflammatory process is there at least delayed, and the rest of the lymphatic system is in this way shielded, as it were, against infection.

The lymphangitis, therefore, cannot any longer be considered the principal and essentially pernicious alteration. It is an accidental change, which usually remains limited to the spot of inflammation, and may heal. Very rarely the thrombosis extends further upwards towards the thoracic duct; if so, there are always other very considerable changes. We have seen a healing process going on in the thrombosed lymphatics in the body of a puerperal woman who died after a general peritonitis had begun to subside. The contents of the dilatation were no longer purulent, but formed an inspissated yellow ball.

In more intense parametritis the ovaries also, which are in direct continuity with the connective tissue, participate in the inflammation, and oophoritis developes. It is, however, almost always of minor importance, an affection of secondary consideration, in so far as it does not materially determine the character of the danger. In cases of putrid decomposition of the subserous tissue the whole ovarian stroma is sometimes seen to be entirely disintegrated, and on incision of its coat the whole contents escape as discoloured serum from a cyst. In such cases all the other changes are also very considerable, and the affection of the ovaries is then, if not less in intensity, at any rate less in extent. Only very rarely abscesses form in the ovaries, which perforate either early or acquire an enormous size and later perforate externally or into adhering neighbouring organs.

The neighbouring layers of the uterus also become in considerable parametritis the seat of an inflammatory œdema. This is distinguished from that of endometritis by the pathological changes being more pronounced in the external layers.

As in parametritis the connective tissue close beneath the peritoneum is affected, it is easily intelligible why the serous membrane itself becomes implicated in the inflammation. In a less virulent infection, and in slight and very gradual swelling, pain, the surest sign of the implication of the peritoneum, may be absent. If the swelling is more considerable the dragging upon and the changes in the position of the peritoneum cause an irritation. Symptoms of perimetritis follow, or, more correctly, of pelvic peritonitis, as the peritoneum covering the uterus is less affected than the other portions lining the pelvis. In fatal cases the inflammation rapidly extends to the peritoneum, and soon a general peritonitis is established.

Pelvic peritonitis, as a rule, consists only in an inflammatory irritation of the serous membrane without much exudation, or pseudomembranes are formed, which lead to adhesions between the organs contained in the pelvis. Their cicatricial shrinking may cause a change of position of the organs, and give rise to a variety of complaints. More intense inflammation may cause intra-peritoneal suppuration, which, when encapsuled, is only very gradually reabsorbed, or the inflammation extends to the whole peritoneum.

General peritonitis arises most frequently in the course of parametritis or pelvic peritonitis in the way described above, and is rarely a consequence of ichorrhæmia. It rarely follows endometritis, so that the inflammation is propagated to the peritoneum through the parenchyma of the uterus or the Fallopian tubes.

In recent and relatively mild cases the whole peritoneum, and especially that enveloping the intestinal coils, is finely injected, and the contents of the abdominal cavity are loosely adherent to one another by means of pseudo-membrane. The exudation is sometimes small in quantity, almost purely serous, and containing few pus-corpuscles. In other cases, again, there are limited patches of pus, or thick yellow membranes of coagulated fibrine have formed, which everywhere cover the organs. The liver has usually a thick coating, and the uterus, except where it is in close apposition to the intestinal coils, is covered by the exudation. The intestines are distended on account of the meteorism and the diaphragm is pushed upwards.

In the most fatal cases due to ichorous parametritis there is no fibrinous exudation. The abdominal cavity contains a thin, brownish, discoloured fluid, of a very putrid odour. The intestinal coils have a dark, brownish-red appearance, the same as in incarcerated hernia. These cases always terminate in death. If the exudation, however, is serous, purely fibrinous, or purulent, recovery may take place. The exudation is then either reabsorbed or encapsuled, or it is gradually inspissated; but the pus may sometimes perforate the intestine, and, from the subsequent escape of fæcal matter into the abscess, a putrid peritonitis follows, or, whilst the pathological changes undergo a retrogressive metamorphosis tending to recovery, a fresh exacerbation causes death.

The peritonitis may also extend through the diaphragm to the pleura. More frequently, however, the pleurisy, as well as the peritonitis itself, is the consequence of ichorrhæmia.

We must now consider the way in which the whole organism is infected, that is, the changes which take place generally, and are not the result of the inflammation spreading by continuity of tissue. It must be borne in mind that the inflammation of more distant organs in puerperal fever arises in the same way as it does in infection after injuries of other parts of the body, and as in surgical diseases.

Experience has shown that in cases of intense septic infection death may take place in a very short time, and that the post-mortem examination reveals no other distinct macroscopical changes than that the blood is dark and non-coagulable, and ecchymoses into various tissues. By an examination of the elementary structure of any organ the commencement of an acute inflammatory process is seen, viz. fine granular infiltration (the so-called cloudy swelling), fatty degeneration, or even disintegration of cells. The experiments of C. O. Weber, Billroth, and many others, have shown that septic matter has pyrogenetic and phlogogeneous properties, that is, it is capable of producing fever and local inflammation. The theory, therefore, is that in cases of acute septicæmia such a quantity of septic matter has been absorbed that the blood has received phlogogeneous properties, and that it is able to produce inflammatory changes wherever it goes. Such a general inflammation of the whole organism, and especially of those organs, the undisturbed function of which is necessary for the preservation of life, must be able to destroy life before more palpable changes have developed in individual organs. Accordingly, in such cases, functional disturbances of the organs are alone observed during life, and after death only the commencement of parenchymatous inflammation of those organs, the cloudy swelling of the cells.

In other cases the infection of the blood is not so intense; fever is the sole symptom of the general disturbance; the functions of the organs important to the maintenance of life are not so disturbed that death must inevitably follow. If infecting matter has only once been absorbed into the blood the disturbances caused by it soon pass off, as shown in numerous experiments on animals; the poison is rendered innocuous within the organism or eliminated from it. Such is the case when putrid matter has been once injected into the blood. By infection from a wound the absorbed matter has still another effect; locally, around the wound, it sets up an inflammation progressive in character, the acute inflammatory œdema, with a tendency to extend along the connective tissue. In this inflamed spot, again, materials are produced by the disintegration of tissues, equally possessed of pyrogenetic and phlogogeneous properties. Continually small quantities of these materials are absorbed into the blood, and thus the fever sustained. At the same time the blood, now possessed, although in a slight degree, of phlogogeneous properties, may also cause inflammation in other organs, predisposed to inflammation, either on account of their anatomical condition or on account of the idiosyncrasy of the patient. Such organs are chiefly the large abdominal glands and the serous membranes, also the striped muscles and the connective tissue. Whilst the process previously described as consisting of a uniformly acute degeneration of all the organs has been called septicæmia, the one just mentioned, where the process is more chronic and limited to individual organs, has been called ichorrhæmia. A specific difference between the two does not exist, it is only one of degree; for where septicæmia has not quite an acute course it ceases to be a pure intoxication with the originally infecting agent, but the infection of the blood is now aided by the absorption of the products of the local inflammation, which products, however, are not specifically different from the original agent. On the whole, septicæmia may be considered the acute, ichorrhæmia the chronic, or rather subacute septic infection.

There is still another way in which the whole organism can be affected from the local disease. It may happen that pieces of a thrombus which had formed in a vein may enter the circulation, and be arrested in an artery and there cause inflammation. Thrombosis of the veins, however, has properly nothing to do with septic infection, and apart from any infection a piece may be torn off a thrombus, and, entering the circulation, may be arrested in the pulmonary artery as an embolus. Embolism, therefore, is not a disease du · to nfection. But, in reality, the case is different. It certainly rarely occurs that particles are torn off a healthy parietal thrombus, which occludes a venous branch and projects somewhat into the main vein. If it occurs, and the embolus is arrested in the lung, under normal conditions it usually becomes encapsuled, and the area supplied by the occluded blood-vessels is in a collateral way again brought within reach of the circulation. But the case is different if the tissue around the thrombosed vein is the seat of a phlegmonous inflammation. Under the influence of the surrounding inflammation the thrombus disintegrates, and the products of disintegration easily enter the circulation. Recklinghausen and Babbnoff have shown that wandering cells pass through the walls of the vessel into the thrombus. If such a particle is arrested in the lung under the altered conditions of the diseased organism, a hæmorrhagic infarct is formed, that is, a stasis in the vessels at the periphery of the occluded spot; and the embolus itself now produces inflammation in its vicinity, which is like to that which caused its own disintegration, and therefore either purulent or ichorous. In point of fact, infection is highly conducive to the origin of embolic centres, but the latter are always complications only of ichorrhæmia, and embolism itself is an accidental affection, not directly due to infection.

A priori it may appear that in puerperal women embolic centres ought to be of frequent occurrence, since there is always thrombosis of the blood-vessels at the placental insertion, and, as Virchow has shown, all the conditions are present which favour the formation of thrombi by compression and dilatation. But this is by no means the case. They appear to be more frequent in surgical cases; and even in diseased puerperal women, in whom embolic centres occur, these appear not to have arisen in the great majority of instances from the sources above mentioned, but from veins which became secondarily thrombosed within a phlegmonous centre.

There is also another important question, whether the circumscribed inflammation of individual organs is exclusively due to embolism or whether they may be a consequence of ichorrhæmia. It must be decidedly admitted that circumscribed lobular inflammation may also arise independently of embolism. Virchow also, although with a certain reserve, has given it as his opinion that he did not consider it improbable; whilst Billroth and Waldeyer derive all circumscribed metastatic inflammations from embolism, although they themselves confess that an exact demonstration of an embolus can very frequently not be made. According to our own observations, circumscribed inflammation of various organs undoubtedly arises independently of embolism, and we ascribe to ichorrhæmia also the power of producing lobular inflammation. It may certainly be very difficult to distinguish between them and embolic centres. In doubtful cases the presumption is in favour of embolic origin in inflammation of the lung, and in that of other organs in favour of ichorrhæmia.

The anatomical appearances in the dead body after infection of the whole organism are, therefore, very variable. In extremely virulent and rapidly acting infection—acute septicæmia—the parenchyma of the most vulnerable organs is found in the stage of cloudy swelling, the commencement of parenchymatous inflammation. There are often also numerous extravasations of blood, especially beneath the endocardium and the mucous membrane of the intestinal canal, together with catarrh of the latter; both of them are regularly met with in septicæmia artificially produced in animals. In the less intense and more chronic action of the virus—in ichorrhæmia—there is especially diffuse or circumscribed parenchymatous inflammation of the large abdominal glands and purulent inflammation of the serous membranes. There is also acute œdema of or abscesses in the connective tissue, partial inflammation of some muscles, and abscesses of the lymphatic glands. In fatal cases of septicæmia circumscribed centres of embolic nature are rarely found in the lungs, and in other organs they are never found if it has run a very rapid course. They are very frequently met with in ichorrhæmia.

We must, however, limit ourselves to mentioning briefly the most important changes that are met with in some organs.

Amongst the serous membrane the pleura is, with the exception of the peritoneum, most frequently the seat of inflammation. Of peritonitis, and the possibility of its origin by ichorrhæmia, we have already spoken. Pleurisy is uncommonly frequent, but does not always arise from ichorrhæmia. It is not infrequently due to perforation, or to a mere propagation of inflammation from an embolic centre in the lung or from pneumonia of another kind. But it may arise also, as shown by the inflammatory œdema of the intermediate layers, by continuity of tissue, the inflammation of the peritoneum being propagated through the diaphragm. We have seen it once caused by the perforation of a purulent abscess of the spleen into the pleural cavity. Pericarditis externa is often connected with pleurisy of the left side. The pleurisy may be of an adhesive character, and thus thick adhesions will be formed, which are sometimes gelatinous from serous transudations, or there is a fibrino-purulent coating, or very frequently a copious serous or purulent exudation, sometimes even discoloured and putrid.

Inflammation of the cerebral membranes is comparatively rare. The internal surface of the dura mater is covered with a gelatinous fibrino-purulent exudation, the pia mater variously altered, from a slight injection and slight œdematous opacity to a purulent meningitis. Virchow found in one case a purulent opacity of the posterior cornu of the left lateral ventricle.

Inflammations of the joints are far more frequent. The shoulder- and the knee-joint are most frequently attacked, but the wrist, the elbow and the hip may be the seat of a purulent inflammation as well as the others. In one case there was pus around the shoulder-joint, without the joint itself being implicated. Pus accumulated in the joint may perforate and undermine the surrounding soft parts to a great extent.

Endocarditis has often been seen in ichorrhæmia and septicæmia. The endocardium is finely vascular, and beneath it are extravasations of blood, which extend into the muscular coat. It may lead also to papillary proliferation of the epithelium, ulceration of the valves, and consecutive embolism. We have once seen circumscribed endocarditis arise from a putrid thrombus, which had been arrested under the inner segment of the mitral valve.

We have already mentioned that the lungs are the most frequent seat of embolic infarction and of abscesses of the known cuneiform shape; for under the influence of the infection the thrombi, as a rule, disintegrate into very small particles, which enter the circulation, and are arrested in the finer branches of the pulmonary artery; in it are often found embolic centres, not very numerous, and seldom large. Thrombi of such a size that they are arrested in the larger branches of the pulmonary artery are a rare occurrence in septic infection. We have, however, seen it once in the body of a puerperal woman, in whom the general peritonitis had subsided, and where the thrombosis of the inguinal and crural veins extended up towards the third lumbar vertebra; very large emboli had been arrested in most of the larger branches of the pulmonary artery, although not completely blocking up the vessels. One branch only was pervious, that leading to the left upper lobe. At the place of division of the artery which supplied the left lower lobe a large discoloured old embolus was found. The same was the case in the pulmonary branch of the right lung, yet those emboli did not perfectly occlude the arteries. Perfect obstruction was only found in some small branches, but in the areas supplied by them infarcts were by no means always found, so that the latter were not numerous, and the greatest part of the lung was only the seat of very intense œdema.

Besides embolism there is often lobar and lobular pneumonia of ichorrhæmic origin. The exudation is rarely of a purely croupous character, mostly somewhat serous and discoloured, and seldom occupies only one lobe and this exclusively; more frequently the greater part of one lobe is infiltrated, and in the other are found lobular infiltrations of the same kind. The greatest tendency to pulmonary gangrene is caused by the presence of putrid emboli, but in discoloured pneumonia also the pulmonary tissue may disintegrate.

The spleen is most frequently enlarged, the pulp soft, greasy, of chocolate colour, and rarely perfectly liquescent; in other cases there are lobular infiltrations of an icorrhæmic nature. Embolic centres in the spleen are equally rare.

The liver is seldom perfectly unaltered. Besides embolism all stages of commencing cloudy swelling of the liver-cells to their perfect disintegration—acute yellow atrophy—are observed. These changes are rarely uniform throughout the whole organ. There is chiefly a far-advanced fatty infiltration, or the already perfect disintegration of cells with relatively or quite intact portions. The fatty infiltration is recognised on section by the clearer spots, which contrast with the brown parenchyma, but the transition into the latter is usually indistinct.

In the kidneys also embolic centres are met with, as well as other circumscribed and diffuse inflammations. The epithelium of the uriniferous tubuli is infiltrated with fat and disintegrated. Cloudy swelling is very commonly met with, and sometimes also, in the cortical portion, degenerative processes.

Buhl saw in one case a parenchymatous inflammation of the pancreas with disintegrated glandular cells as in the liver.

There are also suppurative and putrid inflammations of the parotid glands and of the mammæ, and in strumous cases of the thyroid also. These probably are only exceptionally due to embolism, and they contain either pure pus or a thin putrid secretion.

Inflammation of the eye is most frequently caused by embolism.

It begins with swelling of the eyelids, hyperæmia, and hæmorrhage; the cornea and the iris become opaque, pus is formed, the cornea ruptures, and the process terminates in destruction of the eyeball.

The lymphatic glands may begin to suppurate later on in the course of the disease. The inguinal and the axillary glands are the most commonly implicated.

Purulent inflammations of the muscles and the connective tissue of the extremities are also met with. The cause may sometimes be embolism, but not always so. These abscesses usually contain pure pus, but sometimes the primitive fasciculi of the muscles (also of the heart) are broken down into a molecular detritus, and in the connective tissue circumscribed softened spots also appear, which break and discharge a thin matter with necrotic shreds. The acute swelling of the connective tissue may quite subside.

The intestinal tract is in many cases (even where no calomel has been administered) the seat of a catarrhal inflammation with hæmorrhage, and occasionally also of ulcerations of the mucous membrane, due to hæmorrhagic infiltrations. The same kind of ulcer, due to the same cause, is also found in the bladder. Diphtheritic enteritis is rarely seen.

If we add to the above the inflammations of the skin which occur as circumscribed hyperæmia or as pustules, we shall have mentioned all the more important metastatic centres which, either of ichorrhæmic or embolic nature, occur in the puerperal state, and also in all the organs of the body. (Erysipelas is always due to direct infection, either at the originally infected place at the vulva, or by inoculation at a place of injury. We have seen it start twice from an excoriation of the left nostril.)

It has already been mentioned that at times, when puerperal fever is epidemic, the new-born child also dies of septic infection. This, as a rule, starts from the umbilical wound (doubtless by manual inoculation), and has a similar course to that of puerperal fever in the mother. There is either erysipelas of the abdominal walls, or the inflammation extends from the connective tissue of the umbilicus to the subperitoneal tissue, and causes a secondary and fatal peritonitis. The vessels and the structures in their immediate vicinity may suppurate or mortify. In the umbilical arteries disintegrated thrombi are found. Embolic centres in other organs are very rare, but ichorrhæmic pneumonia is rather frequent.

c.—*Symptoms and Course of Puerperal Fever*

The outbreak of the disease essentially depends upon the period at which infection has taken place. This, as a rule, occurs during the expulsion of the fœtus or in the stage of the afterbirth, but sometimes also at the commencement of labour, and even during pregnancy; for at an early period of labour the os uteri may be already lacerated, and even during pregnancy, when the internal os is patulous, the examining finger may open some of the absorbent vessels by the separation of the decidua, which is shown by the blood covering the finger when withdrawn after an examination.

If infection takes place during gestation this is, as a rule, interrupted. Infection is a rare occurrence a few days after delivery, when the small lacerations have cicatrized, or have begun to form granulations; but it is, nevertheless, possible, because the granulations can be easily destroyed and the recent cicatrices irritated. If, as is usual, the woman is infected in the last stage of her delivery, the first part of the puerperal state passes quite normally. The temperature during and after birth depends exclusively upon the process of labour; even in infected women, if labour has not given rise to any considerable disturbance, the temperature is normal, that is, it rises within the first twelve hours, and in the second twelve hours it again falls very commonly to below 37° C. The pulse may also be slow, not infrequently from 60 to 70 in the minute, but sometimes it is frequent from the commencement.

The first signs of the outbreak of the disease are observed in from thirty to forty hours after infection; usually, however, on the second or third day after delivery, whilst in cases in which infection has taken place earlier the disease begins on the first day or even during labour itself, and in the exceptional cases where infection takes place later the first appearance of the disease may be at a much later period.

The disease is sometimes, but by no means regularly, ushered in by a very pronounced rigor. More frequently the temperature rises with, at least, a subjective sensation of cold and slight shivering; in other cases the fever begins very gradually.

Chilliness or rigor is by no means a symptom of great importance. They are often absent in the most fatal forms of puerperal fever, and, on the other hand, puerperal women so easily feel chilly, that even a very severe rigor, unaccompanied by other symptoms, need not be considered as the precursor of grave pathological changes.

Since almost all the organs of the body may be attacked in puerperal fever, the symptoms of the disease are, therefore, very variable. The following description will be limited to the symptoms peculiar to puerperal fever.

Puerperal ulcers are, as a rule, attended by only very insignificant symptoms; the most constant is a burning sensation during micturition, which, however, is also observed in non-ulcerating lacerations. Usually the inflammation extends to the connective tissue in the neighbourhood of the ulcer, and in consequence of it the labia become very œdematous. Since the ulcers are frequently larger on one side than on the other, it very often happens that one labium is principally the seat of the œdematous swelling. Besides the burning pain on micturition, the discomfort arising from the œdema, and the intense pain on touch, there are no other symptoms which proceed from the ulcers. Fever almost always accompanies their occurrence, on account of the extension of the inflammatory processes to the neighbouring connective tissue; but the ulcers themselves are not directly concerned in the elevation of the temperature, because the temperature often remains rather low when the ulcers are very large, and thickly covered. They heal very slowly; their covering is then thrown off, and healthy granulations appear at the base. Sometimes, even at a late period, they cause very intense pain on walking, and hæmorrhage after the febrile disease has ceased for weeks;

the induration also of the labia often persists in a moderate degree for a long time.

In the chapter on the physiology of the puerperal state we have already called attention to the fact that, in a recently delivered woman, the mucous membrane of the vagina, and especially that of the uterus, is the seat of changes which elsewhere are considered as constituting catarrhal inflammation. It is, therefore, impossible to describe the symptoms of a simple catarrhal endometritis. Usually fever, scanty secretion, and disagreeable odour of the lochia are mentioned as such. From the fever alone we cannot diagnose an endometritis, and the fever itself, if it at all reaches a considerable degree, is in its turn the cause of a scanty secretion, and, consequently, also the cause of the scanty discharge of the lochia. The disagreeable odour is due to the decomposition of the retained shreds of the decidua; but such decomposition may also take place in women in whom the puerperal state is completely within physiological limits.

The commencement, therefore, of an inflammation of the endometrium has no reliable symptoms, whilst the more intense and especially the ichorous inflammation, cannot escape detection. Here the discharge is of a brownish colour, as a rule, thick; sometimes quite serous, and also intensely ichorous and fetid.

The other symptoms due to such changes in the lining membrane of the uterus cannot be accurately followed, because the symptoms of septicæmic and ichorrhæmic infection of the whole organism, as a rule, greatly preponderate.

The symptoms which accompany the acute inflammatory œdema of the pelvic connective tissue are of great importance. The pain which regularly accompanies such swelling shows that the serous covering is implicated in the inflammation, and so far parametritis and perimetritis cannot clinically be distinguished from each other. We, therefore, prefer to describe under parametritis those cases in which the pain is absolutely or relatively inconsiderable in proportion to the extra-peritoneal exudation. Under perimetritis and pelvic peritonitis all those cases will be described in which the symptoms of a partial peritonitis are most marked.

Parametritis is usually accompanied by fever, the degree of which is very variable. It often begins with, but sometimes without, a rigor, most frequently on the second day, and reaches its height either at once on the first day, or at least on the day following the commencement of the disease. The fever is never a purely continuous one, but always shows remissions—as a rule very considerable ones;—very frequently even complete intermissions may occur. Exceptionally, the temperature may remain low; in two cases of very distinct exudation we could not discover an abnormal elevation of temperature, although regular observations were taken morning and evening. But, as a rule, it is very high, and may reach the highest degree with which life is at all compatible. Generally the elevation of temperature corresponds to the extent of the exudation, so that with a considerable tumour we have a high and continuous fever. Sometimes the temperature completely falls, and rises again after a short time with a new and a greater exudation. Almost all the cases, in which at the later period of the puerperal state considerable fever with pains in the abdomen and exudation are observed, have been preceded by a slight parametritis, which, tending to recovery, became more severe through a fresh external injury.

The pulse is usually frequent, corresponding to the temperature; but some cases are distinguished by a very frequent pulse and only a slight elevation of temperature. These cases are always suspicious, because they are easily followed by ichorrhæmic and septicæmic affections. With the beginning of the fever the subjective symptoms also appear, and besides the initial and not infrequently repeated rigors there are heat, thirst, and headache.

Pain is the most important of all the subjective symptoms. It is certainly not caused by the parametritis itself, but is always due to the simultaneous irritation of the peritoneum. The latter lying close above the connective tissue, spontaneous pains are rarely absent, whilst tenderness on pressure is always observed. This is limited to the sides of the uterus, either to one or both of them, and at the commencement of the disease it often changes its place, so that it is found sometimes to the left, sometimes more to the right.

In the progress of the disease swellings form in some places, and most readily between the folds of the broad ligaments. These latter are of the greatest importance, because in combined external and internal examination they are easily accessible to the examining finger, and from the infiltration of that region the affection of the pelvic connective tissue can easily be diagnosed. It may sometimes be found that at the side of the uterus only an increased resistance and swelling exists, so that, without there being a circumscribed and limited tumour, the finger cannot be brought so closely to the uterus as is usually the case, but a sensation is obtained as if a thick layer was situated at the side of the uterus and between the fingers. Of course, the infiltration is limited by the superior margin of the broad ligaments, so that this can be distinctly felt if the thickening extends so far. Sometimes, however, the diffused exudation is situated only in the region of the internal os, and extends thence backwards, so that the sides of fundus uteri are completely free. The exudation can be most distinctly felt if it has originally been circumscribed, or if at a later period the general serous infiltration has decreased, so that only a thick exudation has remained between the layers of the broad ligament. Then a thick tumour is found by the side of the uterus with a broad base, so that only from its abnormal shape, and from the considerable hardness at a later period, the exudation can be dintinguished from the uterus itself. But more frequently a distinct furrow is felt between the uterus and the tumour. Such tumours are not infrequently found on both sides, although, as a rule, the tumour on one side is larger than that on the other. Sometimes, also, one side is entirely free, or is only the seat of a diffused infiltration, whilst at the other side a circumscribed tumour is felt. These tumours are, however, as a rule so high up that they cannot be felt by an examination per vaginam. This explains also why for so long a time they have been overlooked, or at least their frequency of occurrence has been underrated. Sometimes they extend so far downward that they project from the side of the uterus into the vagina, like hard thick semi-circular tumours. Under certain conditions they may become very large—almost the size of a child's head, but commonly they are as small as a hen's egg. Their shape is often irregular, and this may sometimes be due to the ovary, which lies close to or within the tumour. The extent of the exudation is rarely so considerable that it fills up the whole pelvic inlet and encases the uterus so completely that its body cannot be felt separately. In very rare instances the tumour is situate on the anterior or the posterior wall of the uterus.

Somewhat more frequently the infiltration extends from the side of the uterus to the iliac fossa. In such cases the combined examination shows either no tumour at all at the side of the uterus, or only a thin diffused swelling; whilst, by external palpation alone, a tumour can be distinguished situated in the iliac fossa. It, as a rule, gives rise to very decided symptoms by pressure upon the nerves of the lower extremity. Whilst in the tumours resulting from parametritis, lameness or neuralgia is rarely produced, this frequently occurs when the tumour is situated more externally. The infiltration may also extend to the nerve itself. After these tumours have persisted for some time, the contents become gradually more and more inspissated, and in the course of some weeks or months they are completely reabsorbed without leaving any trace of their previous existence. As soon as the inflammatory state and the tenderness have somewhat abated, the tumour begins to contract and becomes somewhat smaller, thicker and harder, and its circumference is able to be distinctly defined. Under favorable conditions it then rapidly becomes smaller. The reabsorption of the exudation takes place with the symptoms of distinct hectic fever, which is especially marked when the tumour is large; the patient begins to feel pretty well, the appetite returns, but the temperature, normal and occasionally also abnormally low in the morning, rises towards the evening without previous shivering to a considerable degree, sometimes up to 40° C., or even more. The tumour diminishes considerably during the fever, so that finally only a slight resistance remains. The uterus is drawn towards that side and fixed there, but the adhesion disappears at a later period, the uterus again returns to its normal position, and no trace remains of the infiltration.

In other cases, especially where external irritations are continuously active, the absorption of the exudation does not take place. The tumour contracts somewhat, becomes hard as wood, and persists in that form.

In relatively rare instances the exudation softens and suppurates. Such a termination we have only seen once in ninety-two cases of distinctly demonstrable exudation. The tumour then gradually becomes softer, and it, together with the adjoining structures, becomes sensitive to pressure. At the same time hectic fever sets in, not, however, with morning intermissions, but simply with remissions. The patients feel ill and lose in weight. Perforations into various organs may then follow, viz., into the rectum, the vagina, the bladder, into the abdominal cavity, or into the uterus. But the abscess may pass also through the thyroid foramen beneath the gluteal muscles and open externally. Abscesses in the iliac fossa easily gravitate towards Poupart's ligament, and discharge their contents externally.

In the more intense forms of parametritis, the veins and lymphatic vessels running through the centre of the inflammation are frequently involved in it. From the clinical course it may safely be concluded, that in the milder forms it hardly ever produces thrombosis of the veins. It is still more uncertain if the lymphatics are implicated. However, thrombosis of the lymphatic vessels has not that importance which was formerly generally ascribed to it.

In rare cases the virulent inflammation of the connective tissue extends less to the connective tissue of the pelvis than to that of the thigh. The inflammation may then extend either to the subcutaneous tissue or to the connective tissue which envelops the large vascular and nervous trunks. It leads to a phlegmonous inflammation of the lower extremity which in puerperal women is known under the name of "phlegmasia alba dolens."

Very rarely the disease takes such a course that thrombosis of the femoral vein is the primary affection, and that this is followed by phlebitis and phlegmasia. As a rule, the inflammation is primary and this is followed by thrombosis of the veins and of the lymphatics. Sometimes thrombosis does not take place, and we, consequently, meet with cases which, belonging to this class of affections, are especially characterised by the absence of thrombosis. If the infection is very virulent a thrombus usually forms in the veins; but at the same time this results in ichorrhæmia, and the disintegration of the thrombus causes embolism. Phlegmasia of the lower extremity frequently commences as late as the second week after delivery, after all the signs of an affection of the pelvic cellular tissue, viz. pain in the abdomen and in the lower extremity, and lameness, caused by the pressure of the exudation upon the nerve had preceded. The swelling usually commences at the thigh and simultaneously also œdema round the ankles. The swelling becomes rapidly considerable, and the circumference of the whole extremity is greatly enlarged. The extremity is movable only with difficulty, and numbness and tearing pain are felt in it. The consistency is not soft and doughy as in simple œdema, but hard as wood. On account of the tension the skin assumes a white and livid appearance and may be elevated in vesicles. Not infrequently the inflammatory swelling of the connective tissue attacks also the adjoining skin of the abdomen, and is sometimes propagated to the other thigh. As long as the acute inflammation exists there is rather high fever which shows more or less distinct remissions. If the inflammation becomes limited, and if the exudation is gradually reabsorbed, the temperature decreases. In other cases recovery is delayed by the formation of abscesses which may continue to suppurate for a long time. Death from phlegmasia is an exception, and only takes place when the skin and the subjacent soft parts become gangrenous. It is more frequently the consequence of the accompanying ichorrhæmic process or the immediate result of phlegmonous inflammation (embolism).

It has already been mentioned that the pains, which in rare cases only are absent in parametritis, are always due to an inflammation of the peritoneum. Therefore, a slight degree of perimetritis or pelvic peritonitis is regularly connected with parametritis.

In some cases the symptoms of pelvic peritonitis are most marked. The infection reveals itself by severe pains in the abdomen, which come on either suddenly, or after the symptoms of parametritis have existed for some time; the pain may be so great that even very patient women complain and cry out loudly. The whole lower portion of the abdomen is sensitive to pressure, and usually one corner of the uterus most so. The temperature rapidly rises with or without a rigor, and reaches 40° or 41° C. Tympanitis is always present. These sudden and severe symptoms of partial peritonitis may be easily subdued by suitable treatment. By local bleeding, the application of cold, and the use of purgatives, recovery may be rapidly effected.

In other cases the pains only slightly abate; or after they have ceased for a short time soon commence again, and the circumscribed peritonitis quickly becomes general. More frequently, however, general peritonitis sets in with less acute and severe symptoms. It has been gradually set up by the parametritis, and it appears in consequence of ichorrhæmic intoxication, just as pleurisy and arthritis.

In these cases the disease begins on the second day after delivery with a rigor or slight shivering. But rigors may be entirely absent during the whole course of the disease; in other cases rigors repeatedly recur. The rigor is soon followed by the usual symptoms of parametritis, tenderness at the sides of the uterus, and remitting fever. The pain gradually increases in intensity, and spreads over the whole abdomen. The tympanites becomes very great, so that at last the pressure of the bed-clothes can be no longer borne, and the patient continually cries out with pain. The diaphragm is pushed upward by the tympanitic intestines, which, together with the concomitant pleurisy, produces great dyspnœa. Palpation of the abdomen is unbearable. By means of gentle percussion we are frequently able to discover that exudation has taken place, which, when the patient changes her position, sometimes slowly changes its place. The diaphragm is very much pushed up, sometimes as high as the fourth or even the third rib; the area of the liver dulness is represented by only a small line. There is persistent nausea or vomiting of greenish fluids, and frequent profuse diarrhœa. Sometimes, also, but not always, in very favorable cases there is obstinate constipation.

The fever is usually continuous or slightly remittent, frequently not exceeding 40° C., whilst the pulse almost always shows special characteristics. If at the commencement of the peritonitis it was not very full and frequent, it quickly rises to 120, 140, 160, or even more, beats in a minute as the inflammation spreads. Towards the fatal end the temperature frequently falls, and the pulse becomes still more frequent, which, however, is a very bad symptom. On account of the great tympanites, respiration is almost always very frequent, forcible, and laborious. The face has an uncommonly anxious expression, the forehead is covered with clammy perspiration, the extremities are as cold as ice, and the patient becomes collapsed often within a few hours.

Important deviations from the usual symptoms of general peritonitis may be met with, so that all the symptoms enumerated, though not entirely absent, yet may be but very slightly pronounced. The most constant symptom is pain, and yet there are cases in which it is very inconsiderable. Sometimes the slight sensitiveness is due to the benumbed sensorium, although occasionally, we meet with semi-comatose patients who complain of pain. But we have also seen cases of peritonitis where a post-mortem examination has shown that the inflammation had spread over the whole peritoneum, and yet the patients, without being drowsy, have only occasionally complained of spontaneous pain, and only slight tenderness was evinced by pressure upon the uterus.

But far more frequently there are spontaneous pains in the abdomen, but they do not reach that degree of intensity which may be expected from general peritonitis. After very copious exudation the pain, previously very acute, has gradually subsided, so that a diminution of sensitiveness, the general state continuing to be bad, as well as persistent tympanites and a frequent and small pulse, by no means give a prospect of a favorable termination. Meteorismus is seldom, if ever, completely absent, but in some cases it is so little marked that it cannot be counted upon as a diagnostic sign, since puerperal women, under normal conditions, also are usually subject to meteorismus. This, however, has a greater diagnostic value, indicative of an affection of the peritoneum, if the contour of some of the intestinal coils filled with gases can be seen through the abdominal walls. In one case we have seen very distinct meteorismus arise shortly before death in a patient in whom peritonitis, with a fibrinous exudation, was already tending to recovery.

Vomiting is not altogether a constant symptom, though, as a rule, it is present; it sometimes occurs rather late; and even nausea is not always observed.

By means of percussion only very considerable quantities of liquid exudation can be recognised, for a large quantity of exudation may accumulate in the small pelvis and at the sides of the vertebral column.

The fibrinous exudations which are so frequently seen coating all the abdominal organs also escape detection by percussion.

The fever, and especially the pulse, show great variations in cases of general peritonitis. The temperature, which sometimes rises to above 41° C., or even 42° C., is in other instances strikingly low, so that it only rises to 39° C., and very considerable remissions alternate with complete intermissions. Fever may often be entirely absent in a very acute case, and where there is very copious exudation. The state of the pulse is more constant. It is always small and always more frequent than would be expected from the degree of the temperature. Yet this also is not without exception, and there are cases in which the pulse always corresponds with the temperature, or where it rises in a striking way only towards the fatal end.

The majority of cases terminate in death. This often occurs even in the first week, and sometimes within thirty-six hours. Consciousness may remain perfectly intact up to the last moment, nay, towards the end the patient may feel so comfortable that, though lying pulseless, she congratulates herself on her supposed recovery. In other instances the patient suffers much from dyspnœa, and is in fearful agony towards the end. Happily, the sensorium is frequently benumbed towards the end; delirium occurs, and the patient becomes perfectly comatose, and then dies collapsed, or very freqently from a complication with pulmonary œdema.

More rarely the symptoms of acute peritonitis subside, and with them the pathological processes within the abdominal cavity. Under such conditions the diffused exudation becomes encapsuled, forming tumours, which distinctly differ from extra-peritoneal tumours. By palpation large and limited tumours are felt, situated usually at the sides of the pelvis, and sometimes reaching upwards to the umbilicus. After prolonged palpation, cooing sounds are perceived, and in rare instances a crackling is produced, as in emphysema.

By an internal examination the fundus uteri is found to be agglutinated to the place which it occupied before the inflammation began, so that the organ having in the mean time diminished in size, the cervix is very high up, and the vaginal portion no longer exists, but the os is represented by a small opening in the upper portion of the stretched vagina. As a distinction between extra-peritoneal exudation, Douglas's pouch is found filled up by a solid, hard tumour, whilst the lateral tumours can be reached from the vagina only with great difficulty. In other cases Douglas's pouch is only slightly affected, even by a very intense peritonitis.

In some of these cases the patients do not finally escape the fatal termination, and this is caused by ichorrhæmic or embolic inflammation, by a fresh exacerbation of the peritonitis, by suppuration of the exudation and by perforations into the intestines. Recovery is not quite so rare, but even in favorable cases numerous peritoneal adhesions remain, which cause colicky pains, changes in the position of the uterus, as well as sterility due to the ovaries being encapsuled, to flexions of the uterus, or to occlusion of the Fallopian tubes. The position of the uterus is so altered by pelvic peritonitis alone, that the fundus is drawn over either to one side, forwards, or backwards, or a general peritonitis may cause the fundus to adhere at an abnormally high level, and after its re-formation it remains elevated, the vagina being at the same time drawn out to a great extent.

General peritonitis is not often met with without ichorrhæmic complications of other organs. The symptoms, therefore, may be very varying, since through the ichorrhæmia almost all the organs may be inflamed. As has already been mentioned, ichorrhæmia and septicæmia are not specifically different, and, as a rule, replace one another, so that their clinical aspects cannot be separately given; yet the extremes of series of cases of prolonged ichorrhæmia and of acute septicæmia are considerably different. The fever of ichorrhæmia has the greatest similarity to that of severe parametritis. The unusually high fever and the frequency of the pulse are decidedly of an ichorrhæmic nature; although in most cases in which recovery takes place, and in which affections of more distant organs cannot be demonstrated, a diagnosis of ichorrhæmia cannot be made. The fever regularly begins with a rigor, or at least with a subjective sensation of cold, and these rigors repeatedly recur.

The type of the fever is quite irregular. The temperature varies and often rises in a few hours some degrees. In the more intense cases the patients feel very unwell; sometimes they moan loudly without complaining of pain in any particular organ, and are conscious that death is approaching.

Icterus and profuse hæmorrhage from the genitals not infrequently occur. These are soon followed by symptoms showing that various organs are affected; cough and pain in the chest show the affection of the lung and of the pleura; there is rarely bloody expectoration. The implication of the kidneys is shown by the presence of albumen, pus, and blood in the urine. There are also pains and swelling of the joints, inflammation of the connective tissue, abscesses in the muscles, and sometimes, though rather late, suppuration of lymphatic glands. The disease either tends to recovery with a gradual abatement of the symptoms, or death more frequently takes place with increasing frequency of the pulse, and sometimes with a very low temperature.

Cases of very acute septicæmia may also commence with a rigor. The temperature does not usually rise very considerably (to about 40° C.), but remains at that height or at least only slightly remits, whilst the pulse and the respiration are very frequent, and death may rapidly occur in two or three days. The most marked symptoms are fetid diarrhœa, increasing frequency of the pulse and of the respiration, a low temperature, and a typhoid condition. In many other cases the disease passes unobserved into ichorrhæmia. The process, which had at first been considered as decidedly septicæmic, now becomes ichorrhæmic if its course be not very acute, as soon as there are affections of an organ, for instance, of the lungs or of the joints.

The previous description refers to the general aspect of the more intense forms which puerperal fever assumes. According to the localisation of the disease in individual organs, the symptoms also will greatly vary. It would lead us too far were we to give all the details of the symptoms produced by the inflammatory processes during a puerperal fever. We, therefore, omit the description of the symptoms caused by individual ichorrhæmic or embolic centres, and we shall only point out important peculiarities whilst speaking of the diagnosis and prognosis.

D.—*Diagnosis of Puerperal Fever*

It is of great importance, both as regards the prognosis as well as the prevention, that we should be able to decide whether a disease of a puerperal woman be puerperal fever, *i. e.* whether it be due to infection or not. If the symptoms of ichorrhæmia or septicæmia be very decided a decision is, of course, easily arrived at, but this may be more difficult in the case of general peritonitis. But here the mistake can be avoided if in all cases in which the processes during labour do not sufficiently explain the occurrence of the peritonitis, we consider it of septic origin.

With regard to parametritis we have already given our opinion. Of course, we cannot deny the possibility that parametritis may arise independently of infection but we consider such an occurrence as rare as malignant embolism apart from infection. From the reasons above mentioned, we consider the parametritis a proper criterion that infection has taken place.

The discharge of fetid decomposed lochia is not, and cannot be considered a proof that infection has taken place. We have often had the opportunity of observing that within a few days after delivery large quantities of foul-smelling lochia have been discharged without there being any trace of disease. Decomposition of the lochia almost always takes place when large shreds of the decidua partly separated from their connection with the surface of the uterus, have remained behind in the uterine cavity. A considerable accumulation of the secretions is usually due to great anteflexion of the uterus, by which the internal os is displaced. Auto-infection does not follow in such cases because there is no solution of continuity of the external integuments. The denuded mucous membrane of the uterus does not absorb, and the lacerations of the mucous membrane during labour have either healed or are now protected by granulations. Yet even healthy puerperal women whose lochia are decomposed require our careful attention on account of the epidemic to which they may give rise. At a later period the women may infect themselves if a mucous membrane has, subsequently, been injured, or other puerperal women with recent lacerations may be infected by the decomposed lochia of the former.

It may be difficult to decide whether acute inflammations of the more remote organs are to be considered of an ichorrhæmic origin or not. Purulent inflammations of the joints are probably always due to ichorrhæmia, and only exceptionally at the commencement of the infection, the distinction between it and an acute articular rheumatism may offer some difficulty. But it is much more difficult to decide upon the nature of a pneumonia or pleurisy arising in the course of the puerperal state. If parametritis exists at the time it will greatly aid the diagnosis, and we shall rarely make a mistake if we consider the affection of the respiratory organs as ichorrhæmic. The nature of the pneumonia is especially marked if it deviates from the usual type, if it does not attack an entire lobe, if there are no characteristic sputa and if the fever is markedly remittent.

It is of less value as regards the prognosis and treatment, though it is a subject of great interest to the accoucheur, to ascertain whether the disease has arisen from auto-infection or from without. We frankly declare that we do not believe in the prevalent notion of frequent self-infection as asserted by several authors.

It cannot be denied that at times, when lying-in institutions are free from epidemic diseases, even very considerable injuries as well as the retention of the membranes and of the placenta, are not attended by unfavorable results. In the absence of an epidemic there is hardly a pronounced case of ichorrhæmia or septicæmia, which we could easily attribute to self-infection, whilst acute peritonitis which often follows severe injuries, cannot be considered as belonging to puerperal fever. The isolated cases of puerperal fever, especially in private practice, show that self-infection is not very frequent; and in lying-in hospitals also in which the material is relatively little used for teaching, a long time may pass before a case of puerperal fever occurs, whilst an exhaustive use of the material for study and the frequent examinations of puerperal women as practised especially in maternities connected with medical schools, soon lead to a deterioration of the sanitary condition. It is easily understood why the accoucheur prefers self-infection to infection by means of the examining finger.

It is also of importance to know whether after infection has taken place the process is still localised or whether it has already extended over the whole organism.

It may be concluded that the process has extended only by continuity of the tissues as long as the general symptoms, especially the temperature, the pulse, the respiration, and the cerebral functions, correspond to the local inflammation, and that inflammatory changes in organs distant from the place of infection cannot be recognised.

But if any slight local inflammation, or a slightly painful parametritis with an inconsiderable exudation, is attended by cerebral disturbance, if the fever is very high, and if especially the temperature does not correspond with the pulse, so that with a temperature of 39·5 the pulse beats 160 in the minute, the fever has already assumed an ichorrhæmic character.

For the diagnosis of the different local diseases everything depends upon an accurate examination performed according to the rules of gynæcology, medicine, and surgery.

We shall only point out the more important facts.

A slight degree of parametritis cannot be recognised without an accurate combined internal and external examination, because in no other way are we able to detect a small infiltration at the side of the uterus.

Partial and general peritonitis as a rule arrest the attention by the severe pain which the inflamed peritoneum gives rise to, and thereby, as well as by the great tenderness on pressure, the meteorismus, the vomiting, and the presence of free exudation, they are easily recognisable. Yet all these symptoms even in acute diffuse peritonitis may be very little pronounced, and in such cases the diagnosis may be attended with very great difficulty. We do not even hesitate to state, that a great many cases which, on account of the paucity of the symptoms, have been considered as pelvic peritonitis, and have ended in recovery, were actually cases of general peritonitis. We base this assertion on some cases we have observed, where, in spite of the slight symptoms, death occurred in consequence of the inflammation; the autopsy showed that the inflammation had extended over the whole peritoneum; and in one case especially, where a year after delivery the patient died of tuberculosis, the remains of general adhesive peritonitis were found on post-mortem examination, whilst during the puerperal state the disease only appeared to be a slight perimetritis occurring in a debilitated subject. For the diagnosis of such cases of general peritonitis with very indefinite symptoms, we attach much value to the general state and the condition of the pulse, which always give an unfavorable impression. Of the distinction between intra- and extra-peritoneal exudation, we have already spoken above.

Pleurisy comes on with pain, and is recognised by its physical signs, especially by a friction murmur and the presence of an exudation.

The diagnosis of lobular pneumonia is much more difficult, because almost always the characteristic sputa are absent, and there is either no dulness whatever on percussion, or it is only very slight. But the inflamed portion can be distinguished by the small vesicular crackling, whilst the smaller centres depending upon embolism usually escape detection.

The diagnosis of meningitis must be made with caution. Whilst cerebral symptoms are very frequent in puerperal women, inflammation of the meninges is so rare, that there must be paralysis before a diagnosis can be made.

It is more easy to recognise inflammation of the joints. The restricted movements and the often very acute pain in the attacked joints are highly characteristic. But the pain may be slight, or even completely absent, if the sensorium is benumbed. There are occasionally very severe pains in some of the joints, and on post-mortem examination we fail to detect any pathological changes. Swellings round the joint are not always due to the presence of pus, but if the capsule of the joint be swollen the tumour always contains pus.

Inflammation of the endocardium can rarely be diagnosed with absolute certainty during life, since functional disturbances of the heart are very frequent, and in many cases the post-mortem examination shows no anatomical changes in the endocardium. Slight degrees of enlargement of the spleen are found in almost all puerperal diseases of a more serious character; but whether the enlargement is due to hyperæmia or to infarctions or abscesses cannot as a rule be decided. We must not omit, however, to state, that in one case, with a very high and rapidly intermittent fever and unusually frequent rigors, we found an abscess of the spleen. When there is peritonitis it is impossible to define an enlargement of the spleen by means of percussion.

The affections of the liver are generally overlooked. Icterus is very frequently met with, even when at the post-mortem examination the liver is found to be normal, or in the first stage of a parenchymatous inflammation. Pain in the region of the liver is rare.

Diseases of the kidneys are indicated by the state of the urine. If in previously healthy women, blood, albumen, and granular casts are found in the urine, the kidneys are implicated in the puerperal disease. But we must also remember that pus and blood may be derived from the bladder. To examine the urine of puerperal women it must be drawn off by means of the catheter.

E.—*The Prognosis of Puerperal Fever*

In all the forms of puerperal diseases due to infection a prognosis is to be made with great caution. For even the mildest form of parametritis may pass into a grave ichorrhæmic affection. If the disease remains stationary as a simple parametritis, the prognosis is decidedly favorable, for in the great majority of instances the contents of the tumour are reabsorbed. It rarely terminates in the formation of abscesses. This is more unfavorable, because the persistent suppuration causes a considerable loss of strength, and the healing of the abscess produces changes in the position of the genital organs. Affections of the peritoneum always render the prognosis doubtful, although the extension of the inflammation to the serous membrane of the pelvis may in many cases be arrested. All intra-peritoneal affections are, therefore, of practical importance because they terminate with the formation of pseudo-membrane and lead to stricture and atresia of the tubes and change of position of the organs of the true pelvis.

In inflammation of the whole peritoneum the prognosis is very unfavorable. Exceptionally only it terminates in recovery, and even in cases where the acute symptoms disappear, and the deposits of pus become encapsuled, many patients die of exhaustion, perforation, or fresh exacerbations of the peritonitis.

Of sixteen patients who had undoubtedly general peritonitis, only four lived, whilst the remaining twelve, in four of whom the inflammatory processes in the abdominal cavity had begun to subside, died.

We have already stated that, according to our conviction, many cases of general peritonitis which were supposed to be only partial pelvic peritonitis may end in recovery. When symptoms of ichorrhæmia and septicæmia appear the prognosis is very grave. In such cases the fatal end must always be expected, although happily it does not often take place. So long as the disease is not localised the prognosis is only unfavorable when the general symptoms are very grave. Inflammations of the respiratory organs are attended with great danger, yet they also may end in complete recovery. In purulent inflammation of the joints death is very frequent; in the great majority of cases it is the result of the intense ichorrhæmic affection, but sometimes also it is the consequence of the suppuration. Recovery is at times only partial, with the usual terminations of a purulent articular inflammation.

Most of the other diseases offer considerable difficulties in both diagnosis and prognosis. Embolism may be suspected, but by no means with absolute certainty, when after a rigor the temperature rises very high, and immediately afterwards falls considerably. The prognosis is always grave, but we have seen a case in which repeated and severe rigors, together with very high temperature, occurred without there being any other disease than peritonitis with exudation.

The treatment consisted in the production of very profuse diarrhœa. On the thirteenth day of the disease the fever which was markedly remittent decreased, and in the third week of her illness the woman was discharged from the hospital cured, although the exudation had not been reabsorbed.

F.—*The Treatment of Puerperal Fever*

Prophylactic measures ought to occupy the first place. Since infection is brought about by septic material, parturient and puerperal women must be carefully watched in order to prevent the reabsorption of decomposed organic matter. Let us therefore, first of all, consider the precautions which we must take in order to guard against self-infection. We have already given it as our opinion that self-infection is far more rare than infection from external sources. We believe that usually the bruised soft parts are encapsuled by inflammation, and are thereby rendered innocuous to the whole organism. We further believe that, as a rule, retained pieces of the ovum only then acquire infective properties when the place at which infection is possible is no longer capable of absorbing. Accordingly we consider that self-infection occurs chiefly in those cases in which septic material has formed before delivery is completed and has passed through the vagina. This is especially the case where the dead fœtus decomposes under the influence of the atmospheric air, or where neoplasms of the maternal organs disintegrate during labour. Pressure bruising the soft parts then only favors infection when delivery takes place after the soft parts have already become gangrenous, so that the gangrenous secretion passes over the recent lacerations of the mucous membrane.

The danger of infection therefore depends less upon the tedious labour attended by great pressure upon the soft parts, but rather upon a tedious and prolonged labour. It is therefore the duty of the accoucheur to shorten the labour. This is not the place to detail in what way this can be done. We will only remark that it is highly important to terminate labour in every case before putrid discharges appear. In all such cases, then, and also when after perforation or embryotomy the parts of the dead fœtal body come in contact with the genital canal, we should immediately after the expulsion of the ovum wash the parts by means of very careful injections with dilute carbolic acid in order to disinfect the small lacerations of the mucous membrane.

Again, when the lochia are decomposed, the accoucheur himself may produce a condition favorable to auto-infection. This is done by destroying with the examining finger the still recently produced mucous membrane lining the uterus, or the granulating wound at the vaginal outlet. In this way a solution of continuity is produced. The septic matter of the lochia is then easily absorbed.

Our chief aim, therefore, should be to prevent decomposed organic matter being brought to the puerperal women. Such septic substances are brought to the genitals of the recently delivered women principally in three different ways. First, by sheets, sponges, &c.; secondly, by the hand or instruments of the accoucheur; and, thirdly, by the hand or instruments of the midwife.

The difficulty with regard to the first point is easily avoided in private practice. Everything that is brought in contact with puerperal women should be very clean, and this is most necessary during the first days after labour. As soon as that period is past the anatomical conditions are, as a rule, unfavorable for infection.

Often enough we have seen quite healthy puerperal women lying on sheets soiled with the discharge; it is evident that scrupulous cleanliness in this respect is necessary.

Secondly. The hands of the accoucheur and his instruments must be absolutely clean, lest the hand which ought only to help, and not harm, should bring disease or death.

Though it is easy to state this, yet it is extremely difficult to carry out fully. It is, of course, well understood that every accoucheur, before examining a woman, should thoroughly well wash his hands; but this alone is not sufficient. Let the accoucheur accustom himself, each time before examining a woman, during or immediately after labour, to think of everything which he has handled in his avocation during the last few days. Washing simply with soap and water is only sufficient if the practitioner knows with certainty that latterly he has not come in contact with decomposed organic matter. We are, of course, not able to mention all the modifications under which septic matter may prove the source of infection. We shall mention only some of the most frequent. The accoucheur must consider his hands not quite free from poison if he has handled any portion of a dead body, if he has seen a patient suffering from phlegmonous erysipelas or any kind of pyæmia, if he has dressed suppurating or diphtheritic wounds, if he has come in contact with decomposing new growths, if, in a case of abortion, he has extracted a decomposed ovum or portions of it, if he has examined women with badly smelling lochia, or suffering from puerperal fever. He must then clean his hands most scrupulously before he undertakes the examination of a parturient or puerperal woman.

Now, simple cleansing with soap and water is not sufficient; he should use a nailbrush or rough cloth, so as to remove the superficial layers of the epidermis. It is also desirable to add disinfecting fluid to the water. We recommend for that purpose chlorinated water, permanganate of potash, concentrated acids, and especially carbolic acid, which we consider the best disinfecting agent.

The instruments should be cleaned with the same precautions. This is best done by means of boiling water, and although this does not destroy, according to Bergmann, the efficacy of the sepsin, it, nevertheless, dissolves the septic material, and most certainly washes it off the instruments. If the accoucheur has for some time been in contact with septic matter it is urgently necessary that he should change his linen and his coat. If, for instance, he had examined, in the usual way, a puerperal woman suffering from puerperal fever, he can never be sure that infecting matter does not adhere to the sleeve of his coat which was brought under the bedclothes.

If the accoucheur takes all these precautions he will not be forced to give up his obstetric practice for a time, if puerperal fever happens to occur.

The third agent which may be the means of bringing infection is the nurse. She requires very careful supervision. She is not acquainted with the danger of the manual transfer of septic matter, and though her attention be called to it, she is only able imperfectly to appreciate the character of the danger. It is, therefore, urgently necessary not to admit nurses or midwives who have attended a diseased puerperal woman to a fresh case of labour.

It is our firm conviction that in this way only the transfer of the disease can be effectually prevented. The proposal which we make may, in the mean time, be received with difficulty by the public, and especially by nurses; but that opposition will soon subside, since the time is not far distant when the bulk of the public will be acquainted with the danger which menaces them through incautious accoucheurs and midwives.

The most careful attention to the prophylactic measures, therefore, is of the greatest importance, for only thus are we able to prevent an outbreak of puerperal fever, and if the disease has once appeared in a puerperal woman there is little prospect of staying its course, although the treatment is by no means so powerless as has been assumed by some.

We do not know any specific remedy with which we are able to neutralise the action of the septic matter in the tissues and in the blood. Polli has lately recommended the sulphites for this purpose. Bernatzik and G. Braun have administered them in some cases, and have found them inefficacious. But we think that their opinion is too unfavorable. We have used the sulphites of soda and potash (about 4 gram. a day) as well as sulphurous acid (2 gram. to 1 lb. of sugar-water per diem). They produced profuse diarrhœa and had a very favorable effect, although we had soon to discontinue them on account of the diarrhœa. We urgently recommend further trials of those preparations. From experiments on rabbits, Hüter and Tommasi recommend carbonate of soda and permanganate of potash. Tyler Smith has seen a desperate case recover after the injection of liquor ammoniæ and water (1 to 3) into the veins.

In the last century purgatives were used in the treatment of puerperal diseases, and recently they have again been strongly recommended by de Latour, Seyfert, and Breslau.

A priori there is a good deal in their favour. If we may expect to remove the absorbed septic material from the blood, there can be no doubt that the best way is through the intestinal canal. Observation also has taught us, that dogs poisoned by septic matter recover after profuse and fetid diarrhœa. The recommendation, therefore, to produce artificially diarrhœa rests on actual observation. In fact, we are inclined to think that we have obtained decidedly favorable results from purgatives, and we therefore strongly recommend them in mild cases.

A decrease of temperature is most commonly seen after their use; the affection of the peritoneum becomes much less painful, and the general symptoms are less troublesome. We have never seen any bad effects resulting from copious diarrhœa; even if it be very profuse the strength does not decline more than may be expected independently of them. This, however, does not mean that we recommend the use of purgatives in every case of puerperal disease without distinction. It appears that some epidemics are distinguished by copious diarrhœa, and we readily admit that the diarrhœa may be so great that it would be more than audacity to increase it still further. In the milder affections we give castor oil, which does not irritate the intestines, and in the graver ones we prescribe the preparations of senna (which, however, cause slight griping) or calomel.

As a last and desperate measure to substitute healthy for the poisoned blood, transfusion may be recommended, which Hüter has performed with at least a transient good result.

All the attention of the accoucheur must be directed to the fever, especially if it assumes the continuous type. A long continuation of high temperature brings the system into the greatest danger on account of the consumption of the tissues. We possess a very valuable remedy against this danger in the methodical application of cold water.

The indication for it cannot be determined by the degree of the temperature alone. The temperature may transiently reach 40° C. or 41° C. or even more, and may so rapidly sink again that a fever which has once reached 41° C. by no means necessarily indicates the application of cold water. But matters are different if the temperature continually remains at such a height or only slightly remits, when cerebral symptoms occur, and when the strength urgently requires to be supported.

The water is most suitably applied in the form of the cold or gradually cooled bath. It is also very useful to wrap the patient in cold wet sheets, which may be especially recommended where the circumstances, under which the patient lives, render the use of cold baths impossible.

The best mode of applying the wet sheets is as follows:—Two beds, each containing a cold wet sheet, are placed side by side. The patient is to be wrapped up in one of these sheets, and after the lapse of a few moments she is taken from this bed to the other, and again wrapped in the same way in the second wet sheet; she must then be again brought into the first bed and wrapped in the first sheet. This procedure is to be repeated from twelve to twenty-four times or more successively; since the above-mentioned method gives great trouble, it suffices, in mild cases, to apply the cold water only to one side of the trunk. For that purpose large sheets are folded so as to be about the size of the trunk; they are then dipped into cold water and placed over the anterior surface of the patient's trunk, and repeatedly replaced by fresh ones. The result of the treatment is generally very evident, but variable in character. If before the application of the water the temperature had reached its greatest height, or if it had already begun to fall, a considerable decrease, down to the normal temperature, or even lower, can be obtained with certainty. But if the temperature was rapidly rising the result is, as a rule, only transient, for the elevation, interrupted for a short time, is soon afterwards continued, and not infrequently with a rigor, but in such cases also a favorable influence on the course of the temperature (especially after repeated applications) cannot be denied. On the following days the fever does not rise to so high a degree, and it becomes more remittent or intermittent. The frequency of the pulse also quite suddenly diminishes after one application of the cold water.

The most beneficial result of the reduction of the temperature consists in the improvement in the general state. Patients with previously quite benumbed sensorium become conscious again for the first time in many days, and by the energetic use of cold water they are enabled to give indubitable signs of the comfort which they derive.

In other cases, when the sensorium is not affected, the subjective state evidently ameliorates, the intense headache diminishes, the tormenting thirst decreases, and the excruciating anxiety and depression under which the patient suffered give way to sensations of general well-feeling.

Other remedies recommended for the reduction of the fever are much less certain. From quinine, which we have given in doses of one and two grammes, we could not see any constant effect. Barker, Elliot, and v. Grünewaldt recommend veratrine. The very great frequency of the pulse indicates the use of digitalis.

The local diseases also must be carefully attended to. The puerperal lacerations must be suitably dressed with carbolic acid. If this is used early the further extension of the laceration is prevented, and the healing process is greatly favoured. There is no necessity to resort to the very painful operation of touching the wounds with solid nitrate of silver. In putrid endometritis injections are to be repeatedly made for the purpose of cleansing and disinfecting; but these, as well as the local treatment of puerperal fever, require every precaution, in order to prevent the transfer of the disease. The ordinary parametritis almost always terminates favorably without complications; it, therefore, does not require active treatment. Moist heat, in the shape of the so-called Priessnitz's embrocation, acts on the local bruises, mitigating and diminishing the pain with great certainty. The exudation, which is often very great, demands special attention. Under favorable conditions, consisting chiefly in maintaining a quiet posture in bed, or at least on the bed, is is quickly and completely reabsorbed. This may be favoured by warm hip-baths, with or without alkalies, and very efficaciously also by iodide of potassium (ten grains dissolved in sixty of water, twenty drops twice daily). Under its continuous use very extensive and hard tumours rapidly diminish in size and disappear almost completely, whilst the general state and the nutrition improve.

The dangers of parametritis, viz. the extension of the inflammation to the peritoneum or the complication with ichorrhæmia are most certainly obviated by the earliest possible use of laxatives. On the first appearance of pain we give a few tablespoonfuls of castor oil at short intervals, until they are followed by a few thin stools; if, as rarely happens, the oil does not act, watery stools are obtained by the use of salines or the preparations of senna.

If the intense pain indicates the appearance of partial peritonitis, the treatment must be energetic, in order to prevent the spread of the inflammation to the whole of the peritoneum. In such cases we give immediately five grains of calomel, and continue every two hours with 2 grains until very copious diarrhœa follows or the gums are affected. At the same time we apply at least twelve leeches to the painful spot, and allow the leech-bites to bleed sufficiently. Olshausen recommends the application to the vaginal portion of from two to four leeches, which bite easily, in spite of the lochia. To prevent the animals from creeping into the uterus a few pieces of cotton wool may be put into the cervix. After the bleeding we place ice-bags on the painful parts of the abdomen. In almost all cases the previously insupportable pains are mitigated by this treatment, and, as a rule, the danger is entirely removed by it. In either case the pains may return, and a renewed application of the leeches and ice may become necessary.

If general peritonitis has occurred the condition is decidedly grave in the highest degree, but not entirely hopeless. If there is no diarrhœa this must be produced as early as possible by means of calomel. The application of ice to the abdomen ought to be continued and the vomiting stopped by pieces of ice or champagne. The more intense pains require subcutaneous injections of morphia. Turpentine enemata render the greatest service against the distressing meteorismus. Tarnoffsky and Dohrn saw very good results from painting the anterior abdominal wall with collodion, whilst other observers have not been successful with it.

A future time will have to decide to what extent the removal of the fluid exudation by puncture will find a place in the treatment of general peritonitis. Spencer Wells removed the fluid which accumulated in Douglas's pouch after ovariotomy by means of the trocar, and made use subsequently of drainage tubes. Thompson and Storer have even washed out the abdominal cavity with greatly diluted carbolic acid. In all cases in which a retro-uterine tumour is attended by grave symptoms of peritonitis, or by those of blood-poisoning, Douglas's pouch may be punctured through the vagina, and the exudation thus removed.

Puncture through the abdominal walls must remain limited to extreme cases, in which the exudation is evidently very considerable and Douglas's pouch not very prominent.

When collapse comes on the fatal end may sometimes be prevented by the copious use of strong wine. We have seen four cases in which with such treatment death did not occur at least at the height of the peritonitis, but only after it had begun to subside, and in four other cases the patients recovered. In one case the patient was already moribund and comatose, but by the free use of wine she was kept alive until she finally and completely recovered.

On the whole, good food and an abundant supply of wine should be given, even when the fever is considerable.

Of the treatment of metastatic inflammations we cannot speak in detail. This must be varied according to the principles of medicine and surgery.

Appendix.—Tetanus Puerperalis

Happily it is only in very rare cases (at least, in Europe) that tetanus occurs after labour at term; it occurs, however, somewhat more frequently after premature deliveries. Its etiology is still quite obscure. We mention it in the Appendix to Infectious Diseases because it is very probably due to infection, or at least to a peculiar irritation proceeding from a puerperal wound.

The causes also which favour the occurrence of tetanus are not yet distinctly known, and the influence of cold is the most uncertain of all of them; relatively speaking, tetanus occurs more frequently after profuse hæmorrhage, especially if this has rendered plugging of the vagina necessary. The symptoms do not differ from those which are found in tetanus following any other injury.

The prognosis is very unfavorable. Of twenty-seven cases collated by Simpson, five only terminated in recovery. The treatment, therefore, can show very little success. In accordance with the views of Simpson, we recommend the use of chloroform so as to produce absolute anæsthesia.

Literature.—R. Lee, Researches on the Pathology, &c., London, 1833, übers von Schneemann. Hanover, 1834.—Eisenmann, Die Kindbettfieber, Erlangen, 1834, and Wund- u. Kindbettfieber., Erlangen, 1837.—Helm, Monographie der Puerperalkrankh. Zürich, 1840.—Kiwisch, D. Krankh. d. Wöchnerinnen. Prag, 1840-41, and Klin. Vorträge, 4 Aufl., I B. Prag, 1854, p. 600.—Litzmann, Das Kindbettfieber. Halle, 1844.—Berndt, Die Krankh. d. Wöchnerinnen. Erlangen, 1846.—Meckel, Charitéannalen, 1854, V, p. 290.—C. Braun, Chiari, Br. u. Sp., Kl. d. Geb., p. 423.—Silberschmidt, Darst. d. Path, d. Kindbettfiebers. Erlangen, 1859.—Hugenberger, D. Puerperalfieber. Petersb. Med. Zeitschr. Sep.-Abdr., 1862.—Leyden, Charitéannalen, 1862, X, H. 2, p. 22.—Fischer, e. l. 1864, B. XII, p. 52.—Hildebrandt, M. f. G., B. 25, p. 262.—Veit, Puerperalkrankheiten, 2 Aufl. Erlangen, 1867, aus d. Handb. d. spec. Path. u. Ther. von Virchow.—Le Fort, Des maternités. Paris, 1866.—Winkel, D. Path. u. Ther. d. Wochenbettes. Berlin, 1866, 2 Aufl., 1869.—Schroeder, Schw., Geb. u. W., p. 197.—Discuss. der geb. Section d. Petersb. Aerzte. Petersb. Med. Z., 1868, H. 6, p. 313.—Hervieux, L'Union Méd., 1869, Nr. 139, and Traité des Mal. puerp., &c., I. Paris, 1870.—Evory Kennedy, Dublin Quarterly Journal, May, 1869, p. 269.—Spiegelberg, Ueber d. Wesen des Puerperalfiebers in Volkmann's Samml. klin. Vortr. Leipzig, 1870, No. 3.—Semmelweiss, Die Aetiologie, d. Begr. u. d. Prophyl. d. Kindbettfiebers, 1861, und Offner Brief an sämmtl. Prof. d. Geb. Ofen, 1862.—Hirsch, Historisch-geograph. Pathol. Erlangen, 1862—1864, B. II, p. 433.—Veit, M. f. G., B. 26, p. 173.—Ferber, Schmidt's Jahrb., B. 139, Nr. 9.—Boehr, M. f. G., B. 32, p. 401.—Stage, Under sögelser, &c. Kjöbenhavn, 1868; s. Virchow-Hirsch'scher Jahresbericht über 1868, B. II, Abth. 3, p. 637.—Martin, Berl. klin. W., 1871, No. 32.—Virchow, Ges. Abh., p. 597, and Virchow's Archiv, B. 23, p. 415.—Buhl, Hecker u. Buhl. Klinik d. Geb., B. I, p. 231.—Erichsen, Bericht, &c. Petersburger Med. Z., B. VIII, p. 257 and 359.—Klob, Pathol. Anat. d. weibl. Sex., p. 235 sequ. —Maier, Virchow's Archiv, 1864, B. 29, p. 526.—Veit, M. f. G., B. 26, p. 127.—König, Archiv d. Heilkunde, 1862, 3 Jahrg, p. 481.—Schroeder, M. f. G., B. 27, p. 108.—Baumfelder, Beiträge zu d. Beob. über Körperwärme. Leipzig, 1867.—v. Gruenewaldt, Petersb. Med. Z., 1868, H. 9, p. 152.—Olshausen, Volkmann's Samml. klin. Vorträge. Leipzig, 1872, No. 28.—Martin, M. f. F., B. 25, p. 82.—Dohrn, M. f. G., B. 25, p. 382.—Breslau, Archiv der Heilkunde, 1863, IV, p. 97 and 481, and Deutsche Klinik, 1865, No. 5.—Ferber, Schmidt's Jahrb., B. 139, H. 10, p. 327.—Simpson, Edinburgh Monthly Journal, February, 1854, p. 97, and Select Obstetric Works, Edinburgh, 1871, p. 569.—Craig, Edinburgh Medical Journal, July, 1870, p. 24.

CHAPTER II

DISEASES NOT DUE TO INFECTION

A.—Febrile Diseases apart from Inflammation of the Genital Organs

Puerperal women are, of course, liable to the various febrile diseases which only form an accidental complication of the puerperal state.

Those diseases do not generally offer any special characteristics. The diagnosis may be difficult when the symptoms are closely allied to those of grave puerperal processes. It may, therefore, be very difficult to distinguish a severe typhoid fever from puerperal infection of a septic character. The explanation also of the inflammations of many organs, especially that of the lungs, may be difficult, inasmuch as they occur, either as accidental complications of the puerperal state or as ichorrhæmic inflammations of puerperal fever. The latter, however, are distinguished by the absence of the typical course characteristic of the former. A similar mistake has frequently occurred between true scarlatina and a scarlatina-like exanthem, which often occurs in puerperal fever.

Although there can be little doubt that occasionally puerperal women are affected by the real contagion, and suffer from true scarlatina, yet they are by no means especially predisposed to infection by that poison, and the large majority of grave diseases termed puerperal scarlatina, with an erythematous dermatitis extending over the surface, is nothing but puerperal fever with such a cutaneous affection.

There are also other causes which, independently of infection or local inflammation of the genitals, very frequently produce fever in puerperal women. These are, first of all, very great distension of the intestine with fæcal matter. We have already stated that, during pregnancy, the action of the intestines is so sluggish that not infrequently enormous quantities of fæcal matter accumulate, and that before or during labour only a very small part is evacuated. The pregnant woman quite commonly enters the puerperal state with a greatly distended intestinal canal, and within the first few days after her delivery the bowels continue to be sluggish.

The retained fæces may produce very decided symptoms of intestinal irritation. In the milder cases there is fulness and swelling of the intestine, with slight inflammatory irritation of the mucous membrane; but certainly, in rare instances, the irritation is continued to the serous coat, and it may assume the proportions of peritoneal irritation, and even that of acute peritonitis.

The abdomen, already somewhat swollen and hard, swells still more. A circumscribed, or more frequently a more diffused, tenderness, which may reach a very high degree, is observed (chiefly in the cæcal region), and prolonged and even persistent vomiting may follow.

The due evacuation of the intestine causes a rapid subsidence of all the symptoms. The diagnosis must rest chiefly upon feeling the fæcal masses. A greatly distended rectum has less value, but more so the distension of Douglas's pouch with a very distended intestinal coil, which can be felt through the posterior wall of the vagina; but the uniform distension of the whole intestinal canal is most distinctly observable by means of palpation of the abdomen, for although in normal puerperal women the abdomen is quite commonly somewhat distended, yet it is soft and easily compressed; but by the retention of large masses of fæces it becomes hard, as a rule uniformly unyielding, so that the uterus can be felt only with difficulty, even in the first days of the puerperal state. In other cases diffused swellings are felt. If also, by a combined examination, the sides of the uterus are found to be free from pain, and if there is tenderness, especially in the cæcal region, we have not to do with peritonitis in consequence of infection, but intestinal irritation through koprostasis. And in spite of the woman's assertion that the bowels have been opened regularly, that diagnosis must be persisted in, for it sometimes (and not rarely) occurs that, in spite of daily evacuations, the intestine is enormously distended. The line of treatment to be adopted is evident.

It is advisable to pay attention to the intestinal functions in all puerperal women immediately after delivery, and especially in cases in which the bowels appear to be greatly overfilled. Aperient medicines ought to be given on the first or at the least on the second day after delivery. They are, as a rule, always indicated when the woman has had no stool until about the fourth day. Castor oil is the mildest aperient, it irritates the bowels the least, and yet causes with great certainty a free evacuation. In rare instances only, repeated doses of it do not act, and preparations of senna or rhubarb or calomel should then be administered. If castor oil is refused on account of its nauseous taste salines may be given, which, of course, usually produce watery stools. But diarrhœa need not particularly be feared in the puerperal state. Often very large quantities of fæces are evacuated by means of purgatives, but Poppel relates a case of koprostasis which was relieved in the course of four days by forty-four very copious stools.

Profuse hæmorrhage is another frequent cause of the elevation of the temperature of puerperal women. There is pretty constantly increased frequency of pulse, lasting for a long time. The temperature, as a rule, rises up to 39° C. (102·2° F.), and remains, with less regular morning remissions, between 38° C. and 39° C.

B.—Change of Position of the Uterus and Vagina

1. *Flexions and Versions of the Uterus*

We have already mentioned in the chapter on the physiology of the puerperal state that an empty uterus is pushed forwards by the pressure of the abdominal muscles acting on its posterior surface, so that it falls upon the symphysis, and that it remains in that position during the first few days of the puerperal state unless other causes (most frequently a distended bladder) press it backwards.

In the first days after delivery the cervix hangs down in the vagina like a loose sail. On the next day or two it is so far formed that its direction can be determined in relation to the axis of the uterus. The cervix is again distinguishable, as a rule, and forms an obtuse, sometimes a right, but often also an acute angle with the axis of the uterus. According to this variation anteflexion is more or less distinctly pronounced. It is most so when the body of the uterus is strongly inclined forwards, and the cervix is directed from behind and above forwards and downwards. The flexion may then become so considerable that the anterior wall of the uterus and that of the cervix almost touch each other. It is rare for the direction of the cervix to be so much inclined backwards that there is, whilst the body of the uterus is bent forwards, exclusively an anteversion, and not also a flexion.

The uterus is usually found directed forwards in the first few days of the puerperal state; it is rarely found inclined backwards. This can be easily explained. In the first few days of the puerperal state the uterus is still too large to fall down beneath the promontory, and in the usual posture the pressure of the intestine is directed upon the posterior wall of the still greatly enlarged organ.

The altered position of the uterus does not give rise to any special symptoms in the early days of the puerperal state. Women with either anteflexion or retroflexion do not complain of it as long as they remain quietly in bed, but we have found in some puerperal women with anteflexion that the discharge of the lochia was prevented by the flexion at the internal os, and when the fundus has been raised some ounces of a brownish fetid liquid flowed off.

But not even that has given rise to complaint. Treatment of the anteflexion is unnecessary because it always gradually disappears of itself.

Retroflexion should be attended to in cases in which the altered position had previously existed and where there is reason to fear that it will recur. We therefore keep the woman in bed for some time.

2. *Prolapsus of the Uterus and Vagina*

During pregnancy the vagina hypertrophies so greatly that its anterior wall, as a rule, projects into the vaginal outlet, and not infrequently prolapses to a slight degree beyond it.

In the puerperal state this latter appearance is quite common. But this prolapsus gradually disappears in consequence of the changes in the vagina. The posterior wall rarely projects beyond the outlet. It most frequently occurs in pluriparæ in whom previously prolapsus of the vagina had existed, and a laceration of the perinæum had healed by granulation. If the prolapsus of the posterior vaginal wall becomes more considerable the uterus itself may be drawn downwards.

The uterus may also primarily descend in the puerperal state, or even be prolapsed. At an early period a true prolapsus scarcely ever occurs, on account of the great size of the organ; but as soon as the uterus has commenced to undergo its retrogressive formation it may prolapse suddenly under certain circumstances. This occurs most easily if the vulva is very large and if there had previously been a prolapsus. Much more frequently it occurs gradually some weeks after delivery.

The treatment of prolapsus consists in maintaining a dorsal posture in the first few days, but afterwards it is the same as in the non-puerperal state.

We may also state that the prolapsus which had previously existed, is in rare cases cured during the puerperal state; the body of the uterus is then fixed by adhesion to the pelvis, or the vagina becomes strictured in consequence of gangrene.

C.—Solutions of Continuity in the Genital Organs

Lacerations of the genital organs almost exclusively occur during parturition. They have been sufficiently spoken of in the pathology of parturition. Here we shall only say a few words on the febrile disturbances which occur in consequence of these lacerations. We omit, however, the perforating ruptures, in which the symptoms of peritonitis very soon predominate.

Traumatic fever, though not regularly, yet very frequently occurs after severe lacerations of the vagina or vaginal outlet, and especially in those of the perinæum. It comes on soon after delivery, within the first few days. It very rarely begins with a rigor, more frequently with slight shivering, and, as a rule, the temperature is not much elevated; exceptionally only it reaches 40° C.

The fever is more or less remittent, exceptionally only intermittent, and at times lasts for five days. Sometimes the fever occurs somewhat later.

D.—New Growths in the Puerperal State

Fibroid tumours of the uterus are a very rare but important complication of the puerperal state. These may give rise to danger during the puerperal state in a variety of ways. By preventing the uniform contraction of the uterus they may cause profuse hæmorrhage; then they undergo changes which are of great interest. The surface of the tumour may putrefy, and the putrid substance absorbed by the smaller lacerations may lead to septic infection. It appears that the acute softening of fibroids, a participation, on account of the similarity of their structure, in that retrogressive metamorphosis which the uterine parenchyma passes through, may cause considerable danger. It is probable, also, that the acute disintegration of such considerable masses of tissue and the re-absorption of the disintegrated parts may produce ichorrhæmia. If the organism is able to bear the involution of the fibroid tumour a complete recovery from the new growth may take place during the puerperal state.

On account of the dangers attending fibroid tumours it appears to be necessary in all cases where an operation can be performed, and especially if there is a fibrous polypus, to remove the new formation by means of the écraseur or the scissors.

We have already spoken of the sad prognosis which carcinoma of the cervix offers in the puerperal state, even if it be not attended by rupture of the uterus during parturition.

E.—Hæmorrhage in the Puerperal State

Hæmorrhage in the puerperal state is of great practical importance. Normally the puerperal woman loses, during the first days, almost pure blood; then, for one or two days more, there is a dark reddish discharge; after this, for about the same time, she has a clear discharge, containing only a little blood; but very frequently these times are not kept.

The bleeding either lasts a longer time or it returns on slight causes, and especially when the woman first leaves her bed. Yet hæmorrhage occurring after the fifth day, as long as only streaks of blood appear in the lochia or at times a small quantity of blood, is not of great importance.

But the cases are not so rare in which a profuse and dangerous hæmorrhage occurs either immediately after delivery or where at a later period protracted and debilitating hæmorrhage sets in. We omit here hæmorrhage which has arisen from laceration either during or immediately after the expulsion of the child, because it has been already mentioned in the pathology of parturition. We shall here consider chiefly all hæmorrhages from the free surface of the mucous membrane, as the blood escapes externally.

The most frequent cause of considerable hæmorrhage immediately after the expulsion of the placenta is atony of the uterus. Its causes have already been pointed out, and we have only to remark that the greatly distended bladder, which presses the uterus upwards, is not a rare cause of metrorrhagia.

The treatment is first of all directed to the production of powerful uterine contractions, since by these alone can the bleeding vessels be made to close. For that purpose the fundus of the uterus should be energetically rubbed by the hand placed on the abdominal walls, and, at the same time, the uterus should be compressed in the way recommended by Credé. By the last manœuvre coagula of blood which remain in the uterus are pressed out with great certainty, and the introduction of the hand for that purpose is rendered unnecessary.

If in spite of energetic friction of the uterus no tendency to contraction is shown we may endeavour to bring it about by introducing one hand into the uterus and rubbing the uterus between the two hands.

Great irritation of the inner surface of the uterus is produced also by the injection of cold water or vinegar and water, and very suitably also by the introduction of large pieces of ice. It is, besides, also advisable, in a case of atony of the uterus, to administer immediately ergot of rye, at least thirty grains, in two doses, at a short interval.

Too much reliance is never to be placed upon that remedy, and the treatment must be continued as if no ergot had been given.

If satisfactory uterine contractions cannot be obtained in this way, we may try to stop the bleeding by compression, *i. e.* by pressing the anterior and posterior walls of the uterus against each other. Deneux accomplishes this by pressing the uterus from the outside against the vertebral column, whilst Hamilton pressed the two surfaces of the uterus against each other with one hand applied externally and the other internally. Hubbard recommends compression of the uterus in the half-lateral position against the iliac fossa, the muscles in the lateral abdominal wall, and the column of the lumbar vertebræ. Fasbender's method is as follows:—He introduces one hand high up along the posterior wall of the vagina, and the other hand he places externally upon the anterior surface; he compresses thus both walls of the uterus.

We think that the pressure would be more practically exerted if one hand were placed against the anterior wall of the vagina whilst the other hand externally pressed the posterior surface of the uterus, and thus the uterus would be pressed against the symphysis, and also against the hand placed in the vagina. In cases in which there is no absolute want of contraction of the uterus, but where this is required only in order to prevent the escape of blood from the placental insertion, bleeding may be arrested by the injection of liquor ferri perchloridi, in order to favour the formation of thrombosis, a method of treatment which Barnes has lately recommended extensively. The liquor is to be so much diluted that the fluid to be injected should have the yellow colour of wine. The proportion of four ounces to twelve ounces of water, as recommended by Barnes, is not necessary. In thus injecting into the uterus great attention must be paid to prevent the admission of any air at the same time. Wynn Williams recommends a more rapid way of applying a solution of perchloride of iron. He dips a sponge in it, introduces the sponge into the uterus, and swabs its inner surface.

If we succeed in causing contraction of the uterus, the next most important thing is to guard against its relaxation, and thus prevent the recurrence of the bleeding. For that purpose the ergot already administered is very useful, even though it is by no means to be trusted absolutely.

The further state of the uterus is best controlled by the hand applied to the abdominal walls; the uterus must be constantly watched, and as soon as it begins to be soft the hand must cause it to contract again by means of gentle friction.

Danger is only quite over when the uterus remains hard spontaneously. If it shows a tendency to relax and to bleed again, the hand must remain there to watch over it during the relaxation. If the bleeding again stops and the uterus becomes softer only from time to time, the attention to the uterus, since a contracted uterus can easily be felt through the abdominal walls, may, in case of need, be left to the nurse, or even to the puerperal woman herself. Their attention, however, must be directed simply to the fact that as soon as the hard body can no longer be felt in the belly they must rub the whole inferior region of the abdomen until the circumscribed hard body of the contracted uterus is again felt. We prefer to keep up the pressure by the hand to the employment of a sand bag or a thick book.

It will be readily understood that the momentary as well as the lasting consequences of the bleeding must be obviated in a suitable way. Consequently, in order to avoid syncope, the head must be placed low, and wine, punch or brandy given in sufficient quantity. If the anæmia becomes extreme transfusion may be necessary. The more remote consequences of anæmia may be removed by good food and large doses of iron.

If the uterus has contracted well hæmorrhage occurs only in extremely rare cases. Kiwisch noticed very considerable hæmorrhages occurring in disease of the heart and other disturbances of the circulation, in consequence of which there was congestion of the veins of the lower half of the body, and attributes it to abnormally large venous openings at the placental insertion. Hecker saw a fatal hæmorrhage from a varicose vein of the cervix. Mikschik observed it in a case where there was an ulcer of the cervix.

Graily Hewitt reported a fatal case in consequence of repeated hæmorrhage from an exposed aneurismal sac of a uterine artery; and Johnson communicates a case in which the fatal bleeding was caused by rupture of a thrombus of the cervix. Hæmorrhage from such causes are very rare, but when they occur they are extremely dangerous, since the blood persistently flows from the vessels, even when the uterus is well contracted.

There is nothing more to be done in these cases but to favour the formation of thrombi in the uterine cavity by various injections. The diagnosis can, as a rule, only be made by way of exclusion.

Another kind of very dangerous hæmorrhage occurs in the otherwise well-contracted uterus in cases where the placental insertion does not participate in the contraction. This constitutes the paralysis of the placental insertion. The contracting parenchyma pushes that portion into the cavity of the uterus, where it projects as a nodular tumour, whilst externally a funnel-shaped depression of the uterine wall can be felt. The diagnosis is easily made by a careful examination as long as the cervix is permeable. If the finger of one hand be introduced into the uterine cavity whilst the uterus is pressed against it with the other hand, the projecting piece of the placental insertion can be felt internally, and the depression with the other hand externally. In the highest degree of paralysis the prognosis is very unfavorable, since the continuance of the bleeding causes death. We must endeavour to overcome the paralysis by large doses of ergot, as well as by continued friction of the uterus externally, Or in severe cases an active injection, such as the liquor ferri perchloridi diluted with water, must be used to stay the bleeding from the paralysed part.

Deficient contraction of the uterus also gives rise to some other kinds of hæmorrhage which have already been mentioned, as in inversion and in folding-in, and in fibroid tumours of the uterus.

Besides the hæmorrhage immediately following delivery others may occur a few days after or even later, coming on suddenly and in more or less abundance; or the sanguineous lochia may continue for an unusual time, and may thereby (at least temporarily) become so profuse that the patient is weakened. Such hæmorrhage is, as a rule, due to something being retained within the uterus; this is for the most part coagulated blood or a piece of the placenta.

The portions of the ovum retained in the uterus undergo various changes. If they are not too large, as only a small cotyledon of the placenta, they are thrown off, as a rule, with the piece of the decidua which remains in the uterus, and are discharged with the lochia, without giving rise to any other symptom than that which the thick fetid lochia produce.

In other cases, very frequently, but by no means always, after operations in which air has entered the uterine cavity, as, for instance, in artificial detachment of the placenta, the pieces left putrefy, and may then cause an intensely putrid discharge, giving rise to septicæmia and endometritis, and thus may occasion all the complex symptoms known as puerperal fever. The portion of the placenta left behind may continue perfectly fresh after premature labour and labour at term, as it at times does after abortion, and may give rise, even at a late period, to profuse and obstinate hæmorrhage.

By the retention of such pieces at the placental insertion polypoid tumours may form in the puerperal uterus, which acquire considerable practical importance.

Small fibrinous coagula intimately blended with the projecting thrombi at the placental insertion are quite commonly found in the bodies of puerperal women. But larger coagula also, the result of repeated hæmorrhages, either of a round shape the size of a walnut, or flat and lobulated, which may also project into the uterine cavity like a cockscomb, are by no means rare. We ourselves had lately an opportunity of observing such a case. Those cases only are very rare where large fibrinous coagula of a polypoid shape are seated at the normal placental insertion, and project with their obtuse end into the cervix or into the vagina. A fibrinous polypus of that kind, the free polypous hæmatoma of the uterus (Virchow), consists of coagulated fibrine, including a nucleus of coagulated blood.

It is not quite known what circumstances favour the formation of fibrinous polypi. They do not exclusively occur in an excessively large uterine cavity, and, on the other hand, there are very large uterine cavities which do not contain large coagula at the placental insertion. A peculiar roughness or too great projection of the placental insertion into the uterine cavity appears chiefly to predispose to their formation (therefore greatly prominent thrombi and paralysis of the placental insertion are associated). We have, however, lately seen two cases with very considerable coagula, and there was no question of an excessive roughness.

The fibrinous polypus occurs after premature labour and after labour at term. If it forms after abortion it appears to favour the return of hæmorrhage at a much later period (after weeks or even months), whilst after labour at or near term it produces hæmorrhage in the first or second week of the puerperal state.

Polypoid formations are far more frequently observed when one or more pieces of the placenta have been retained. Upon these pieces blood is deposited in the way just mentioned, and a fibrinous polypus is formed with a pedicle of placental tissue, or the retained cotyledon becomes bloodless, firm and hard, and assumes a shape corresponding to that of the uterine cavity. This forms the so-called placental polypus.

Under such circumstances hæmorrhage takes place at an early period of the puerperal state. Sometimes, however, an abundant flooding does not occur before the end of the first week, more frequently in the second, and often in the third; and this may be repeated unless the polypus be removed.

The diagnosis of such polypi is not difficult, because the internal os, normally closed at the twelfth day, in such cases remains patulous for a much longer time, or reopens with the repeated hæmorrhage. The tumour very rarely projects into the vagina; sometimes, if the uterus is approached to the examining finger by pressure externally, it can be felt in the external os, but, as a rule, in the internal os, surrounded on all sides, and attached to the uterine wall. The uterus is usually slightly anteflexed, and, apart from the increase in its contents, appears to have undergone the normal changes.

When the disease has been recognised early and the proper treatment instituted, the prognosis is not unfavorable. In other cases death may follow from the hæmorrhage, from the decomposition of the polypus, or from septic endometritis.

The cause of the hæmorrhage can only be obviated by the removal of the polypus. This may be done, as a rule, by a little manipulation, the soft polypus easily yielding to the pressure of the finger, or the hard pedicle is nipped off by the finger-nail.

If two fingers can be introduced, the removal of the polypus is much easier. If the pedicle cannot be separated in this way, the uterus must be fixed in a suitable position from without by an assistant, a forceps is introduced, guided by the finger in order to grasp the polypus, and it is now removed entire or in pieces by pinching or torsion.

After the removal of the polypus the bleeding stops, as a rule, and thin, slightly coloured, sometimes also thick streaky, lochia are discharged. If the bleeding still persists after the removal, injections of diluted liquor ferri perchloridi should be made. The internal os contracts only slowly, and the woman requires much care and attention for a long time.

A thrombus or hæmatoma of the vagina or vulva consists in an effusion of blood into the submucous connective tissue of those parts.

Larger blood tumours of that region are not frequent. According to Winkel hæmatoma occurs once in 1600 births. It appears to be somewhat more frequent in pluriparæ. Greatly varicous dilatations of the veins do not predispose to it. If the mucous membrane over the bleeding vessel tears at the same time, the blood escapes externally, otherwise a tumour is formed in the surrounding connective tissue. Effusion of blood externally and into the tissue may both occur together.

As a rare exception, the rupture occurs in the latter part of pregnancy, as in the case communicated by C. Braun. As a rule, the passage of the head causes the rupture. The tumour is not always seen immediately after the expulsion of the child, and this is partly due to the slow bleeding, partly also to the gradual necrosis caused in some cases by the pressure upon the venous walls and the consequent escape of blood. Such ruptures very rarely occur in a later period of the puerperal state after great bodily exertions, as in the case of Helfer, on the twenty-first day.

The seat of the tumour varies according to the place of injury. If the latter is beneath the pelvic fascia, the blood descends into the greater labium, more rarely into the smaller or towards the perinæum. If the bleeding vessels are situated between the pelvic fascia and the peritoneum, the tumour may at first extend upwards, and may infiltrate extensively the subserous connective tissue on the iliac bones, towards the kidneys, and also towards the umbilical region, but it may also gravitate downwards. Usually the tumour is situated only on one side, but on account of its large size it may extend also to the other side. Very rarely hæmatomata occur on both sides independently of each other.

The somewhat larger tumours are always attended by pain, which, apart from the size of the tumour, is the greater the more rapidly the bleeding occurred. The tumour may contain such a quantity of blood as to give rise to symptoms of excessive anæmia, though hardly a case of death is met with without external hæmorrhage. In very rapid extravasation the pressure of the blood may cause the superjacent skin or mucous membrane to burst, and the consequent external hæmorrhage will certainly then prove fatal.

If the size of the tumour is not very considerable, it usually happens that the liquid constituents are reabsorbed, and the thickened coagula of blood are encapsuled. If the tumour is very large its covering (sometimes the recto-vaginal wall) mortifies under the pressure, and the effused blood now escapes, partly coagulated, partly dark and liquid. Repeated bleeding may again bring imminent danger, or death may follow from septic infection, due to the putrefaction of the greatly compressed sinuous walls. But the sac may also gradually cicatrize with suppuration, sometimes forming perineal or rectal fistulæ.

The diagnosis offers no difficulties. The rapid formation, with the attending anæmia, the blue colour of the mucous membrane covering it, and frequently also of the external skin, and the liquid contents (the tumour feels uniformly elastic, sometimes even fluctuating), exclude all other kinds of tumour which may arise here.

The prognosis is the more unfavorable the larger the tumour. After rupture of the covering the effusion of the blood or the subsequent suppuration and putrefaction may cause death. But most cases of thrombus terminate in recovery. According to Winkel, death occurred six times in fifty cases (three times by bleeding).

Prophylactic measures can scarcely be taken, since hæmatomata mostly arise quite unexpectedly, and do not occur in cases where, from the presence of large varicose veins, they might be suspected. If the tumour is observed early, and whilst it is still increasing in size, its growth must be diminished and coagulation and thrombosis favoured by the use of compression and cold. Both indications are best met by the introduction of a caoutchouc tampon filled with ice-water. If the enlargement of the tumour has stopped there need not be any active treatment as long as particular symptoms do not demand it. Even now the formation of thrombi may be very suitably favoured by the application of ice. The most favorable termination is always that of absorption without opening the tumour. If the coats of the tumour do not mortify, and if parts around do not become the seat of a more intense inflammation, the tumour, on the contrary, becoming harder and smaller, the treatment ought to be merely expectant. Thus tumours, even larger than a fist, may heal without accidents, leaving behind only an inconsiderable swelling.

If the skin, however, becomes discoloured and threatens to burst, an incision ought to be made into the tumour, the cavity emptied, and a suitable tampon used, in order to prevent any further bleeding. If this does not succeed the cavity may then be syringed out with vinegar or diluted liquor ferri perchl., or the tampon again may become necessary. Injections, however, always delay the healing process for a long time. The incision is to be made at the most dependent part, and this may also be found advisable, simply on account of the size of the tumour, without there being any urgent symptoms. It is best to wait three or four days (the longer it is possible to wait the less easily dangerous after-hæmorrhages occur), then a large incision is made, all coagula are removed, and the cavity treated like a large abscess. Purifying injections are made, and by a careful and gradual compression from above we shall favour union of the walls of the abscess.

F. Diseases of the Mammæ

1. *Diseases of the Mammary Glands*

(*a*) *Anomalies of Secretion.*—The quantity of milk secreted by the mammary gland varies greatly in different individuals. Some women have for years a great abundance of milk, whilst others are after a short time no longer able to satisfy the child.

The secretion of milk is very rarely completely stopped during the puerperal state—agalactia—without sufficient reasons to account for it. More frequently the secretion is only imperfect and ceases entirely after a short time. This may be due to an imperfect development of the gland, which sometimes is congenital, but also occurs in very young and very old mothers. It is more frequently found in very feeble women, whose state of nutrition is faulty and unsuitable, and in great general development of the panniculus adiposus. Great mental excitement also may cause a sudden suppression of the lacteal secretion. A gradual diminution of the secretion occurs especially frequently from unsuitable or unaccustomed nourishment.

Intense fever and severe watery discharges may rapidly diminish the quantity of the milk.

Cases where the want of milk is due to an atrophied gland or to a general adiposis are, of course, not accessible to treatment. In the other cases the regulation of the diet is of the first moment. A nursing woman must generally be left to her customary habits of life. She has to avoid overwork and great fatigue, and her customary food must be of good quality and in sufficient quantity. The existing secretion of the milk is best maintained in this way. Milk, farinaceous food, and beer, favorably influence the lacteal secretion.

If menstruation returns in suckling women the nursing need not, as a rule, be interrupted, for in the great majority of instances the secretion of the milk does not diminish and the child thrives.

In very rare cases the secretion of milk becomes so abundant that the general health suffers under it—polygalactia; or after the weaning of the child a secretion of milk, which weakens greatly, continues in the majority of cases from both breasts—galactorrhœa. The quantity of milk secreted may amount to five pounds daily.

In such cases symptoms appear very similar to long-continued bleeding or suppuration. The nutrition of the woman suffers; great weakness and emaciation follow a too long-continued nursing. Menstruation thereby often becomes very abundant, and profuse hæmorrhage sets in, which can only be stopped by the weaning of the child. Prolonged nursing has also a very unfavorable influence on women suffering from osteomalacia.

Disturbances of vision (hemeralopia, amblyopia, amaurosis) also occur in puerperal women who do not suffer from anomalies of the lacteal secretion; they consist in a hyperæmia of the conjunctivæ and no changes can be detected by means of the ophthalmoscope. The prognosis is always good. The disturbances of vision disappear in a few days or weeks.

The great abundance of the secretion of milk can be rapidly diminished by copious watery stools, which are obtained easily and without harm by sulphate of magnesia. The quantity of food also must be somewhat restricted.

Galactorrhœa may be diminished by pressure. Of internal medicines iodide of potassium is principally to be recommended. With the return of menstruation galactorrhœa, as a rule, stops. Abegg cured two cases by producing uterine hæmorrhage by the application of the douche.

Very little is known of the qualitative changes of the milk. Change of food appears to have more influence on the quantity, and secondarily only on the quality. Experience has shown that mental emotions of the mother may cause qualitative changes of the milk, but that influence has frequently been overrated.

(*b*) *Inflammation of the Mammary Gland; Milk Fever; Mammary Abscess.*—The parenchymatous inflammation of the mammary gland—mastitis—in the puerperal state, may pass off in a few days without any consequences, or may go on to suppuration.

The former and milder form is a gradual transition from the physiological development of the lacteal secretion in the puerperal state.

Even in the perfectly normal course of the puerperal state, the temperature is highest on the third or fourth day, and this elevation, as has been shown elsewhere, is not due to the genital organs proper, but is undoubtedly connected with the secretion of milk.

If the mammæ are not very sensitive and the temperature is only slightly elevated, the function then strictly remains within physiological limits. Not infrequently, however, the secretion begins with somewhat more marked symptoms. The gland is greatly swollen, the whole or at least parts of it are extremely sensitive, and there are spontaneous pains; the skin is tense and inflamed, the lymphatics are often seen to proceed to the swollen lymphatic glands in the axilla. The temperature rises on the third or fourth day to 39° C., or even 40° C., and may remain three to four days at an abnormal height with morning remissions.

Such an inflammation most frequently, but not exclusively, occurs in cases where the child has been applied to the breasts too late, or where (on account of the weakness or the abnormal condition of the nipples) not much milk has been drawn from the gland, or where the infant is not nursed by its mother. In most of these cases the inflammatory symptoms and the fever abate after the first or, more rarely, after the second day, and the function of the gland then becomes normal.

As there are unmistakable signs of a local inflammation, and as, doubtless. the fever is due to that inflammation, the name of mastitis may be reserved, as it is usually done, for the suppurative inflammation, and the symptoms in connection with the commencement of the secretion of milk may be designated as "milk fever."

Although the inflammatory state just described disappears in a great majority of instances, yet there are exceptional cases where, under unfavorable conditions, suppuration sets in.

Abscesses of the mammary gland, however, develop at a later period, after the slight inflammation connected with the commencement of the secretion has long since disappeared. Exceptionally the abscess begins during pregnancy, and continues into the puerperal state.

Suppurative inflammation of the mammary gland commonly arises through the small chaps of the nipple, and through the scab formed on it occluding the excretory ducts of some of the lobes of the gland. The secretion is pent up, the lacteal ducts and gland-vessels are dilated and inflame, and suppuration begins in their walls. It is known to occur from an accumulation of milk in the breasts from the sudden weaning of the child. It only exceptionally occurs that the suppuration is caused by an injury, as a blow.

As a rule, only one lobe of the gland of one side is inflamed, yet through improper treatment almost the whole gland may suppurate; more frequently the inflammation commences somewhat later in another lobe or in the whole gland.

The first signs of suppurative mastitis are the general characters of inflammation. The gland becomes swollen, spontaneously painful and tender on pressure, and the skin reddened. It is accompanied by a great elevation of temperature, even to 40° C., and often preceded by a rigor. The more superficial the inflammation is the easier the doughy feeling is obtained which indicates the subjacent pus. There is fluctuation in large abscesses only, but when a greater portion of the gland has suppurated fluctuation becomes very distinct. The fever, which was very high at the onset, usually falls again on the second or third day, or at least there are great remissions, and then passes into a prolonged suppurative fever. The most prominent symptom is pain. It may become so severe that it entirely deprives the patient of her sleep. It only ceases when the abscess is artificially opened, or when it spontaneously bursts. Gradually the connective tissue above the abscess is more and more thinned, the skin becomes bluish, the abscess points, and perforation follows through a small opening, which frequently closes again. Other openings may then form, and the skin between them becomes necrosed. The bursting usually takes place after several weeks.

Sometimes the abscess very early communicates with a perforated lacteal duct, through which the pus escapes with the milk. The symptoms just described are then far less severe, and spontaneous pains may be entirely absent. It is in such cases that a diagnosis can easily be made by pressing pus out of the nipple.

After the spontaneous bursting also suppuration commonly lasts a long time, and months may pass before the abscess has closed. If it communicates with a lacteal duct at the place of perforation a lacteal fistula may remain, which closes spontaneously when the secretion of milk has ceased. It is extremely rare that in otherwise healthy puerperal women the gland mortifies, and that death follows from ichorrhæmia. Sometimes the abscess remains so long open, that gradually the whole gland suppurates. By means of an incision an enormous quantity of pus will be discharged in a thick stream, and the skin of the whole gland will collapse like an empty bag.

The diagnosis of mammary abscess is easy. The sensitiveness of the swollen lobe, the change of the original hardness into a soft doughy consistency, together with the symptoms of the general state, make the diagnosis certain.

The prophylactic treatment consists, first of all, in careful attention to the chaps. Though the first symptoms of inflammation may already be present, suppuration can be prevented by energetic treatment. The breast is to be supported by a suitably arranged suspensory bandage, and sulphate of magnesia should be given until a few watery stools follow and the breasts become thinner. This treatment, early applied, is just as efficacious as the often-praised starch bandage, which is said to prevent suppuration by uniform compression. Simple and useful also is the painting with collodion, which may, however, easily cause excoriations. If suppuration has begun, incisions must be made as early as possible. As soon as a soft spot is felt an incision, not too small and rather deep, must be made in a line radiating from the nipple, and the pus evacuated as completely as possible.

If the incision is kept open the process does not extend any further, and the cavity of the abscess heals up spontaneously if it is always sufficiently emptied. It takes a long time before the cicatrization is complete. To accelerate that process various remedies have been recommended. Winkel strongly recommends the starch bandage. This is very troublesome, as it has to be renewed frequently; if that effect is desired, collodion exerts excellent compression on a tense breast before the abscess is opened, but is not applicable in a flaccid breast. Most simple of all is the compression by suitably arranged napkins and the daily emptying of the cavity of the abscess. If the abscess is in the lower lobe a simple suspensory bandage suffices; if it is higher, the bandage must be applied round the thorax.

If the pus is thin and the granulations flabby it is well to stimulate the healing process by the injection of a solution of nitrate of silver.

An abscess does not render the weaning of the child absolutely necessary; but if an abscess communicates with the lacteal duct it is advisable to wean the child, at least from the diseased breast. The sinuses and lacteal fistulæ which remain commonly heal only when lactation is entirely stopped.

(c) *Galactocele.*—It very rarely happens that when the excretory duct is occluded the secretion still continues without an abscess developing. The lacteal duct becomes more and more dilated, and even forms a large cavity, or one of its walls ruptures, and the milk remains in the newly formed abscess-like cavity. Usually these milk-abscesses are small, yet they may acquire an excessive size, as in the case of Scarpa, where the tumour descended to the left flank, and contained ten pounds of milk.

The tumour at first contains pure milk; later on, the serum separates, and the solid constituents are inspissated, or hæmorrhage into the cyst produces a mixture with blood and of a great variety of colours.

The diagnosis may be very difficult. If the wall of the lacteal duct has not ruptured, from the hard consistency and the distinct fluctuation it may be easily mistaken for a cyst, and after rupture into the surrounding tissue for an abscess.

The treatment is not materially altered by the uncertainty in the diagnosis. After puncture alone the contents again accumulate, and inflammation of the walls of the duct must be produced by injections of iodine. If the cyst does not close in this way it must be laid open by an incision, and allowed to close by suppuration.

2. *Inflammation of the Breast*

The phlegmonous inflammation of the connective tissue of the breast may be limited to circumscribed spots of the areola. The inflammation then usually proceeds from the fissures of the nipple or from the glands of the areola to the connective tissue in the immediate vicinity. Around the nipple small furunculous abscesses form.

Besides the implication of the connective tissue, which regularly occurs in parenchymatous mastitis, there are also during the puerperal state inflammations of the connective tissue, with a tendency to suppuration, most frequently due to an injury. More frequently this is phlegmonous erysipelas in consequence of septic infection from slight excoriations of the nipple. They are either limited to the breast, and may then terminate in an abscess, or they may end in mortification, and death may follow in consequence of an infection of the whole organism. When abscesses form, the pus must be evacuated as early as possible, and this is done by multiple, large, and deep incisions.

In rare cases the submammary connective tissue between the gland and the thorax inflames in the puerperal state. This is the consequence of a blow or of the implication of the connective tissue in the inflammation of the deeper glandular lobes.

The breast is swollen and feels as if it rested on a water-cushion. The base of the breast becomes œdematous. Unless incisions are early made, it may go on to tedious and dangerous burrowing of the pus.

3. *Diseases of the Nipples.—Chaps*

The nipples, covered by a delicate epidermis, are in women who nurse easily exposed to disease. Various conditions predispose to it. By nursing the child or the flow of the milk, the nipple is softened, and the epidermis is raised in the shape of vesicles, which form scabs, under which the epidermis is replaced. Also at the tip and the base of the nipple folds are found, where the epidermis is still more tender and more closely adherent. If the folds have been bridged over by scabs formed of discharged colostrum and dirt, and if the scabs are torn asunder by sucking, the folds are opened up in their entire length, and chaps result. In women in whom those folds are little developed, so that their nipples have quite a smooth appearance, chaps are not easily formed. Others, on the contrary, have deeply fissured nipples, and on separating the small papillæ, even before the child is applied, red, somewhat moist and painful spots are seen. Chaps in these cases are seldom absent when the child begins to suck. Nipples also which freely project, so that the child can grasp them without trouble, are much less disposed to the formation of chaps than those which the child must continually drag upon in order to retain its hold on them.

These chaps are a great source of annoyance to nursing women. Nursing the child causes very severe pain, whilst the simple excoriations with small scabs and ecchymoses are much less sensitive, but may, of course, lead to actual chaps. The chaps also may cause intense fever, though this is absent in most cases. The occurrence or absence of the fever depends partly upon the size of the chaps, partly also upon their depth, and the sensitiveness and irritability of the individual. In some cases the fever may rise to above 40° C., but with the cessation of the irritation it rapidly abates.

If the chaps are neglected they become deeper and more sensitive. The women are in a continual state of anxiety, and greatly dread each fresh application of the child to the breast. There is want of sleep, the appetite is diminished, and fever consumes their strength. In suckling the child the nipple is involuntarily drawn back for fear of the pain, so that the chaps are still more torn open, and, at last, when the pains become too severe and the secretion of the milk ceases, the child has to be weaned. Under certain circumstances the chaps may extend so much in depth that the nipple is almost entirely separated from the breast, and the connection is maintained only by the lacteal ducts.

In other cases the resulting scabs occlude some of the larger lacteal ducts, and in the corresponding glandular lobes the secretion is retained, which is followed by a parenchymatous inflammation, often terminating in suppuration.

Prophylactic measures are required even during pregnancy. Care must be taken that the nipples sufficiently project and that the epidermis is somewhat hardened. The former is obtained by frequent traction with the fingers or with sucking glasses; the latter by scrupulous cleanliness of the nipples and frequent washing with cold water, spirits of wine, or if the skin is very tender with weak solutions of tannic acid.

If the woman begins to suckle with prominent, healthy nipples, chaps do not easily form. If, nevertheless, the nipples remain small, so that the child can only grasp them with difficulty, the milk pump must be used before the child is put to the breast, and the child will then find the whole nipple full of milk.

When there are chaps it is best to cauterize them immediately with the solid nitrate of silver, so that the whole base of the chap is covered over by an eschar. The child, of course, cannot be put to the breast until the surface beneath the scab is healed, else the scab would be torn off and the chap again irritated.

If the child cannot be weaned for so long a time, the base of the chap must be frequently cleansed, or the base is cleared by cauterization, and a few threads of charpie soaked in a solution of tannic acid (gram j in 30—50 grammes of water) are placed into the chap. If this is kept clean, or treated as recommended, it rapidly fills from below upwards and heals.

G. Mental Diseases of Puerperal Women

The mental diseases of puerperal women have by no means particular and distinctive characters. It suffices to say that gestation, labour, and the puerperal state are very powerful etiological factors in certain mental disturbances.

Daily observation shows what a great *rôle* the sexual apparatus plays in the psychical life of women. Witness hysteria, and its intimate connection with diseases of the genital organs. Already in the physiology of parturition attention has been drawn to its influence upon the mental state of women. Sometimes seriously disposed women become excessively merry, whilst, on the contrary, gay young women easily assume a serious, even shy and melancholy disposition. But in rare instances this latter goes so far as to develop into a real disease. Most frequently the gloomy frame of mind passes off with the puerperal state.

The influence of pregnancy upon the cerebral organs may partly be explained on physiological principles. The congestions to the head which accompany pregnancy may just as well give rise to nutritive disturbances of the brain as they do to the formation of osteophytes in its bony envelope. And not only is the great quantity, but also the abnormal quality, of the blood concerned in those nutritive disturbances. The not infrequent acute hæmorrhages during pregnancy and parturition are undoubtedly in some causal connection with the psychical state, which arises from the puerperal condition.

The fact also must be borne in mind, that often pregnant and parturient women are extremely sensitive to psychical affections, and this is especially the case with those illegitimately pregnant. In a great number of the mental diseases of women, the origin of the disease can easily be deduced from the puerperal state. It is, of course, easily intelligible that the predisposing causes which are to be taken into consideration in other causes of mental diseases are of great moment as regards the origin of those diseases after child-bed.

Corresponding to the state of mind displayed during pregnancy or parturition, the psychical disturbances are either those of depression or of exaltation.

The slight forms of melancholy usually originate during pregnancy, and are continued into child-bed, where they may develop into the graver forms of melancholy.

The attacks of mania most frequently occur during parturition. At the moment when the head passes the vulva, and when the suffering reaches the highest degree, acute mania is not rarely seen. Otherwise patient and intelligent women suddenly cry out loudly, wildly stare around them, and menace and strike those who are near them. In the great majority of instances this state disappears immediately after the expulsion of the child, but exceptionally it may also continue after delivery.

Attacks of mania accompanying puerperal diseases are also symptoms of cerebral irritation.

The prognosis of mental diseases originating during the puerperal state is, on the whole, relatively favorable.

As regards the treatment, it is best to watch the patients carefully until their conditions will allow them to be removed to an asylum. Melancholic lying-in women must never be left alone on account of the possibility of their committing suicide.

Appendix.—Sudden Death in the Puerperal State

We shall have to consider those sad cases where death quite suddenly and unexpectedly occurs in lying-in women who have been perfectly well or only slightly indisposed.

The more frequent causes of sudden death, such as very profuse hæmorrhage, the very acute form of septic infection, dangerous apoplexies, or rupture of the heart in consequence of acute myocarditis, of which Spiegelberg has observed a very interesting case, have already been mentioned in the previous chapters. It only remains to speak of two not very rare causes of sudden death during or after delivery, namely, embolism of the pulmonary artery and entrance of air into the uterine veins.

1. *Embolism of the Pulmonary Artery*

As already stated above, embolism occurs most frequently in the course of puerperal fever. Under the influence of the septic infection the thrombi which have formed in the veins, either as a physiological process or in consequence of an extension of the inflammation of the connective tissue around the veins, become disintegrated. However, in the great majority of instances only the smaller branches of the pulmonary artery are the seat of those embolisms, because the thrombi usually break down into very small fragments.

Embolism is happily very rare apart from septic infection. Only, if a physiological thrombosis continues into a larger vein, a piece of the thrombus may be washed away by the current of the blood. Such a large thrombus may then pass through the right heart into the pulmonary artery and obstruct it or, at least, one of its larger branches. Death occurs then either suddenly or in a few days with symptoms of increasing dyspnœa, cyanosis, and a low temperature. The carefully observed case of Rutter is very instructive in this respect. After the observations of Playfair and others, it is not improbable that recovery may take place under favorable conditions.

2. *Entrance of Air into the Uterine Veins*

Experience has taught us that air entering into the veins may give rise to very alarming symptoms or even cause sudden death. This has been frequently seen in operations about the neck, when the veins which are situated between the fasciæ, and therefore unable to collapse, have been wounded.

But air may also enter the uterine veins during or shortly after delivery, and may cause sudden collapse or even death. Now, many facts go to prove that in parturient and lying-in women air may and not infrequently does enter the vagina and uterus, not only in operative procedures, but also in simple examinations. But it is doubtful whether the air spontaneously enters the veins, or whether it is pressed into them by the contractions of the uterus, whilst the os uteri is displaced. The cases of Olshausen and Litzmann unmistakably show that such an entrance can take place, when by means of an injection apparatus air is forced with the water into the uterus under a certain pressure.

Every precaution should, therefore, be taken in making injections into the uterus of a parturient and lying-in woman. The piston must accurately fit into the syringe, and every bubble of air must be previously removed from it.

Literature.—Deneux, Mém. sur les Tumeurs sang. de la vulve et du vagin. Paris, 1830.—Blot, Des Tumeurs sang., &c. Paris, 1853.—Chiari, Braun and Spaeth, Kl. d. Geb., p. 219.—Hecker, Kl. d. Geb., B I, p. 158.—v. Franque, Wiener Med. Presse, 1865, Nos. 47, 48.—Hugenberger, Petersb. Med. Z., 1865, H. 11, p. 257.—Winkel, Pathol. d. Wochenb., 2 Aufl., p. 129.—Kiwisch, Krankh. d. Wöchnerinnen, II, p. 160, 167, and 170.—Scanzoni, Kiwisch's Klin. Vortr., B. III. Prag, 1855, p. 108.—Veit, Frauenkrankh., p. 612.—Winkel, Path. u. Ther. des Wochenbettes, p. 353.—Winkel, M. f. G., B. 22, p. 348, and Path. d. Wochenb., 2 Aufl., p. 405.—Schroeder, M. f. G., B. 27, p. 114, and Schw., Geb. u. W., p. 194.—Wolf, M. f. G., B. 27, p. 241.—Schramm, Scanzoni's Beiträge, B. V, H. 1, p. 1.—Veit, Frauenkrankh., p. 606.—Virchow, Geschwülste I, p. 283.—Veit. Frauenkrankh., 2 Aufl, p. 610.—Winkel, M. f. G., B. 22, p. 345, and Berl. klin. Woch., 1864, No. 2.—Scharlau, Berl. klin. Woch., 1864, Nos. 19 and 20.—Berndt, Krankh. d. Wöchn. Erlangen, 1846.—Leubuscher, Verh. d. Ges. f. Geb. in Berlin., III, p. 94.—Ideler, Charitéannalen, 1852, I.—Scanzoni, Kl. Vorträge von Kiwisch., B. III, p. 520.—Veit, Frauenkrankh., 2 Aufl., p. 705.—Tuke, Edinburgh Medical Journal, January, 1867 (s. Virchow-Hirsch'scher Jahresb. über 1867, B. II, Abth. 3, p. 605.—Winkel, Path. u. Ther. d. Wochenbettes, 2 Aufl., p. 449.—Weber, Allg. Med. Centralzeitung,

1870, Nos. 87, 88. Ausserdem s. die fachwissenschaftl. Werke über Psychosen.—Mordret, De la mort subite dans l'état puerperal. Mém. de l'Acad. de Méd., Tome XXII.—Hecker, Deutsche Klinik, 1855, No. 36.—Charcot and Ball, Gaz. Hebdom., 1868, V, 44, &c. (Schmidt's Jahrb., B. 104, p. 187).—Mackinder, Obstetrical Transactions, Vol. I, p. 213.—Hervieux, Gaz. des Hôp., 1864, No. 8.—v. Franque, Wiener Med. Halle, 1864, Nos. 33 and 34 (s. M. f. G., B. 25, Suppl., p. 335).—Barnes, Obstetrical Transactions, IV, p. 30.—Frankenhäuser, Jenaische Z. f. M. u. N., B. III, p. 74.—Steele, British Medical Journal, April 7, 1866 (Virchow-Hirsch'scher Jahresb. üb., 1866, p. 542).—Chantreuil, Gaz. des Hôp., 1866, No. 59 (ebend.).—Playfair, Lancet, July and August, 1867 (Virchow-Hirsch'scher Jahresb. über 1867, p. 603), and London Obstetrical Transactions, Vol. X, p. 21, nebst der Discussion.—Duncan, Results in Obstetrics, p. 399.—Ritter, M. f. G., B. 27, p. 138.—Worley, British Medical Journal, May 7, 1870.—Olshausen, M. f. G., B. 27, p. 353 (mit Angabe der einschlägigen Literatur).—Litzmann, Arch. f. Gynaec., B. II, p. 176.—Winkel, Path. u. Ther. d. Wochenb., 2 Aufl., p. 88.—Schroeder, Schw., Geb. u. Wochenb., p. 187.—Bidder, Petersb. Med. Z., 1869, B. 17, H. 4 and 5.—Credé, Arch. f. Gyn., B. I, p. 84.—Hecker, Kl. d. Geb., B. II, p. 124, and M. f. G., B. 26, p. 446.—Horwitz, Petersb. Med. Z., 1868, H. 5, p. 249.—Kiwisch, Klin. Vorträge, 4 Aufl., B. I, p. 243.—Chiari, Braun u. Spaeth, Klinik der Geburtskunde, p. 201 and 218.—Elliot, Obstetrical Clinic, p. 223.—Winkel, Pathol. u. Ther. d. Wochenb., p. 108.—R. Barnes, Obstetrical Transactions, Vol. XI, p. 219, and Obstetrical Operations, 2 ed., p. 464.—Breisky, Volkmann's Samml. klin. Vortr. Leipzig, 1871, No. 14.

INDEX

ERRATA.

Page 63, 9th line from top, *omit* "scarcely."

" 176, 9th line from bottom, *for* "a cross bed," *read* "lying across the bed."

" 177, 3rd line from top, *for* "on the cross bed," *read* "with the patient lying across the bed."

" 182, 7th line from bottom, *for* "upon a cross bed," *read* "across the bed."

www.ingramcontent.com/pod-product-compliance
Lightning Source LLC
Chambersburg PA
CBHW021352210326
41599CB00011B/842
9783743357518